D1171321

Animal
Cell Culture
CONCEPT AND APPLICATION

Animal Cell Culture

CONCEPT AND APPLICATION

Sheelendra M. Bhatt

Alpha Science International Ltd.
Oxford, U.K.

Animal Cell Culture
Concept and Application
330 pgs. | 159 figs. | 43 tbls.

Sheelendra M. Bhatt
Assistant Professor
Department of Biotechnology
Lovely Professional University
Jalandhar, Phagwara
Punjab 144402

ALPHA SCIENCE INTERNATIONAL LTD.
7200 The Quorum, Oxford Business Park North
Garsington Road, Oxford OX4 2JZ, U.K.

www.alphasci.com

ISBN 978-1-84265-488-0

Printed in India

Dedicated to my esteemed teacher Professor S. K. SRIVASTAVA whose guidance enabled me to present this book

PREFACE

The Author present this book on "Animal Cell Culture" with great pleasure. The main credit of the work goes to my students of Biotechnology, whose questions inspired me to write a conceptual book. Currently, tremendous research is going on in various field of Biotechnology such as Gene therapy, livestock as Bioreactor, and stem cell culture of various mammalian cells. Therefore, this applied area has great potential to fight with various unsolved problems related to health.

This book is intended to serve as logical and conceptual introduction to animal cell culture. The chapter starts with breif concept, historical aspects and then content has been developed analytically with recent technique and developments and finally applications has been discussed.

Since highly aceptic environment is required before and after starting the cell culture, so chapter 2, 3, 4 and 5 is dedicated for wide discussion on concepts and related problems and method to cop up the real solution. Since culture condition predicted in one laboratory can't be replicated in same way in other laboratory therefore, emthasis is given on concept and limit of culture in Chapter 6 and 7. After culturing the cell various assay is required which has been discussed in Chapter 9. Many time after culturing cells undergoes various changes has been discussed in Chapter 10, 11 and 12. Chapter 13, 14 and 15 emphasis on cloning and maintaining the cells while application has been discussed in Chapter 16.

The present book is inspired from a famous proberb "The worth of a book is to be measured by what you can carry away from it." It is hoped that student will be able to use this book without a teachers help, since book has been conceptualise after long experiences of working and teaching of the subject understanding the need of the student belong to various discipline. Therefore book emphasis on learning the concept rather than protocol and also application in related field such as diagnostic kits and drugs developed.

Therefore, author is grateful to many scientific works. Various figures in the book have been redrawn in the Coral Draw, which may be similar but not a copy of any book. The text where say has been written by the author and few important issues have been adopted from original work of Biotechnology. Though, I have taken great care to avoid mistakes in the book, still if one finds any anomly these would be rectified if brought to knowledge.

Chapters in the book has been arranged to meet in the requirement of the undergraduate and postgraduate courses in most of the universities and will also be helpful in various competative exams such as ICMR, NET, GATE, JNU, IISc, and TIFR. This book is not going to cover every topic in the animal cell culture and is also not the authors real work.

Author is indebted to various teachers who helped directly or indirectly during writing of the manuscript Prof. S. K. Srivastava (School of Biochemical Engineering), Dr. Arun K. Pandey

(Agrasen P.G. College, Varanasi), Dr. Ajay Pandey (IIVR, Varanasi) and Prof. Ashish S. Verma and Prof S. M. P. Khurana (Amity University) for inspiring, guiding and reading the manuscript. I am grateful to my teachers of Biotechnology Department, BHU, Varanasi for giving me the ground concept of the subject. Author is grateful to: Prof. Subir Kundu, Prof. Pradeep Srivastava, Dr. Banik, Dr. Abha, Dr. M.D. Das of Biochemical Engg, IIT, BHU. Author is grateful all the colleague teachers specially to Dr. Arya of Punjab University, Dr. Bhuminath Tripathi of Banasthali Vidyapeeeth, Dr. Anil of Gorakhpur University, Dr. Alok of Balia PG College, Dr. Rajesh of Biotechnology department, Purvanchal University, and Prof. Anita singh (AKPG College), who directly or indirectly inspired me at every stage.

I am indebted to my mother, father, wife, my son Anshuman and my brother who always inspired me in every moment, during preparation of this book. Last but not the least, I am sincerely thankful to Mr. N. K. Mehra (Narosa Publishing House Pvt. Ltd.) who took every pain in publishing this book.

Sheelendra M. Bhatt

CONTENTS

ABBREVIATIONS

A1AD	Alpha1-Antitrypsin Deficiency
AAV	Adeno-Associated Virus
ADA	Adenosine Deaminase
ALV	Avian Leukosis Virus
AM	Calcein-Acetoxymethyl
BAC	Bacterial artificial chromosomes
B-CLL	Chronic Lymphocytic Leukemia
BSS	Balance Salt Solution
β*hCG*	β-human chorionic gonadotropin
BVDV	Bovine Viral Diarrhea Virus
CAG	Cytosine-Adenine-Guanine
CAM	Calcium Dependent Cadherins Cell Adhesion Molecules Cell-Adhesion Molecules
c-AMP	Cyclic Adenosine Monophosphate
CDR	Complementarity Determining Regions
CFC	Colony-forming cell
CFU-DC	Colony-forming-unit dendritic cells
CFU-GM	Colony forming-unit granulocyte macrophage
CHO	Chinese Hamster Ovary Fibroblast
CIG	Cold Insoluble Globulin
CMC	Carboxy Methyl Cellulose
CNTF	Ciliary Neurotropic Factor
CSTR	Continuous flow stirred tank reactor
D	Decimal Reduction Times
DAF	Decay Accelerating Factor
DMD	Duchenne muscular dystrophy
DMEM	Dulbecco's modification of Membrane
DMSO	Dimethylsulphoxide
EBV	Epstein-Barr virus
ECM	Extracellular Matrix
EGFP	Enhanced green fluorescent protein
ELISA	Enzyme-Linked Immunosorbant
FBS	Fetal Bovine Serum
FKBP	FK506- binding protein
GCV	Gancyclovir
GIFT	Gamete Intrafallopian Transfer
GnRH	Gonadotropin Releasing Hormone
GVHD	Graft-versus-host disease
HAC	Human Artificial Chromosome
HAMA	Human Antimouse Antibody
HAR	Hyperacute Vascular Rejection
HEPA	High Efficiency Particle Filter
HEPES	N-2-hydroxyl ethyl piperazine-N'-2-ethane sulphonic acid
HSC	Hematopoietic Stem Cells
HSPG	Heparan sulfate proteoglycans
HSV-1	Herpes Simplex Virus Type-1
HTT	Human Huntingtin
ICM	Inner Cell Mass
ICSI	Cytoplasmic sperm injection
IDDM	Insulin-Dependent Diabetes Mellitus
IGF	Insulin-Like Growth Factor
IP	Intraperitoneal Injection
IVF	In-Vitro Fertilization
JIVET	Juvenile In Vitro Embryo Transfer
KSM	Keratinocyte serum free media
LIF	Leukemia Inhibitory Factor
LTC-IC	Long-Term-Culture Initiating-Cell

MAB	Monoclonal antibodies
MAC	Mammalian Artificial hromosomes
MAS	Marker Assisted Selection
MBC	Minimum Bactericidal Concentration
MCB	Master cell banks
MDCK	Madin-Darby canine kidney
MIC	Minimum Inhibitory Concentration
MOET	Multiple Ovulation and Embryo Transfer
MR	Batch membrane reactor
MT	Metallothionein
MWCB	Master Working Cell Banks
NT	Nuclear Transfer
OV	Ovalbumin Gene
PBL	Peripheral blood lymphocytes
PBR	Packed bed reactor
PGD	Preimplantation genetic diagnosis
PID	Pelvic Inflammatory Disease
PKC	Protein Kinase C
PVP	Polyvinyl Pyrolidone
PWM	Pokeweed mitogen
RIA	Radioimmunoassay
RISC	RNA-induced silencing complex
SB	Sleeping Beauty
SCNT	Somatic-Cell Nuclear Transfer
SMGT	Sperm-mediated gene transfer
SNP	Single Nucleotide Polymorphism
STR	Stirred tank batch reactor
SUZI	Subzonal Sperm Insertion
T3	5-tri-iodo-tyrosin
TCA	Trichloroacetic Acid
TDT	Thermal Death Time
TET	Tubal Embryo Transfer
TK	Thymidine Kinase
TNF	Tumor Necrosis Factor
TPA	Tissue Plasminogen Activator
X-SCID	X-linked severe combined immuno deficiency
YAC	Yeast artificial chromosomes

DEVELOPMENTS OF THE ANIMAL CELL CULTURE

1

1.1 HISTORY

Year of Discover	*Workers*	*Discoveries*
1950's		Emphasis was given on development of viral vaccines, Anchorage dependent vessels like Glass T-Flasks and Roller bottles (for culture of the normal cells).
1960'S		Emphasis was given on the development of *Microcarriers for cell culture*, utilization of Stirred tank bioreactors.
1970's		Emphasis was given on the development of *Hollow fiber bioreactors, Cell* immobilization and encapsulation techniques development of suspension cultures, utilization of Air-Lift bioreactors, development of serum free media, development of plastic ware, utilization of antibiotics in cell culture *media*, Improved systems for scale up, Increased emphasis on MAB production, development of several cell culture products like TPA (Tissue Plasminogen Activator), Human Growth Hormones, and Interferon.
1878	**CLAUDE BERNARD**	Established the fact that physiological systems of an organism can be maintained in a living system even after the death of the organism.
1885	**ROUX**	Established the fact that the specific developmental process occurs independently of the other parts of the embryo by maintaining a portion of chicken embryo in vitro in a warm physiological solution
1907	**HARRISON**	Star ted animal cell culture and cultivated amphibian spinal cord in a lymph clot, thereby demonstrating that axon produced by extension of single nerve cell. He maintained embryonic frog neural crest fragments in culture for several weeks and observed the growth of nerve fibers in vitro. They utilized a drop of coagulated frog lymph as the medium for the growth. He took frog as a source of tissue because it is a cold blooded animal and incubation is not required. Later they were considered as the **father of animal cell culture**.
1910	**ROUS**	Induced the tumor cell by using a filtered extract of chicken tumor cell.
1910	**BURROWS**	Tried first long term successful cultivation of chicken embryo cell in plasma clots and also described first, detailed observation of mitosis.
1911	**LEWIS & LEWIS**	Observed limited monolayer growth in the media that was first liquid media which consisted of, Sea water, Serum, Embryo extract, Salts and peptones.
1913	**CARREL**	Proved that the cell can grow for longer period if grown under aseptic condition.
1916	**ROUS & JONE**	The first use of the proteolytic enzyme, Trypsin to suspend attached cells in culture.
1923-31	**CARREL & BAKER**	Developed T-Flasks techniques for cell culture, and for their scale-up and Microscopic evaluation of cells in culture.
1927	**CARREL & RIVERA**	Production of the first viral vaccine – Vaccinia against chicken pox.

Contd..

Contd..

1933	**GEY**	Developed the **Roller Tube Technique** for cell culture.
1948	**FISCHER**	Developed a chemically defined medium called as **CMRL 1066**, still used in virology and cytological studies.
1948	**EARLE**	Isolated single cell line from clone of cells in tissue culture.
1952	**DULBECCO**	Developed plaque assay for animal viruses using confluent monolayer of cultured cells.
1952	**GEY**	Established a continuous cell derived from human cervical carcinoma (**He** Lacell).
1954	**ABERCROMBIE**	Observed contact inhibition in monolayer culture (culture ceases when contact is made with adjacent cells); motility of diploid cells in monolayer.
	LEVI MONTALCINI & ASSOCIATES	Proved that NGF (nerve growth factor) stimulates growth of axon in tissue culture.
1955	**EAGLE**	Developed first systematic investigation of essential requirement of cell in culture and found that animal cell could propagate in a defined media with small proportion of serum protein. Media formulated by them were called as **Eagle's media**.
1956	**PUCK & ASSOCIATE**	Observed that selected mutant cell has altered growth requirements for nutrition e.g. He La cells.
1956	**LITTLE FIELD**	Introduced the HAT (Hypoxanthin Aminopterin and Thymidin) medium for the selective growth of fused somatic cell during development of monoclonal antibody from unfused one.
1965	**HAM**	Introduced a defined serum free media that was able to support the growth of certain mammalian cells. The media were later called as **HAMS's** media.
1965	**HARRIS & WATKINS**	Produced first **Heterokaryon** of mammalian cells by the virus induced fusion of Human and mouse cell called as **Hybridoma**.
1973	**KOHLER AND MILSTEIN**	Prepared the **Monoclonal antibody** by fusing the human B-cells with cancerous cell.
1975	**GOSPODARO-WICZ**	Fibroblast growth factor was discovered totipotancy of embryonal stem cell
1976	**ILLMENSEE & MINTZ**	**Hayashi & Sato** serum free media
	SATO and associates	Published first a series of papers showing that different cell lines requires different mixtures of hormones and growth factors to grow in serum free medium.
1977	**NELSON REES & FLANDERM-EYER**	Conformation of Hela cell cross contamination
	RHEINWALD & GREEN	3T3 feeder layer and skin culture
	WIGLER, AXEL	Developed an efficient method for introducing single-copy mammalian genes into cultured cells, adapting an earlier method developed by **Graham** and Van Der Eb.
1978	**HAM & MC KEEHAN**	MCDB selective, serum free media
	GOSPODAR-OWICZ	Matrix interaction
1980	**DARNELL**	Regulation of gene expression
	WEINBERG	Oncogene, malignancy and transformation were discovered
	HASSELL E AL	Matrix from EHS sarcoma (Matrigel)
1980-1987	**PEEHL & HAM**	Many specialized cell line

Contd..

Contd..

1982		Human insulin became the first recombinant protein to be licensed as a therapeutic agent.
1983	**EVANS HUSCHTSCHA & HOLLIDAY**	Regulation of cell cycle Immortalization of SV40
1984	**BELL**	Reconstituted skin cell culture
1985	**COLLEN**	Recombinant TPA (Tissue Plasminogen Activator) in mammalian cell Human growth hormone produced from recombinant bacteria was accepted for therapeutic use.
1986	**MARTIN, EVANS**	Isolated and cultured the Pluripotent embryonic stem cells. Lymphoblastoid ©IFN licensed.
1987		Tissue-type plasminogen activator (tPA) from recombinant animal cells became commercially available.
1989		Recombinant erythropoietin in trial.
1990		Recombinant products in clinical trial (HBsAG, factor VIII, HIVgp120, CD4, GM-CSF, EGF,mAbs, IL-2).
1991	**CAPLAN**	Culture of human adult mesenchymal stem cell
1998	**DENNIS**	Culture of human embryonic stem cell
1996	**WILMUT AND CO-WORKERS**	Successfully produced a transgenic sheep named **Dolly** through nuclear transfer technique. Thereafter, many such animals (like sheep, goat, pigs, fishes, birds etc.) were produced. In 2002, Clonaid, a human genome society of France claimed to produce a cloned human baby named **EVE**.
1998	**THOMSON AND GEARHART**	Isolated human embryonic stem cells.
2000	**HUMAN GENOME PROJECT**	Genomics, proteomics, genetic deficiencies and expression errors
2001	**LEE ET AL**	Spermatids were cultured from newborn bull testes and co-cultured with Sertoli cells
2002-2004	**ATALA & LANZA, VUNJAKNOVAKOVIC & FRESHNEY**	Exploitation of tissue engineering for generation of tissue equivalents by organotypic culture; Isolation and differentiation of human embryonal stem (ES) cells and adult totipotent stem cells such as mesenchymal stem cells (MSCs); Use of gene transfer, materials science, bioreactors, and transplantation technology for implanting normal fetal neurons into patients with Parkinson disease has been demonstrated; In vitro fertilization (IVF) technique were developed from early experiments in embryo cuture.

1.2 TYPES OF CELL CULTURE

1. Cell Culture

Culturing of the tissue explants in vitro is called **cell culture**. The tissue is dispersed in the medium, mostly by using enzymes, into a cell suspension which may then be cultured as a monolayer or suspension culture.

2. Tissue Culture

Fragments of the excised tissue, if placed in fresh culture medium, then its normal functions may be maintained but original organization of tissue is lost. The tissue culture is better than the organ culture.

3. Organ Culture

The maintenance or growth of the organ primordia or the whole parts of an organ in vitro in a way that may allow differentiation and preservation of the architecture or function of the organ is called the **organ culture**. In the organ culture whole tissue is taken out of the body for culture while in the primary culture a fragment of tissue is placed over a glass (or plastic) in liquid interface, where, following attachment, migration is promoted in the plane of the solid substrate which can gain normal physiological functions and cells remain fully differentiated.

4. Organotypic/Histotypic Culture

The Histotypic culture resembles a tissue-like morphology in vivo, usually, a three- dimensional culture is re-created from a dispersed cell culture that attempts to regain a tissue-like structure by cell proliferation. Organ cultures cannot be propagated, whereas the Histotypic cultures can be. Histotypic culture involves more than one cell type.

5. Serial Passaging or Sub Culture (Secondary Culture)

The transfer of cells after one generation is called as the subculturing. Usually, but not necessarily, this promotes the subdivision of a proliferating cell population, enabling the propagation of a cell line or cell strain.

6. Cell Line

Cell line arises from the primary culture after few subculture is called as "finite" cell line.

7. Continuous Cell Line

A cell line which has been "transformed" and thus they divide for indefinite period of time and therefore they were called as "infinite" life span.

8. Split Ratio

The divisor of the dilution ratio of a cell culture during subculture.

9. Passage Number

It is the number of times that the culture has been recultured.

10. Generation Number

It refers to the number of doublings that a cell population has undergone.

11. Primary Culture

Isolated cells, from tissues or organs taken directly from an animal, when grown in culture media, first time is called as Primary Culture.

12. Explant

An excised fragment of an organ which usually retains some degree of tissue architecture.

13. Monolayer

Single layer of cells growing on a surface is called as Monolayer.

1.3 CELL CULTURE AND THEIR SCOPE

Cell Culture is the technique by which cells can be grown in vitro, (i.e. outside the body) in an incubator. Cells present inside the body functions in a coordinated manner where cell-cell interaction, tissue-tissue interaction, organ-organ interaction, system-system interaction takes place but outside the body environment is different since cells are the individual units, and behaves like microbes or unicellular organism. Thus, culturing of animal cells is more difficult in comparison to the culture of microbes as they are less adapted to in vitro condition.

Historically, the earliest cell cultures involved the growth of cells from fragments of tissue embedded in clot of plasma, but it was not suitable for experimental analysis. In the late 1940s, a major advancement was the establishment of cell lines that grew from isolated cells attached to the surface of culture dishes. A widely used human cancer cell line called HeLa cells was initially established in 1952 by growth in a medium consisting of chicken plasma, bovine embryo extract, and human placental cord serum. The use of such complex and undefined culture media, made analysis of the specific growth requirements of the animal cells impossible.

Harry Eagle was first to solve this problem, by carrying out a systematic analysis of the nutrients needed to support the growth of animal cells in culture. Eagle studied the growth of two established cell lines: HeLa cells and a mouse fibroblast line called **L cells**. He was able to grow these cells in a medium consisting of a mixture of salts, carbohydrates, amino acids, and vitamins, supplemented with serum protein. By systematic variation in the components of this medium, Eagle was able to determine the specific nutrients required for cell growth. In addition to salts and glucose, these nutrients included 13 amino acids and several vitamins. A small amount of serum protein was also required.

The medium developed by the Eagle is still the basic medium used for animal cell culture. Its use has enabled scientists to grow a wide variety of cells under defined experimental conditions, which has been critical to studies of animal cell growth and differentiation, including identification of the growth factors present in serum (now known to include polypeptides, that control the behavior of individual cells within intact animals.). Now due to increase in population, there is pressure to utilize the potential for biotechnology to improve the productivity in animal agriculture. **Recombinant DNA technology** now allows us to introduce foreign genes into organisms for the expression of specific new traits. Animals can also be engineered for a variety of purposes, ranging from use as human disease models to introduction of desirable traits into a variety of agronomically important animals. Genetically engineered poultry, swine, goats, cattle and other livestock are now available as generators of pharmaceutical and other products and potential sources for replacement organs for humans. The technology to produce foreign proteins in milk by expressing novel genes in the mammary glands of livestock has already taken the shape with some products currently under clinical trials.

Transgenic animals, thus not only provide invaluable research tools for studying gene regulation and diseases but they may be genetically modified for the production of pharmaceuticals, vaccines and rare chemicals as well as for food production. Now a days animal "bioreactors" has been developed for production of rare pharmaceuticals and other medical compounds. Genetically engineered livestock are yielding important products in milk or blood for treating a variety of human diseases and health needs.

Animal biotechnology is considered to be synonymous to rDNA technology and also to some of the older technologies along with the state-of-the-art cutting-edge technologies such as cell culture, monoclonal antibodies, bioprocess engineering and manipulation of reproduction. Thus, animal biotechnologists not only manipulate the genomes of the targeted animals, but also the process that exist in the organism but are out of reach for manipulation. Animal biotechnology offers supplementation of selective breeding, helping to effect changes at the organism level by the manipulation of cells and genes within an organism.

Animal cell cultures are used to produce **virus vaccines,** as well as a variety of useful biochemicals like enzymes, hormones, cellular biochemicals like interferon, and immunobiological compounds including monoclonal antibodies. Animal cells are also goodhosts for the expression of recombinant DNA molecules and a number of commercial products have been/are being developed. Initially, virus vaccines were the dominant commercial products from cell cultures, but at present monoclonal antibody production is the chief commercial activity. It is expected that recombinant proteins would become the prime product from cell cultures in the near future. Transplantable tissues and organs are another very valuable product from cell cultures.

1.4 PROPERTIES OF THE ANIMAL CELLS

The animal cells are of mostly 10-100 microns size. Animal cells are spherical in shape in suspension medium. Animal cells are devoid of any cell wall, i.e. only Plasma membrane is present which is thin, fragile and shears sensitive and have negative charge over surface, therefore they can be grown on positively charged surfaces.

1.5 ADVANTAGES

1. **Characterization of the Cells.** Specific sample cell can be detected by presence of markers e.g. expression of specific receptors, type of molecule secreted, their Karyotype; e.g. HeLa cells, Dendritic cells.
2. **To Obtain Heterogenous or Homogenous culture.** Heterogeneous culture containing mixture of cell types are called as the primary cell whereas homogenous cultures have only one cell type are called as secondary cell; since they are obtained after selection of single cell from mixed population and their further culture yields clone of single population. Heterogeneous cultures can be transformed into homogenous culture by selective techniques. Homogenous cultures may be called as clones, if it contains genetically identical cells derived from a single parental cell (cloning technique). Homogenous cultures are used to study origin and biology of the cells.
3. **Scaling up and Mechanization.** Micro-culture can be prepared in 96 well plate culture plate which requires low numbers of cells, low amount of media and agents. Mechanization can be done by robotic control over the instruments for scaling up the products from cell culture.
4. **In-Vitro Modeling of In-Vivo Conditions.** Organotypic/Histotypic culture can be done which mimics respective organ/tissue. This is helpful in production of artificial tissues (e.g. Skin), which can be helpful in reduction of sacrificing animals for obtaining the tissue.

5. **Study of Effect of Environment.** Effect of physiochemical properties on cells at different pH, temperature, O_2/CO_2 concentration, osmotic pressure and physiological properties (hormones, cytokines, and nutrients) can be studied. It minimizes batch to batch variation and thus reliability increases.

1.6 LIMITATIONS

1. **Costly equipments and expertise needed.** For the culture of specific type of cells, well equipped lab is required since culture must be carried out under strict aseptic conditions. Therefore, high level of care is needed to avoid the problems of chemical contamination, microbial contamination and cross contamination.
2. **The output is costly.** A major limitation of cell culture is the expenditure of effort and little product from cell culture. Finally, there is production of relatively little tissue e.g. Milligrams of monoclonal antibody are produced from cell culture while in vivo it is produced in bulk (gms to kgs amount). It is expected that recombinant proteins would release the prime product from cell culture in near future. Transplantable tissues and organs are another valuable product obtained from cell cultures.
3. **Cells undergoes Dedifferentiation, and are unstable.** Dedifferentiation describes the loss of the differentiated properties of a tissue (when it becomes malignant or when it is grown in culture) e.g. hybridoma cell differentiates into fibroblast cells which is different from original cells.
4. **Cell Culture produces low quantity of products.** Animal cells have low productivity, (good producers produces only 100 microgram/ml/day while an average producers 20-100 microgram/ml/day, poor producers <20 microgram/ml/day) because of slow growth of the cells, and low expression rate. Most of the small laboratories can produce up to 1–10 g of the cells in the batch culture. With a little more effort (facilities of a larger laboratory), 10–100 g of product is possible; but above 100 g, industrial pilot-plant are required, (a level that is beyond the reach of most laboratories).
   Animal cell culture generally yields lower cell density, commonly around 5×10^6 / ml. This results in lower volumetric productivity (g product formed/l/h), and specific productivity (g product formed/g cell/h). This problem is being addressed by improvements in fermentation conditions to allow the use of dense cultures in the order of 10^8 to 10^9 cells/ mL (Tyo & Spier, 1987).
5. Animal cell typically **exhibit slower growth rate** and metabolic rate compared to microbial cells. The doubling time of animal cells is in the order of 20 to 100 h, with averages ranging from 18 to 24 h, compared to 0.5 to 2 h for most bacteria.
6. Animal cells are **more susceptible to contamination** by microbial cells, especially mycoplasma which are difficult to detect and highly contagious (Arathoon & Birch, 1986).
7. Many animal cell cultures **require more careful handling** to avoid cell breakage. This problem is being addressed by selecting and using more robust cell lines. Thats why, most of the animal cells are grown on flat glass or plastic surfaces as monolayers; or as suspended cells in slowly rotating containers.

1.7 IMPORTANCE OF THE ANIMAL CELL CULTURE

1. **In understanding the physiology and the biochemistry of specific cell types.** Cell culture can be used to trace out the metabolic pathways using C^{14} isotopes to trace out various products of the Krebs cycle. Cell culture can also be used to study the metabolism, of secretion, tissue organization and various interactions among cells (cell to cell and cell to matrix).
2. **In understanding the action of molecules (agents) on cells.** Action of different biochemicals like hormones and cytokines on various cell types, or toxicity of the pharmacological agents (drug) can be studied by cell culture techniques since very less amount of the agents (in nanogram amount) and cells are required.
3. **In production of cells in bulk for various applications.** Production of artificial skin (from epidermal cells) and various other transplantation tissue is possible by tissue engineering applications.
4. **In production of various cellular products.** Cell culture is used for production of various cellular products e.g. cellular glycoprotein (hormones, cytokines), vaccines from the virus culture, and monoclonal antibodies from hybridoma cells, etc.

1.8 APPLICATIONS

1. **Cells as model system.** Cells can be used as model for study of basic cell biology, biochemistry, effect of drugs on cells, and senescence of cells. This model gives idea of in-vivo condition without killing animals.
2. **In toxicity studies.** Toxic effect of an agents on cells can be studied, with very little quantity (nanogram-picogram) of toxic substance. Toxicity can be individually checked by applying variety of parameters such as Survivals of cells, Apoptosis, Proliferation, and Functional capacity etc. Cells of kidney and liver have been used for such toxicity studies.
3. **In cancer research.** Normal and transformed cells can be studied at the same time in presence of carcinogens. Thus drug can be designed based on metabolic difference.
4. **In virology.** The animal cell culture are used now days for production of viruses in aim to get large number of virus proteins, for vaccine production. Cell culture also helps in various clinical detections and isolation of virus from biopsy sample and to understand the basis of infection, and their mechanism for example attenuated viruses are used in production of vaccine against the Polio, Rabies, Influenza, Chickenpox, Hepatitis B, Measles, Mumps, Rubella, HIV vaccines (envelops proteins, GP 120 have been cultured in Chinese Hampster,or Ovary cell culture for making vaccine against the AIDS.)
5. **In large scale production of the Hormones.** Genetically engineered cells have commercial or medicinal application in hormone vaccine production e.g. **Insulin, Hormones, monoclonal antibodies** (can be produced at large scale applying cell culture in bioreactors).
6. **In large scale production of the tissue and organs.** Various tissue and organs like artificial skin and cartilage, can be obtained at large scale e.g. the Skin of whole body can be reproduced within 3 weeks from $2 \times 2\ cm^2$ skin of the body (Keratocytes obtained from

the skin are used). **Chondrocytes cells** can be used to produce cartilage for repair of knee joints. Now a days, development of the artificial organs like pancreas, kidney and and liver etc. is under study.

7. **In genetic counseling.** **Amniocentesis** is the part of genetic counseling that can be done by taking out the cells of the fetus and culturing them to check the chromosomal abnormalities. This may be helpful in checking the chromosome related disease in the children before their birth. (down syndrome)
8. **In genetic engineering.** Transformed and reprogrammed cells can be produced by introduction of foreign DNA and hence the cellular effect of new gene can be introduced. Genetically engineered cells have been used to produce large quantity of desired protein for e.g. Insect cells has been used as miniature cell factories after infection with baculovirus.
9. **In gene therapy.** Gene therapy involve two types of cells **a. Somatic cells:** Patient having defective gene for specific protein can be replaced or repaired by introduction of the missing proteins in the cells by using viral vector (vaccinia vector). **b. Germinal cells:** Embryonic stem cells are modified (showing totipotency) by introducing the gene of choice into selected cells and transferring those to a blastocyst) thus, offspring inherits specific genes to cure the genetic diseases.
10. **In cytotoxicity testing.** Screening of potential compounds as future drug can be done by culture application.
11. **In transgenic animal creation.** Transgenic animal can be produced by collecting the fertilized egg in appropriate stage and then culturing in suitable medium and finally microinjection of the desired DNA to introduce the new character. This cell can be placed inside the uterus for developing the complete animal called as transgenic animal.
12. **In bulk production of monoclonal antibodies.** **Monoclonal antibody** are produced in bulk for use in protein purification, disease diagnosis and treatment, (as they have single affinity for proteins which can be used) in treatment of cancer and for making immunotoxins. Monoclonal antibodies can be tagged with toxin specific for tumor antigens and thus Immunotoxin produced. **In Radioimmuno-diagnosis**, in radiotherapy of cancer, Isotopes have been used which have, very high energy, very low penetrating power, very short half life, and the decay products are inert. **IN111, Y Technetium** is used for tagging antibodies, which produces very low energy radiation, and have very short half life against cancerous cells.

1.9 IN-VITRO AND IN-VIVO CULTURE

There are following differences in in-vitro culture compared to invivo-culture.

1. Cell-cell interaction is reduced in vitro
2. Cell-matrix interactions reduced in vitro
3. Nutritional milieu is changed in vitro
4. Spreading is increased in vitro
5. Migration is increased in vitro
6. Proliferation is increased in vitro

Table 1 Comparison of differet cells in culture

Property	*Bacteria*	*Mold/fungi*	*Animal/plant cells*
1. Doubling time	Shortest	Intermediate	Slowest (days)
2. Viscosity	Normal/low	Often a problem	Low
3. Medium cost	Low	Low to intermediate	High
4. Shear Resistance	High	Often high	Often low
5. Cost of downstream processing	Modest	Modest	High
6. Culturing	Suspension	Suspension	Substrate or
7. Product concentration	High	Intermediate	Suspension can be very low
8. Aggregation	Nil to low	Low to intermediate	May be a problem
9. Product value	Low to intermediate	Low to high	High to very high
10. Cell density	Very high	Intermediate to high	Usually
11. Cleaned by steam	Almost never	Almost never	Very often
12. Contamination	Nil to low	Nil to low	Often
13. Genetic stability	Usually stable	Occasionally	Sometimes

BRAIN QUEST

1. Write the features of an animal cell that make it difficult for culture compared to bacteria or other microorganism.
2. Write the advantages of Animal Cell Culture.
3. Write the Limitation of Animal Cell Culture.
4. Give a brief description of application of Animal Cell Culture in health sector.
5. How a animal cell can be modified to get more product to meet the growing demand in health sector.
6. Who is the father of animal cell culture?
7. Who developed the 'Monoclonal Antibody'?
8. Who discovered 'Abzyme'?

LABORATORY DESIGN

2

In this chapter concept we will discuss regarding basic design of lab that should be aimed to provide sufficient area and complete aseptic environment along with necessary facility of equipments, light, humidity, working area, washing area, culture transfer area, and media preparation area.

2.1 CELL CULTURE LABORATORY

The tissue culture is often done in laboratories in conditions that are not ideal. Therefore, one of the most important aspects of the tissue culture is the need to design a laboratory to ensures the culture in a safe and efficient manner. There are several aspects in the designing of good tissue culture laboratories. Therefore, work is separated into different areas. 1. Main Culture Room; 2. Sterile and aseptic condition; 3. UV irradiated chamber; 4. Ozone for Ozonization; 5. Restricted entry; 6. Positive air pressure; 7. Sink for waste disposal; 8. Air conditioning for microbe's free air. (See Fig. 2.1)

2.1.1 Main Culture Room

A tissue culture room should have favorable conditions to control the temperature, humidity, air circulation, and light quality duration. These factor influence the growth and differentiation process directly during culture or indirectly by affecting their response in subsequent generations. Typically, the culture room should have temperature between 15° and 30° C (±0.5°C); however, a wider range of temperature may be required for specific experiments. The temperature should be constant throughout the culture room (i.e., no hot or cold spots). The culture room should have enough fluorescent lighting (10,000 lux). The lighting should be adjustable in terms of quantity and photoperiod duration. Both light and temperature should be programmable for a 24-hrs. The culture room should have fairly uniform forced-air ventilation, and a humidity in range of 20-98% (±3%). Many incubators, and large growth chambers, should meet these specifications. All new material should be handled as 'quarantine material' (assume all material are infected) until it has been shown to be free of contaminants such as bacteria, fungi and particularly mycoplasma.

2.1.2 Preparatory Room

There are following instruments/facility should be present in the preparatory rooms

1. *Containment Room*
2. *Washing room with division into media preparation area, sterilization area*
3. *Storage room/incubator room*
4. *Microscopic room/observation area*

2.1.3 Containment Room

Used for disposal of waste material which may be potentially hazardous.

2.1.4 Washing Room

Deep sinks with lead lining (resistance to acid) helps in good washing while oven helps in drying the culture materials after washing. Beside this washing area must contain distillation plant for distilled water. Other area include Media preparation area which should have ample storage space for the chemicals, culture vessels, glassware essential for media preparation and dispensing. Bench space should be equipped with hot plate/stirrer, pH meter, balances, water baths and media dispensing equipment. Other equipments include distillation plant, bunsen burner with gas source, refrigerators and freezers for storing stock solution, chemicals, a microwave, a oven and a autoclave for sterilizing media glassware and instruments.

In preparing culture media, analytical grade chemicals should be used. The water used for preparing media should be pyrogen free since tap water contains mostly microorganism, salts and particulate matter. Water used should have an electrical coductivity less than 1.0 μ mho/em

Transfer area is meant for sterile transfer room for culture. Therefore this room should be equipped with UV light and a positive pressure ventilation unit equipeed with HEPA filter. All surface in the room should be thoroughly cleaned and disinfected. All transfer should be done inside laminar flow hood equipped with UV light, and HEPA filter.

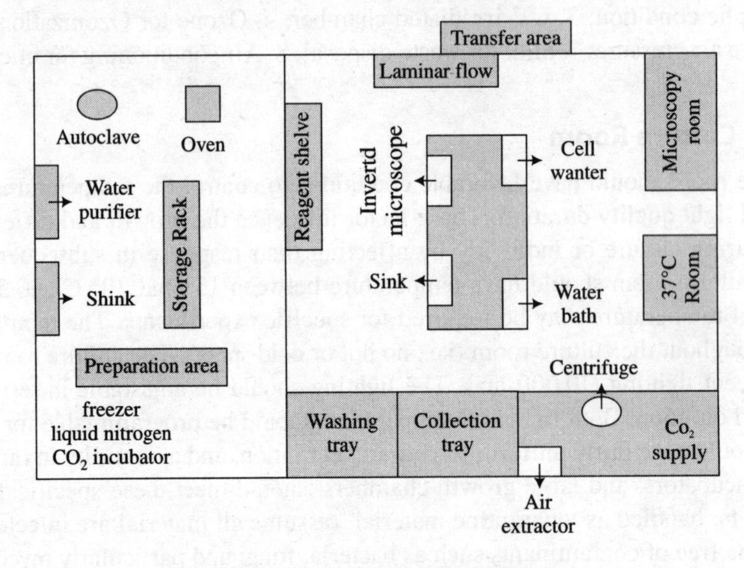

Fig. 2.1 Tissue Culture Lab design

2.1.5 Sitting Room

For sitting of personal in the laboratory

2.1.6 Storage Room

Storage room is used for storing liquid nitrogen containers, CO_2 cylinders, Piped CO_2 and air supply instruments.

2.1.7 Microscopic Room

These rooms are equipped with different type of microscope such as Inverted microscope with camera.

2.1.8 Quarantine Area

This is an specialized area reserved for handling newly received material or for handling of biopsy (dead) materials. They should have proper burning facility and disposal facility without spreading any infection. (See Fig. 2.2)

Fig. 2.2 Quarantine room

2.2 COMMON CULTURE INSTRUMENTS

2.2.1 Laminar Flow Hood

This is a type of closed chamber used commonly for creating aseptic environment during the animal cell culture. Laminar flow hood is available in two formats: (See Fig. 2.3)

a. Horizontal Flow Hood

(Air flows parallel to the ground) Horizontal hoods are designed such that the air flows directly at the operator hence they are not useful for working with hazardous organisms but are the best protection of the cultures. Both types of hoods have continuous flow of air that passes through a HEPA (high efficiency particulate) filter that removes particulates from the air.

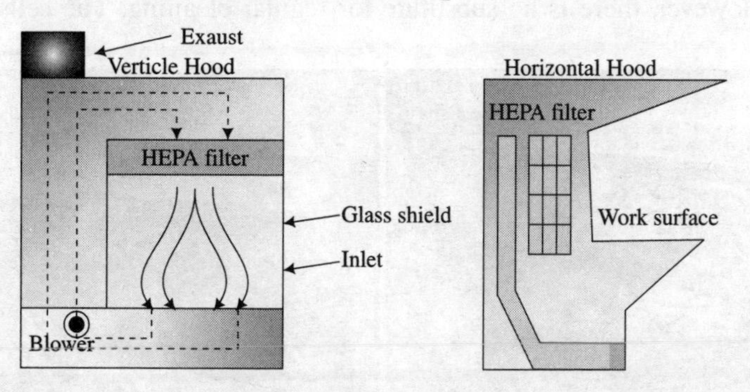

Fig. 2.3 Mechanism of air flow (a) in vertical hood, (b) laminar flow hood (horizontal)

b. Vertical Flow Hood

(Air flows vertical to the ground) It is much safer and best for the animal cell culture labs. The vertical hood is also known as a **safety cabinet**, and is best for working with hazardous organisms, since the aerosols generated in the hood are filtered out before they are released into the environment. In a vertical hood, the filtered air blows down from the top of the cabinet. The hoods are equipped with a short-range UV light that requires exposure for a few minutes (20-30 minutes) prior to start of the work in aim to sterilize the surfaces of the hood and all the material kept inside. But be cautious to never put your hands or face near the hood when the UV light is on, as the short wave light can cause skin and eye damage. The hoods should be turned on for about 10-20 minutes before being used and all surfaces should be wiped with 70% ethanol before and after each use.

2.2.2 Centrifuge

Centrifuge is used for separating the cells. Centrifuge is used routinely in the tissue culture during subculture of cell lines or before cryopreservation. Centrifuge produces aerosols, and thus it is necessary to minimize this risk. This can be achieved by using Centrifuge with sealed buckets. Ideally the centrifuge should have a clear lid so that the condition of the load can be observed without opening the lid. This will reduce the risk of being exposed to hazardous material, if a centrifuge tube breaks during centrifugation. Care should always be taken as not to overfill the tubes and to balance them carefully. These simple steps, will certainly reduce the risk of aerosols being generated. The centrifuge should be situated where it can be easily accessed for cleaning and maintenance. Centrifuges should be checked frequently for any signs of corrosion.

2.2.3 Incubator

Cell culture requires a strictly controlled environment in which cell has to grow and Incubator are better suited to the growing cells. Special incubators are used routinely to provide the correct growth conditions, such as temperature, and CO_2 levels in a controlled and stable manner (see Fig. 2.4). Generally the temperatures is set in the range 28°C (for insect cell lines) to 37°C (for mammalian cell lines) and CO_2 at the required level (e.g. 5-10%). Copper-coated incubators are also now available. These are reported to reduce the risk of microbial contamination within the incubator due to the microbial inhibitory activity of copper. The inclusion of water bath treatment fluid in the incubator water trays will reduce the risk of bacterial and fungal growth in the water trays. However, there is no substitute for regular cleaning. The cells are grown in an

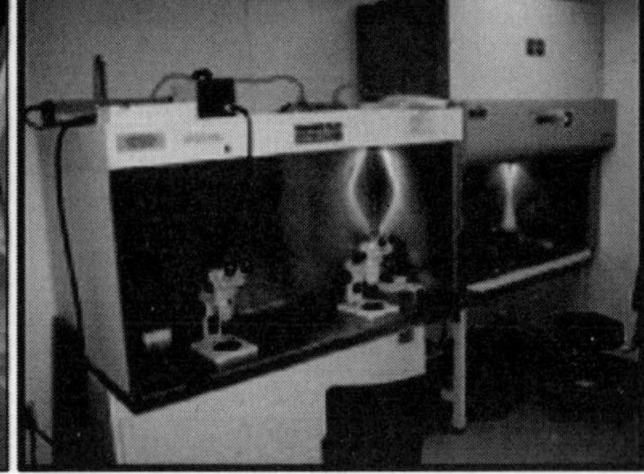

Fig. 2.4 Inside view of CO_2 incubator and Laminar flow

atmosphere of 5-10% CO_2 since water react to form/ carbonic acid (buffer). Culture flasks should have loose caps to allow sufficient gas exchange. In CO_2 incubator, required quantity of CO_2 is mixed with air (2-10% CO_2 and 90-98% air). CO_2 sensor helps in maintaining CO_2 concentrations inside the incubator. (1. Thermal conductivity sensor (Long lived); 2. Infrared sensors (more sensitive, but short lived)). Incubator are available in two formats one with *A. Air jacketed and other is B. Water jacketed. Water jacketed incubator,* hold the temperature for longer times in the events of heater failure or power cut. Heat is evenly distributed thus avoiding the formation of cold spots. The atmosphere is kept humidified by placing a pan of dH_2O containing a little Roccal in the bottom. The incubator should be checked periodically for water level and CO_2 pressure. Temperature and CO_2 levels are regulated by the incubator itself and do not require attention unless their alarms sound. The best way to prevent mold growth in the incubator is to constantly monitor the contents for contaminated cultures. In the event that a contaminated culture is found in the incubator, all of the shelves within the incubator should be removed and washed with Roccal followed by Ethanol. The interior of the incubator should be washed in the same manner. If mold growth becomes a persistent problem, the shelves and supports (sides and top) will have to be removed and autoclaved. (See Fig. 2.5)

Fig. 2.5 Outline of CO_2 incubator

Function of Incubator

Incubator helps in maintaining the constant temperature, concentrations of the gase and humidity. It has cut off temperature around 39°C. Since beyond this cells dies. Fan maintains uniform temperature circulation. Therefore, both heating and cooling device is present, which helps in keeping temperature at a constant level.

2.2.4 Microscope

Microscope helps in detecting the minute organism like bacteria, fungi. Generally compound microscope gives poor visibility of microbes. Therefore inverted microscope, are useful in visualizing the microbes, contamination and animal cell culture well. There are two types of inverted phase contrast microscopes: (See Fig. 2.6)

1. Upright Microscope

It is required for chromosome analysis, mycoplasma detection and autoradiography. It has better resolution.

2. Inverted Microscope

It is required for observing living cultures in the culture vessel. Various companies are manufacturing such microscope like Olympus CM (Japanese), Zeiss Axiovert (Germany), and Leitz Diavert (Germany). Here resolution is low than upright one. Note that microscopes should be kept covered and the lights should be turned down when not in use.

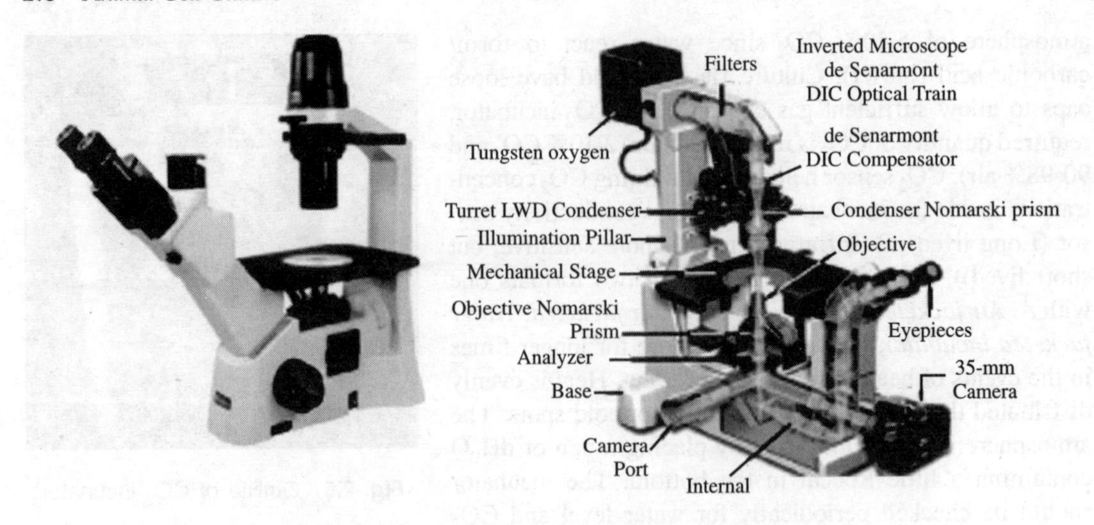

Fig. 2.6 (a) Microscope and its (b) internal parts

2.2.5 Aspirator Assembly

This is a type of suction pump which is used to add the sterile medium in the culture vessel. A peristaltic pump may be used to remove spent medium and other reagents from the culture flask and culture plates. (See Fig. 2.7)

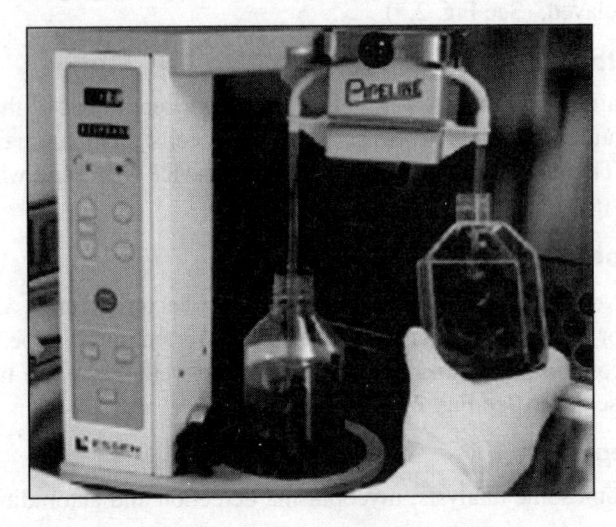

Fig. 2.7 Outline of aspirator assembly

2.3 DRY HEAT STERILIZATION

2.3.1 Oven

It is a tightly closed chamber sealed internally by thick aluminum sheet while outside portion is made up of thick metal sheet of two-three layers. The Maximum temperature that can be raised is about 700°C, at which most of the microbes are killed. The instrument is designed so as to provide controlled heat to the glass/plastic wares. It is often used in laboratory or clinic for sterilizing those substances that cant be sterilize well with moist heat, but they must be heat resistance. RNA can be killed by this procedure which may be attached with the glass tube or pipette released from fingertips. Culture media contains most of the labile substances like proteins, enzymes, hormones, vitamins, and therefore they are difficult to sterilize by this method since labile substances get coagulated at high temperature by a process called as denaturation. Buffer can be sterilized by this technique because they do not contain the any such labile organic component.

2.3.2 Bunsen Burner

It works on the principle of dry heat sterilization. There are three zones having different temperature, uppermost blue, violet in middle and pink zone at lowest. The blue region is the hottest zone and gives temperature about 1870-2000°C. This can kill all type of microorganism spontaneously. Therefore inoculation needle, borocil, or other glassware are sterilized within seconds during the culture, before plating or during the transfer of media. Around 1 cm^2 area of the flame is free of microbes and is the safest method of sterilization for bacterial culture but this can not be used for animal cell culture. This may be used in the open environment or within laminar hood. During the plating the care must be taken that the needle should not be much hot during the transfer of the Inoculums because all the culture would get die. Therefore, best method is to heat the needle in the flame, touch it for sometime in the plate containing agar in the corner. This technique will prevent the burning of the culture organism.

2.4 MOIST HEAT STERILIZATION

2.4.1 Autoclave

In function they resemble the pressure cooker. Framing of the autoclave is such that they can endure high temperature and pressure. Autoclave has a cylindrical metal chamber with an air tight door on one end, and a cylindrical rack to hold the material to be autoclaved, with a steam filled jacket around sterilizing chamber. (See Fig. 2.8). As we know water boils at 100°C at normal pressure, but on increasing pressure, temperature increases. It is observed that at 5 psi above normal pressure, the temperature increased up to 109°C and when pressure is increased up to 10 psi the temperature rises to 115°C and at 15 psi (2 atm.) it will be 121°C. Therefore' both temperature and pressure is essential for the killing of any microorganism especially any spores. Pressure also increases latent heat of the vapour and thus causes more intense killing of organism. This can be checked by the reference spore of *thermobacillus* that is killed exactly at 121°C. The sterilization of all type of germs is possible because steam condense against the object in the chamber and releases a lot of latent heat of evaporation that makes the temperature beyond 1000°C because each drop releases 530 k cal heat and due to longer holding time

Fig. 2.8 An autoclave outside view

(from 10-40 minutes) all the protein get coagulated and membrane structure burns out. At 15 lb pressure the temperature of autoclave water rises to 121°C.

2.5 WORK SURFACE AND FLOORING

In order to maintain a clean working environment the laboratory surfaces such as bench tops, walls and flooring should be smooth and easy to clean. They should also be waterproof and resistant to a variety of chemicals (such as acids, alkalis, solvents and disinfectants). The areas used for the storage of materials in liquid nitrogen, the floors should be resistant to cracking if any liquid nitrogen is spilt. In addition, the floors and walls should be continuous with a covered skirting area to make cleaning easier and reduce the potential for dust to accumulate. Windows should be sealed. Work surfaces should be positioned at a comfortable working height.

2.6 PLASTIC WARES AND CONSUMABLE

These include various petridishes, multiwell plates, microtitre plates, ELISA plate, roller bottles, and screw cap flasks such as - T-25, T-75, T-150 (number denotes surface area in cm^2). Almost every type of animal cell culture vessel, together with consumable such as tubes and pipettes, and other Plasticwares are commercially available for single use from different suppliers like Sigma-Aldrich, Nunc, Greiner, Bibby Sterilin and Corning and are sterile packed. The use of such plastic ware is more cost effective than cleaning and sterilization procedures. Plastic tissue culture flasks are usually treated to provide a hydrophilic surface to facilitate attachment of anchorage dependent cells since, anchorage dependent cells require a nontoxic, biologically inert, and optically transparent surface that will allow cells to attach and allow movement for growth. The most of the commercially available vessels are g-treated polystyrene plastics that are sterile and are disposable.

2.6.1 T-flask

A T-75 cell culture flasks are shown in Fig. 2.9. Its design helps in stacking of several flasks with scale marking areas showing volume and Plug-seal screw cap are provided to avoid any contaminations. The flasks are available in sterilized form which is sterilized usually by gamma irradiation and thus they are nonpyrogenic.

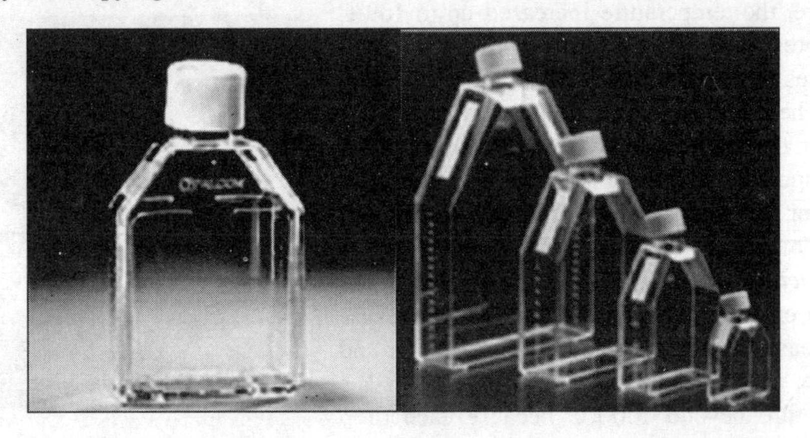

Fig. 2.9 T-75 FLASK (Photograph from lab of immunology)

2.7 CARE AND MAINTENANCE

2.7.1 Laboratory Areas

In order to maintain a clean and safe working environment tidiness and cleanliness are the key factors. Obviously all spills should be dealt with immediately. Routine cleaning should also be undertaken involving the cleaning of all work surfaces both inside and outside of the microbiological safety cabinet, the floors and all other part of equipment e.g. centrifuges. All major part of equipment should be regularly maintained and serviced by qualified engineers for example Microbiological safety cabinets should be checked six monthly to ensure that they are safe to use in terms of product and user protection. These tests confirm that the airflow is correct and that the HEPA filters are functioning properly. The temperature of an incubator should be regularly checked by any authentic agency or equivalent calibrated thermometer and the temperature should be adjusted as necessary CO_2 and O_2 levels should also be regularly checked to ensure the levels are being correctly maintained inside the incubator.

2.7.2 Work Area

The washing area should contain large sinks (lead-lined to resist acids and alkalis) distilled water, and double-distilled water. There should be space for ovens or racks, dishwasher, acid baths, pipette washers and driers, and storage cabinets should also be available in the washing area.

2.7.3 Media Preparation Area

The media preparation area should have sufficient storage space for the chemicals, culture vessels and closures, and glassware required for media preparation and dispensing. Bench space for hot

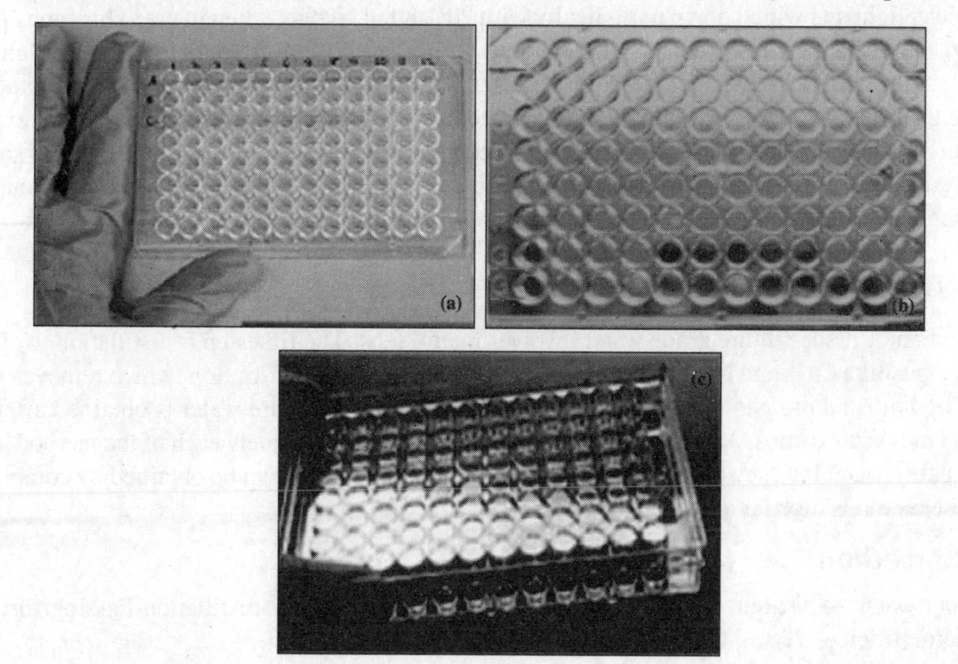

Fig. 2.10 Plastic ware and consumable: 24 well plate (a) and 96 well normal (b) and ELISA plate (c) (Photograph from Lab of Immunology)

plates/stirrers, pH meters, balances, water baths, and media-dispensing equipment should be available. Other necessary equipment may include air and vacuum sources, distilled and double-distilled water, Bunsen burners with a gas source, refrigerators and freezers for storing stock solutions and chemicals, a microwave or a convection oven, and an autoclave or domestic pressure cooker for sterilizing media, with glassware, and other instruments. In preparing culture media, analytical grade chemicals should be used and good weighing habits practiced. The water used in preparing media must be of the utmost purity and highest quality. Tap water is unsuitable because it may contain various cations (ammonium, calcium, iron, magnesium, sodium, etc.), anions (bicarbonates, chlorides, fluorides, phosphates, etc.), microorganisms (algae, fungi, bacterial), gases (oxygen, carbon dioxide, nitrogen), and particulate matter (silt, oils, organic matter, etc.) Water used for tissue culture should meet, at a minimum, the standards for type II reagent grade water, i.e., free of pyrogens, gases, and organic matter.

2.7.4 Transfer Area

All surfaces in the room should be designed and constructed in such a manner that dust and microorganisms do not accumulate and the surfaces can be thoroughly cleaned and disinfected. A room of such design is particularly useful if large numbers of cultures are being manipulated or large pieces of equipment are being utilized. Within the transfer area the most desirable arrangement is a small dust-free room equipped with an overhead ultraviolet light and a positive-pressure ventilation unit. The ventilation should be equipped with a high-efficiency particulate air (HEPA) filter. A 0.3 mm HEPA filter of 99.97-99.99% efficiency works well. Another type of transfer area is a ***laminar flow hood***. The simplest type of transfer area suitable for tissue culture work is an enclosed plastic box commonly called a glove box. This type of culture hood is sterilized by an ultraviolet light and wiped down periodically with 70% ethyl alcohol when in use. This type of unit is used when relatively few transfers are required. Typically, the culture room for growth of animal tissue cultures should have a temperature between 15° and 30° C, with a temperature fluctuation of less than ± 0.5 °C; however, a wider range in temperature may be required for specific experiments. The culture room should have enough fluorescent lighting (10,000 lux) the lighting should be adjustable in terms of quantity and photoperiod duration. Both light and temperature should be programmable for a 24 hrs.

2.8 TISSUE CULTURE GRADE WATER

Preparation of tissue culture grade water involves many steps. The first step is distillation in aim to remove ions like Ca^{+2} and Mg^{+2} and then filtration through Carbon filtration (which removes most of the organic and inorganic component of the media, while ultra pure water is obtained after the process of reverse osmosis) and finally by Micropore filtration.Separately each of the method gives high quality water but they are not pyrogen free. Pyrogen free water can be obtained by combining all these methods together (see Fig. 2.11) (Pyrogen = germ free)

A. Ist method

Take tap water → Deionize (Ion exchange) → Triple glass /quartz distillation Passing through submicron filter → ***Tissue Culture Grade Water.***

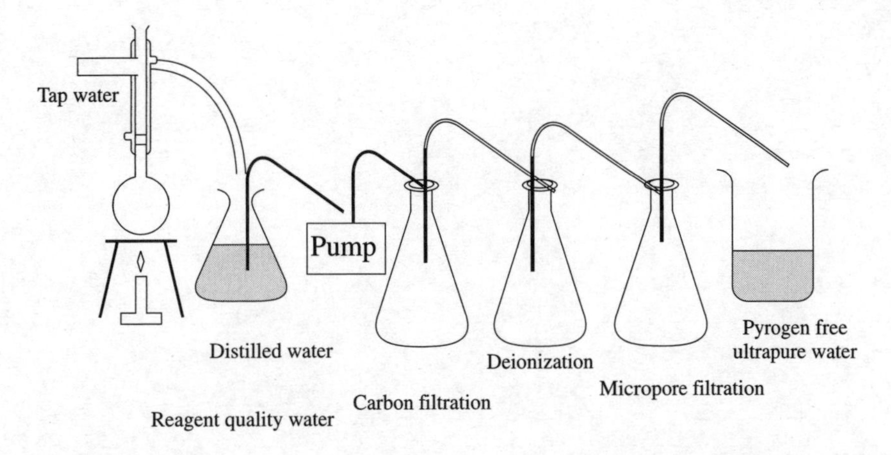

Fig. 2.11 Steps involve in cell culture grade water: 1-Distillation, 2-Carbon filtration, 3-Deionization

B. IInd method

Take tap/ treated water → Filter to remove particulate material → Passed over activated charcoal (to absorb organic containments) → Deionize (Mixed bed ion exchange resin) → Reagent grade water → Passed through submicron filter (Millipore water purification system) → ***Tissue Culture Grade Water.***

C. IIIrd method

Take pre filtered tap water → Reverse osmosis → Deionize (Mixed bed ion exchange resin) → Passing through submicron filter → ***Tissue Culture Grade Wat*er.**

BRAIN QUEST

1. Give a brief description of laboratory maintenances and its importance.
2. Give a detail description of instruments used commonly in animal cell laboratory.
3. Give a brief description of care that should be taken inside the laboratory before and after cell culture.
4. Why tap water is not suitable for animal tissue culture?
5. What are the different methods to make the culture grade water?
6. Define pyrogen free water.
7. How flask used in animal cell culture are different to that of normal cell culture. Write the name and features of different such flask used in the laboratory.
8. Write the mechanism of sterilization of Autoclave.
9. Why HEPA filter is suitable for media sterilization.
10. Write the name of substances that can't be sterilized by the autoclave.

LABORATORY PRACTICES 3

In this Chapter we will learn about different containments safeguard and international guidelines for safe use of modified microorganism so that we and our society don't have any problem in future. We will also learn how to get rid of highly dangerous microorganisms after working

3.1 CONTAINMENTS

Managing infectious agents in the laboratory environment safely, where they are being handled and maintained are referred as containment. The purpose of containment is to minimize or eliminate the risk after exposure of various agents to the lab workers, or other persons. The term "containment" is used to describe the primary and secondary barriers for managing hazardous materials in the laboratory environment. The three elements of containment include laboratory practices and techniques, safety equipment, and facility design.

Containments are the set of guidelines which was proposed in international conference held at Asilomarin February 1975 after request made by famous scientist "Paul Berg" to discuss potential hazards of infectious agents. In that conference the novel concept of Biological containment was given which recommends the use of disabled vector/host systems which can survive only under certain condition. Before this conference the National Institute of Health or in short RAC committee formulated a set of guidelines. These guidelines were issued on June 23, 1976 by the United State Secretary of Health, Educational welfare. These guidelines set up and defined categories of Physical and Biological containment procedures which were to be followed by researchers in the field according to type of vector and origins of DNA to be used. There are three main risks involved in laboratory during working

1. **Physical Risk** which includes intense cold and heat, fire and electrical shock.
2. **Chemical Risk** includes chemical exposure, their toxicity and volatility, release of aerosols; therefore there is need of appropriate training in all these aspects.
3. **Biohazards** includes risk from the culture itself. It is due to the pathogenicity of the culture, and techniques used for containment procedures. Other assay technique like Radioisotopes, (Radioactivity level, their half life, and Disposal of the waste) may present a serious risk to the health. Therefore a special Health consideration includes for Personnel health, Pregnancy, Immunodeficiency and Allergies due to use of latex coat and wear in the laboratory.

3.2 CONTAINMENT SAFEGUARD

There are two means of containment; primary and secondary level.

3.2.1 Primary Containment

Primary containment represents the procedures practices and equipments used to prevent the dissemination of research materials from the location where the are being used. Laboratory practice and technique are the most important elements of primary containment. The protection of personnel from the immediate laboratory environment after exposure to infectious agents can be provided by: (1) Good aseptic techniques (2) Use of appropriate safety equipment (3) The use of vaccines. Primary barriers are offered by 1. Biosafety Cabinets 2. Blender, 3. Centrifuge cup, 4. Variety of Enclosed Containers.

3.2.2 Secondary Containment

Secondary containment consists of the laboratory or facility where the work takes place. It includes all of the elements that prevent research materials from entering the environment. Examples of secondary barriers include floors, walls and ceilings, air locks and self-closing doors, differential pressures between spaces (positive pressure and negative pressure designs to ventilation system), exhaust filtration, as well as devices for treating contaminated air, liquids and solids. The function of these barriers is to prevent both the release of micro-organisms into the environment in the event of a failure in a primary barrier and to prevent environmental organisms from contaminating the workspace.The protection of the environment external to the laboratory from exposure to infectious materials can be provided by the 1. Facility and design of lab, 2. Operational practices (how ones do in the lab). Secondary containments are of two type one is physical and other is biological. There are four levels of Physical containment designated biosafety levels and two Biological containment HV1 (host -vector system 1) and HV2.

3.2.2.1 Physical Containment

Physical containment is achieved through the use of primary barriers (laboratory practices and containment equipment) and special barriers (laboratory and building design). The four physical containment levels ranged from standard microbiological practice (P1) to carefully designed laboratory conditions involving negative pressure, air locks autoclaves and so forth. It includes personnel training in 1. Aseptic Techniques; 2. Biology of the organism used; 3. Potential biohazard of the organism; and 4. Emergency procedures in case of accidents.

Physical Containment Levels

1. **BL1.** Biosafety Level 1 (BL 1) requires no special design features beyond those suitable for a well designed and functional laboratory. Biological safety cabinets are not required. Work may be done on an open bench top and containment is achieved through the use of practices normally employed in a basic microbiology laboratory. This level applies to the basic laboratory for the handling of Risk Group 1 agents.
2. **BL2.** Biosafety Level 2 (BL 2) is suitable for experiments involving agents of moderate potential hazard to personnel and the environment. These organisms are generally not infectious via an aerosol route. Access to the laboratory is limited when experiments are being conducted. Procedures involving large volumes or high concentrations of agents, or in which aerosols are likely to be created, are conducted in biological safety cabinets.
3. **BL3.** At Biosafety Level 3 (BL 3), facility design plays a significant role in safety. BL 3 activity involves organisms or systems which pose a significant risk or represent a

potentially serious threat to health and safety of workers. Such facilities include special engineering design features and containment equipment. These facilities are usually separated from the general traffic flow by controlled access corridors, air locks, locker rooms, or other doubledoor entries. Biosafety cabinets are required for all technical manipulations that involve viable cultures (no work is allowed on an open bench). The surfaces of all walls, floors and ceilings are sealed and, therefore, impervious to liquids that may spill onto them. This means that all penetrations (telephone, lights, plumbed lines for gas, vacuum, electrical lines, electrical switches, etc.) are sealed to prevent leaks. The collars and seals are also made of material which can be cleaned. The ventilation system in the Risk Group 3 facility is designed to exhaust more air than is supplied, resulting in a directional airflow from the outer corridors, which are regarded as clean, into the laboratory which is regarded as contaminated. The air is discharged to the outdoors and not recirculated to other parts of the building without appropriate filtration treatment.

> This laboratory design is suitable for experiments involving:
> (i) Recombinant DNA molecules requiring physical containment at the Risk Group 3 level including animal studies with BL 3 and some BL 2 agents.
> (ii) Microorganisms of moderate biohazards potential such as those in Risk Group 3 or BL3.
> (iii) Oncogenic viruses that have human cells in their host range.
> (iv) The production of large volumes or high concentrations of certain Risk Group 2 and all Risk Group microorganisms or viral infected cells (where the virus is infectious for man and requires BL 3 containment).
> (v) Production activity which involves Risk Group 3 and some Risk Group organisms.

4. **BL4.** Separate building; no windows; exhaust air decontaminated; personnel with positive pressure suits; airtight doors.

3.2.2.2 *Biological Containment*

1. **HV1.** (Host vector system I) which provides a moderate level of containment e.g. *E.coli* K12 and the vectors are non-conjugative plasmids.
2. **HV2.** (Host vector system II) which provides a high level of containment. In EK2 system with plasmid vectors no more than one in 10^8 host cells should be ab!: to perpetuate a cloned fragment in a non-laboratory environment.

3.3 LABORATORY PRACTICES

3.3.1 Containment Equipments

1. Experimental procedures involving organisms that require P4-level physical contain ment shall be conducted either in: (i) a class III cabinet system; or in (ii) class I or class II cabinets that are located in a specially designed area.
2. Laboratory animals involved in experiments requiring P4-level physical containment shall be housed either in cages contained in class III cabinets or in partial-containment caging systems (such as horsfall units, open cages placed in ventilated enclosures, or solid-wall

and bottom cages covered by filter bonnets, or solid-wall and -bottom cages placed on holding racks equipped with ultraviolet irradiation lamps and reflectors) that are located in a specially designed area.

3.3.1.1 Class I (Personnel and Environmental Protection Only)

The Class I biological safety cabinet is designed to provide personnel and environmental protection only. Class I cabinet does not protect the product from contamination because "dirty" room air constantly enters the cabinet front to flow across the work surface. Unlike conventional fume hoods, the HEPA filter in the Class I cabinet protects the environment by filtering air before it is exhausted. Personnel protection is made possible by constant movement of air into the cabinet and away from the user.

3.3.1.2 Class II (Product, Personnel And Environmental Protection)

A Class II cabinet protects product, personnel and the environment. This type of cabinet is widely used in clinical, hospital, life science, research and pharmaceutical laboratories. The Class II biological safety cabinet has following three key features: 1. A front access opening with carefully maintained inward airflow; 2. HEPA-filter, vertical, unidirectional airflow within the work area ; 3. HEPA-filtered exhaust air to the room; 4. Vertical, unidirectional airflow and a front access opening are common to most Class II cabinets. But, since Class II design permits different airflow patterns, velocities, HEPA filter position, ventilation rates and exhaust methods, therfore, a sub-classification of type is needed to differentiate Class II designs. In 2002, the National Sanitation Foundation (NSF International) restructured the Class II classification system to reflect specific performance and installation attributes.

3.3.2 Safety Cabinet

3.3.2.1 Class II, Type A

Class II, Type A hoods are used to protect personnel, product and environment as well from bioaerosols and particulates. These hoods offer personnel protection through negative pressure airflow into the cabinet. To protect the product, the work area in the cabinet is continuously bathed with ultra-clean air provided through the supply HEPA filter. Approximately 70% of the air from each cycle is recirculated through this supply HEPA filter. The remaining air (approximately 30%) is discharged from the hood through the exhaust HEPA filter, protecting the environment.

3.3.2.2 Class II, Type B2

The Class II Type B2 cabinet is equipped with a Vertical Laminar Flow Hood. It protects from diverse types of biohazards material without recirculating gases and fumes in the work area. The conventional fume hood only protects the user. The Class II, Type A (70% recirculating hood) in addition to protecting the user, also protects work in progress and the environment from all particulates. The Class II, type B2 Biological Safety Cabinet carries this protection one step further by ducting (venting) to the outside of the building (100% of the air entering the hood). This total exhaust feature makes the B2 hood suitable for not only handling particulates, but also fumes and gases. It is especially desirable for work with radionuclides, effluents from subliming particulates and noxious odors.

3.3.2.3 *Custom Class II*

The Laminar Flow (Biological Safety Cabinet) was developed to provide protection for personnel and environment from the hazards of aerosols generated by handling of low to moderate risk solids and liquids. It also protects the work in progress with ISO Class 5/Class 100 air. These same protective features which make this hood so valuable with biological and pharmaceuticals, have many potential uses in industry, e.g., for the safe handling of dust and powders. Prefilters can be added in the work area to extend the life of the HEPA filters when working with such materials. Biological Safety Cabinets can be designed and built for any sized equipment.

3.4 P-4 LEVEL LAB REQUIREMENTS

1. Laboratory doors shall be kept closed while experiments are in progress.
2. Work surfaces shall be decontaminated following the completion of the experimental activity and immediately following spills of organisms containing recombinant DNA molecules.
3. All laboratory wastes shall be steam-sterilized (autoclaved) before disposal. Other contaminated materials such as glassware, animal cages, laboratory equipment, and radioactive wastes shall be decontaminated by a method demonstrated to be effective before washing, reuse, or disposal.
4. Mechanical pippetting devices shall be used; pippetting by mouth is prohibited.
5. Eating, drinking, smoking, and storage of food are not permitted in the P4 facility.
6. Persons shall wash their hands after handling organisms containing recombinant DNA molecules and when they leave the laboratory.
7. Care shall be exercised to minimize the creation of aerosols. For example, manipulations such as inserting a hot inoculating loop or needle into a culture, flaming an inoculation loop or needle so that it splatters, and forceful ejection of fluids from pipettes or syringes shall be avoided.
8. Biological materials to be removed from the P4 facility in a viable or intact state shall be transferred to a nonbreakable sealed container, which is then removed from the P4 facility through a pass-through disinfectant dunk tank or fumigation chamber.
9. No materials, except for biological materials that are to remain in a viable or intact state, shall be removed from the P4 facility unless they have been steam-sterilized (autoclaved) or decontaminated by a means demonstrated to be effective as they pass out of the P4 facility. All wastes and other materials as well as equipment not damaged by high temperature or steam shall be steam-sterilized in the double-door autoclave.

   Other materials which may be damaged by temperature or steam shall be removed from the P4 facility through a pass-through fumigation chamber.
10. Materials within the class III cabinets shall be removed from the cabinet system only after being steam-sterilized in an attached double-door autoclave or after being contained in a nonbreakable sealed container, which is then passed through a disinfectant dunk tank or a fumigation chamber.

11. Only persons whose entry into the P4 facility is required to meet program or support needs shall be authorized to enter. Before entering, such persons shall be advised of the nature of the research being conducted and shall be instructed as to the appropriate safeguards to ensure their safety. They shall comply with instructions and all other required procedures.
12. Persons under 18 years of age shall not enter the P4 facility.
13. Street clothing shall be removed in the outer side of the clothing-change area and kept there. Complete laboratory clothing, including undergarments, head cover, shoes, and either pants and shirts or jumpsuits, shall be used by all persons who enter the P4 facility. Upon exit, personnel shall store this clothing in lockers provided for this purpose or discard it into collection hampers before entering the shower area.
14. The universal biohazard sign is required on the P4 facility access doors and on all interior doors to individual laboratory rooms where experiments are conducted. The sign shall also be posted on freezers, refrigerators, or other units used to store organisms containing recombinant DNA molecules.
15. An insect and rodent control program shall be instituted.
16. Animals and plants not related to the experiment shall not be permitted in the laboratory in which the experiment is being conducted.
17. Vacuum outlets shall be protected by filter and liquid disinfectant traps.
18. Use of the hypodermic needle and syringe shall be avoided when alternate methods are available.
19. The laboratory shall be kept neat and clean.
20. If experiments involving other organisms which require lower levels of containment are to be conducted in the P4 facility concurrently with experiments requiring P4-level containment, they shall be conducted in accordance with all P4-level laboratory practices specified in this section.

3.5 HANDLING FROZEN BIOLOGICAL MATERIAL

Storage and retrieval of frozen biological material from liquid nitrogen requires appropriate personal protection equipment. The three major risks associated with liquid nitrogen (-196°C) are frostbite, asphyxiation and exposure. Gloves thick enough to insulate but flexible enough to allow ampoule manipulation should be worn. When liquid nitrogen boils off during routine use, regular ventilation is sufficient to remove the excess, but when nitrogen is being dispensed, or a lot of material is being inserted into the freezer, extra ventilation is necessary.

BRAIN QUEST

1. What are the most important aspects that one should keep in mind during lab design?
2. Write a short description of following instruments used most often in the laboratory.
   (a) Autoclave, (b) Laminar hood, (c) Incubator
3. What should be the quality of water used in the animal tissue culture? Write one method for preparation of such tissue grade water.

4. How one should keep the work area aseptic and what measure should be taken to avoid any contamination in the area.
5. Write down the physical parameter of the laboratory necessary for maintaining animal cell culture laboratory?
6. Write down the importance of containment & also describe its type and classification.

IDENTIFICATION AND CHARACTERIZATION OF CONTAMINATIONS

4

In this Chapter, we will learn how to get rid of different type of contamination that may destroy whole culture. We will also learn technique of its identification, characterization, and its final destruction before and after the culture.

4.1 INTRODUCTION

The unwanted growth of the microorganism in the pure culture is called as contamination and specific class of microorganisms is called as contaminants. The presence of contaminant can inhibit cell growth, kill cells, and lead to inconsistent results. Contamination can occur with both cell culture and by novice and experts. Potential contamination routes are numerous for example, through poor handling, from contaminated media, reagents, and equipments (e.g., pipets), and from microorganisms present in the incubators, refrigerators, and laminar flow hoods, as well as on the skin of the worker and in cultures coming from other laboratories. Bacteria, yeasts, fungi, molds, mycoplasmas, and other cell cultures are common contaminants in animal cell culture. Contamination may occur because of the technique used during the culture, Reagents, and equipments used. Therefore, contaminants may be due to. 1. Material and reagent; or from biological sources.

4.1.1 Materials and Reagents Contamination

Contamination often occurs due to use of non-sterile materials and reagents, dirty storage conditions, inadequate sterilization procedures and poor commercial supply of reagents and media. **Glassware** contamination occurs due to dust and spores from storage, poorly sealed bottles, ineffective sterilization of materials, instruments and pipettes, and invasion by dust or mites (or fruit flies). **Equipments and facilities** may be one of the major causes of contamination often e.g. Laminar flow hoods containing perforated filter allow the growth of molds and fungi because of spores carried during air recirculation.

4.1.2 Biological Contaminations

Biological contamination may occur due to invasion of Bacteria, Yeasts, Molds, Fungi, and Mycoplasma. Therefore, monitoring for contamination is essential during cell culture. Monitoring can be done visually, checking at each handling (microscopically at 100 X), and if suspected, verify with microscope at high magnification (400 X). If contamination occurs, better to discard the media.

Table 4.1 Common source of contaminations

Organism	*Sources*
Bacteria	Clothing, Skin, Hair, Aerosols, Air currents in labs, Humidified incubators, Purified
Fungi	Fruits, Damp wood or cellulose products (cardboard), Humidified incubator, Plants
Yeast	Breads, Humidified incubator, Operators
Mycoplasma	Contaminated cell lines, Serum, Medium, Operators, Viral infection

A. Cross Contamination

Invasion of other cells in the culture is called as cross contamination.

B. Microbial Contaminations

It involves invasion of the bacteria, yeast, fungi, mycoplasma and the viruses.

4.2 MYCOPLASMA

Mycoplasma is the smallest prokaryote capable of self replication and lacks a true cell wall. They are found as pathogens in plants and animals, in normal flora of respiratory and genitourinary tracts and are causative agent of typical pneumonia. They are recognized infection that failed to respond against penicillin. They are very small and hard to detect. Their characteristics feature includes, pleomorphic shape, sensitive to lysis by osmotic shock (because of no cell wall), Immune to penicillin (no cell wall), and unique membrane structure that produces "fried egg" colonies in the culture. Its common species are *M. orale, M. hominis, M. arginine, M. hyorhinis, acholeplasma laidlawii, M. salivarum, M. pneumoniae, Ureaplasma urealyticum.*

Table 4.2 Medium used for Mycoplasma Detection

Broth Base	20 gm
Horse serum	100 ml
Arginine	1 gm
Dextrose	20 gm
Phenol Red	20 mg
Water	850 ml

4.2.1 Detection of Mycoplasma in the Culture

Mycoplasma passes through the filters and due to lack of easy, reliable detection method 15-50% of all cell lines contain mycoplasma infections. The main Effects in the culture are altered nucleic acid synthesis and RNA profiles, altered membrane antigenicity, reduced cell fusion / hybridoma formation, depletion of arginine in media (may reduce viral yield in vaccine production), sometimes (cytopathic effect) causes cell lysis, alters growth rate of cell lines.

Mycoplasma spreads rapidly in the infected cultures. Mycoplasma invasion should be tested at once in newly arrived culture prior to freezing or test when cultures are suspected to get infection. For detecting mycoplasma generally good idea is to use positive and negative controls in **LAL**

assay in 2-4 days after passage or grow cells in antibiotic free media for at least two passages. Therefore, use of at least two different tests is beneficial to detect mycoplasma.

Isolation and detection of mycoplasma have proved to be very difficult because there is no specific and sensitive detection methods exists until now. Unlike bacteria, mycoplasma needs very specific and elaborate culture media, and prolonged incubation periods. Over the years, many methods have been devised for routine testing, including DNA staining using fluorescent dyes such as **Hoechst dye, enzyme-linked immunosorbant assay (ELISA) methods**, and biochemical detection (e.g. using 6-methyl-purine deoxyriboside).

4.2.2 Methods for Detection of Mycoplasma

There are several methods that can be used for detection of microbial contamination

1. Culturing; and 2. Gram staining;

Specific method of detection of mycoplasma

3. Limulus Amoebocyte Lysate assay; 4. Dye; 5. ELISA; 6. Electron microscopic analysis and 7. PCR method

4.2.3 Mycoplasma Detection By Conventional Test Methods

Two test methods are employed to detect Mycoplasma contamination:

4.2.3.1 Detection by Colony Formation on Selective Agar Plates. Samples are inoculated into mycoplasma enrichment broth and onto selective medium agar plates and broths are subcultured on agar plates for up to two weeks. Cultures are incubated aerobically and microaerophilically. Inoculated media are examined for evidence of mycoplasma growth.

4.2.3.2 Fluorescent Staining of DNA. Samples are inoculated onto susceptible mycoplasma-free host cell cultures and incubated for 4 days. The cells are then passaged to fresh cultures and further incubated. After incubation, the cells are stained with **Hoechst stain** and are examined under a fluorescent microscope for signs of any mycoplasma infection.

4.2.4 PCR Test for Mycoplasma Detection

A more rapid detection method employed for mycoplasmas detection is the PCR technique. This has recently been an approved method. The basic principle is the amplification of the Mycoplasmal DNA from within the sample matrix using specifically designed primers. These are then detected under UV illumination in a UV illuminator.

4.2.5 Limulus Amoebocyte Lysate Assay (LAL)

Limulus Amoebocyte Lysate is obtained from horse shoe crabs (*Limulus polyphenius*).

Horse shoe crabs are found naturally in eastern coast of USA. These have one cell type in their body called amoebocyte which provide defense against bacterial infections. Whenever crabs are injured, the amoebocyte releases a clotting enzyme which causes formation of clot.

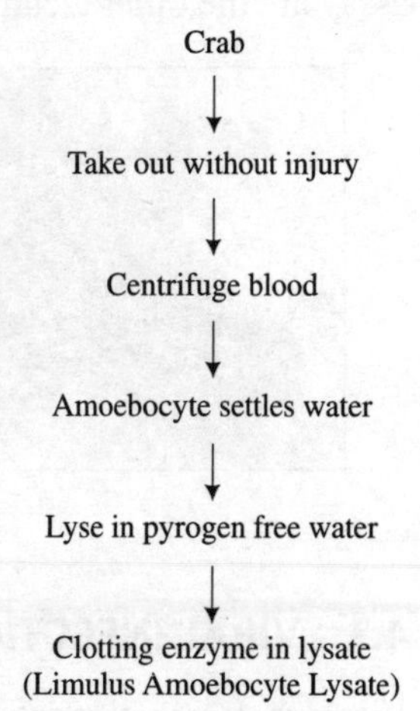
Crab
Take out without injury
Centrifuge blood
Amoebocyte settles water
Lyse in pyrogen free water
Clotting enzyme in lysate
(Limulus Amoebocyte Lysate)

Fig. 4.1 Procedure for LAL Extraction

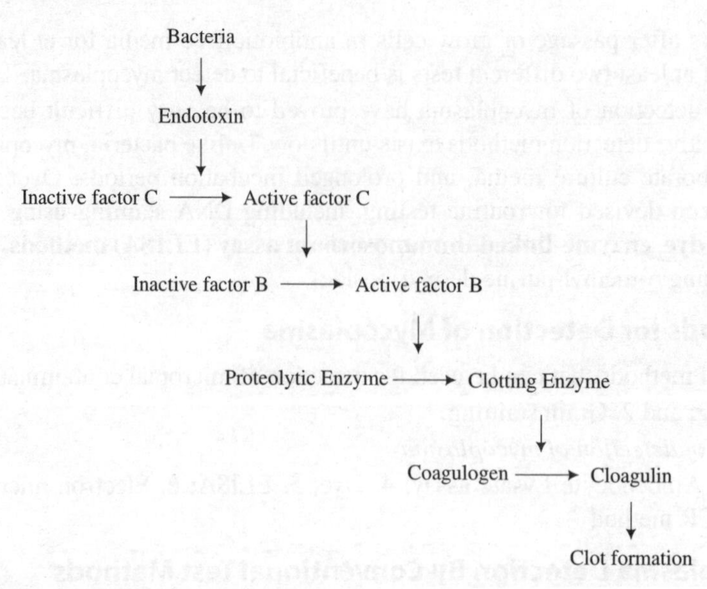

Fig. 4.2 Mechanism of clot formation by using clotting enzyme from crab

4.2.6 Dye

HOECHT company makes a fluorescent dye called 33258, which binds to DNA and emits fluorescence and the amount of fluorescence can be standardize for normal and infected culture with mycoplasma.

4.2.7 Other Techniques for Detection of Mycoplasma

DNA fluorescence, Immunofluorescence, Electron Microscopy, ELISA, DNA Probes and Mycotest assay are the other technique for detection of mycoplasma.

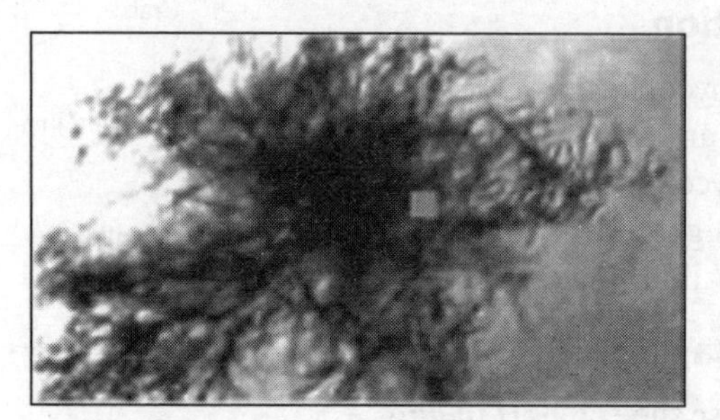

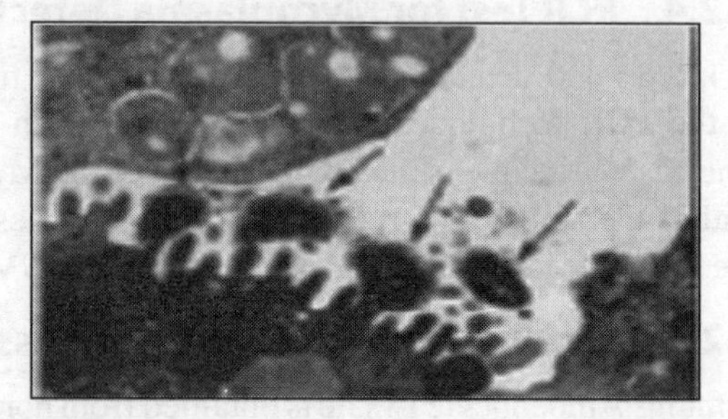

Fig. 4.3 Culture after Mycoplasma infection

4.3 VIRAL INFECTION

Viral contamination needs particular attention because infection may be without cytopathic effect for the cell culture or may be latent (e.g. herpesvirus) and hard to detect. Human and non-human primate cultures are more likely to harbour viruses that are highly pathogenic to humans. Particular concern are the blood-borne viruses such as Hepatitis B Virus (HBV), Human Immunodeficiency Virus (HIV) and others such as Hepatitis C virus (HCV) and Human T-cell lymphotropic viruses (HTLV). However, non-human cell cultures are not without risks as they may contain viruses with a broader host range able to infect humans such as rodent cell culture carrying hantavirus (Lloyd G and Jones N, 1986) or primate cells harbouring Marburg virus.

4.3.1 Detection and Removal

The cells contaminated with the virus can be checked either by a PCR methods such as DNA fingerprinting, or by classical method of Heme absorption methods. Removal of viruses can be done by membrane filtration method, or by adding other metallic compounds in the culture medium.

4.3.2 Symptom of Contamination

Observation is the best option, and symptom like pH (drops with bacterial contamination, yeast cell) medium transparency e.g. infection of fungal cell causes cloudiness in the medium, presence of other structures like cell clumping which can be visualized at 400X to 1000X, and cells may exhibit Cytopathic effect (look funny).

4.3.3 Virus Free Culture

Viral infections are virtually impossible to remove from cultures since they do not respond to antibiotic treatment. Also, since they are intra cellular parasites, it is not possible to remove them by the centrifugation or other separation techniques. If virus free stocks or a virus free alternative is not available, then a thorough risk assessment should be undertaken prior to continuing work with the infected cell line. Appearance of vacuoles, decrease in confluency, and rounding.

4.4 FUNGAL CONTAMINATIONS

Typical infections in the cultures are fungus and mold spores that are ubiquitous in the environment and generally infect culture via an air borne route. Heating and air-conditioning systems are notorious for having high Concentrations of spores. Therefore, the sea -sonal changes like fall spring usually results in an increase in this type of contamination in cultures as heating or A/C systems are mostly switched off. Also, particularly in the spring, the higher bio-burden in the air from pollen particles can carry fungal spores into air handling systems and into labs on the clothes of lab personnel. Fungi mostly grow in the medium unattached to the cells or growth vessel, but can become attached to either. The spores that give rise to the mycelia formation are often hard to detect in cultures. In early stages of contamination, they do not typically cause pH change in the medium nor do they have significant toxic effects on animal cells. Cell cultures can often be cured of fungus contamination when detected early by treatment with certain antibiotics (actually antimycotics).

4.4.1 Antifungal and Antimycotic

The two most common antimycotic agents used in the cell culture that is effective against fungi or molds are **Amphotericin B** (Fungizone) and **mycostatin** (Nystatin). Routinely used antibiotics such as **penicillin/streptomycin**, but **gentamicin**, and **kanamycin** are not effective against fungi or molds. Fungizone can be used in media at final working concentrations between 0.25 micro. g/ml and 2.5 micro g/ml. The fungizone is typically very toxic in cell culture systems and should be used conservatively. Nystatin can be used at final working concentrations between 100 U/ml and 250 U/ml. Nystatin is a colloidal suspension rather than a solution and should be mixed thoroughly before it is added to the cell culture media. When nystatin is in the medium and viewed under a microscope, it will appear as small crystal-like particles. The decision to use the antibiotics to prevent contamination should be based on the individual researcher's needs and experience.

4.5 ANTIBIOTICS

One hugely underestimated problem in tissue culture is the routine use of the antibiotics. Continuous use of the antibiotics leads to the development of resistant strains and may be toxic to the cell cultures. In addition the use of antibiotics and antimycotics at high concentrations can be toxic to some cell lines; therefore, it is necessary to perform a dose response test to determine the level at which an antibiotic or antimycotic becomes toxic. This is particularly important before using an antimycotic such as fungizone or an antibiotic such as tylosin. Therefore,

1. Culture the cells for two to three passages using the antibiotic at a concentration one to two fold lower than the toxic concentration and see the growth and diameter of the cells.
2. Culture the cells for one passaging in antibiotic-free media. or
3. Culture the cells in antibiotic-free medium for 4 to 6 passages to determine, if the contamination has been eliminated.

4.5.1 MIC Determination Methods

The MIC is the lowest or minimum concentration of the antibiotic that results in inhibition of visible growth (i.e. colonies on a plate or turbidity in broth culture) under standard conditions. The MBC is the lowest concentration of the antibiotic that kills 99.9% of the original inoculum in a given time. MICs can be determined by agar or broth dilution techniques by following the reference standards established by various authorities such as the Clinical and Laboratory Standards Institute (CLSI, USA), British Society for Antimicrobial Chemotherapy (BSAC, UK), AFFSAPS (France), Deutsches Institute für Normung e.V. (DIN, Germany) and ISC/WHO.

The basic quantitative measures for in vitro activity of antibiotics are the minimum inhibitory concentration (MIC) and the minimum bactericidal concentration (MBC). The disc diffusion method employed for antibiotic susceptibility testing is the Kirby-Bauer method. It is relatively inexpensive, versatile, and easy to perform. In addition, it requires media, reagents, equipment and supplies that are readily accessible to most clinical laboratories.

4.5.1.1 *Kirby-Bauer Method*

A suspension of the isolate is prepared and, then is spread evenly onto an appropriate agar (such as Muller-Hinton or for a more defined media Iso-Sensitest™ agar) in a petri dish, disks impregnated with various defined concentrations of antibiotics are placed onto the surface of the agar. A multichannel disk dispenser can speed up placement of the disks. After the incubation, a clear circular zone of no growth in the immediate vicinity of a disk indicates susceptibility to that antimicrobial. Using reference tables the size of zone can be related to the MIC and results recorded as whether the organism is susceptible (S), intermediately susceptible (I), or resistant (R) to that antibiotic. A variation on this approach is to use a strip impregnated along its length, with a gradient of different concentrations of antimicrobial, after incubation this creates an ellipse shaped zone of no growth, where the ellipse meets the strip, the MIC can read from the concentration markings on the strip. These are easy to read, no tables need to be referenced to get an MIC value and the test requires less manipulations, as one strip will cover the whole concentration range. These again can be manually or instrument read. If the zone of inhibition is equal to or greater than the standard, the organism is considered to be sensitive to the antibiotic. If the zone of inhibition is less than the standard, the organism is considered to be resistant.

Clinically effective antimicrobial agents exhibit selective toxicity toward the bacterium rather than the host. It is this characteristic that distinguishes antibiotics from disinfectants. The basis for selectivity will vary depending on the particular antibiotic. When selectivity is high the antibiotics are normally not toxic. However, even highly selective antibiotics can have side effects.

The therapeutic index is defined as the ratio of the dose toxic to the host to the effective therapeutic dose. The higher the therapeutic index the better the antibiotic. The Antibiotics are categorized as bactericidal if they kill the susceptible bacteria or bacteriostatic if they reversibly inhibit the growth of the bacteria.

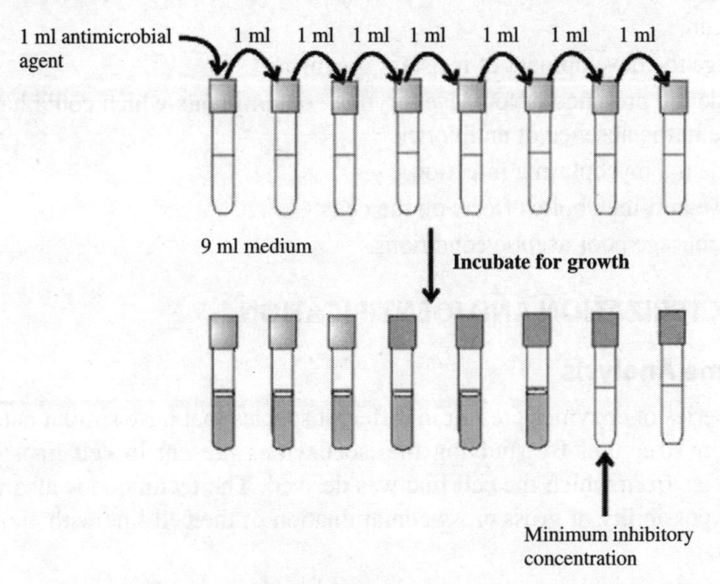

Fig. 4.4 Method of MIC determination

4.5.1.2 The Broth Dilution Method

The broth dilution method depends upon the inoculation at a specific density in broth (in tubes or microtitre plates) containing antibiotics at varying levels - usually doubling dilutions are used and after incubation, turbidity is recorded either visually or with an automated reader, and the breakpoint concentration established. Microtitre plates or ready-to-use strips are commercially available with antibiotics ready prepared in the wells. A variation on this approach is the agar dilution method where a small volume of suspension is inoculated onto agar containing a particular concentration of antibiotic, when the inoculum has dried the plate is incubated and again examined for zones of growth.

4.5.2 Criteria for Using Antibiotics

1. They must eliminate the microbial contaminants entry.
2. They must not affect the growth and metabolism of cells and tissue culture.
3. It must provide protection for the entire period of experiments.
4. It should not affect the ultimate use intended for the cell culture.
5. It should be non toxic in state for handling by laboratory person.
6. They should be compatible with other component of culture medium.
7. Antibiotic should be inexpensive.
8. It should be broad spectrum.

4.5.3 Disadvantage of Using Antibiotics

Their wide use can:

1. Encourage the development of resistant organism.
2. They hide the presence of low level cryptic contaminants which could have become fully operative in the absence of antibiotics.
3. They hide the mycoplasma infection.
4. They have anti-metabolic effects on the cells.
5. They encourage poor aseptic conditions.

4.6 CHARACTERIZATION AND IDENTIFICATION

4.6.1 Isozyme Analysis

Isozymes are a series of enzymes present in different species that have similar catalytic properties but differ in their structure. By studying the isoenzymes present in cell lines it is possible to identify the species from which the cell line was derived. The technique is also used as a means of excluding the possibility of gross cross-contamination of the cell line with another culture of a different species.

The principles upon which isoenzyme analysis is based are:

1. Each isoenzyme has multiple gene loci coding for different polypeptides with identical enzyme activity (e.g. lactate dehydrogenase, LD has 5 loci)
2. Electrophoretic migration rates change dependent on subunit composition e.g. LD has five possible iso-forms (LD 1-5)
3. Different species have different combinations of these iso-forms: Using a typical panel of 4 isoenzymes a composite picture is built up enabling the species of origin to be determined by the use of reference tables.

4.6.2 DNA Fingerprinting

DNA fingerprinting is a unique band appear in the gel after cutting and loading of DNA of different species due to presence of unique restriction sites that enables the following features: 1. Identification of individual cell lines from the same species; 2. Confirmation of the identity of cell banks compared to reference master stocks; 3. Detection of cross-contamination.

4.6.3 Multi Locus DNA Fingerprinting

Multi locus DNA fingerprinting and multiplex - PCR DNA profiling are the methods used routinely as part of many routine cell banking procedures. For multi locus DNA fingerprinting multi locus Jeffrey's probes can be used along with Southern blotting technology and produces a complex banding pattern. This probes cross-hybridize with most common species and other disadvantage is that the profiles require visual interpretation and comparison with other samples.

4.6.4 Multiplex - PCR (STR) DNA Profiling

It uses a set of multiple primers for recognizing micro-satellites a small repetative DNA sequence specifically present in each species, using PCR and automated DNA sequencing techniques. Primers are species specific and are used only for human cell lines that Produces a color-coded banding pattern that translates into a digital code that can easily be stored on a database and compared to other stored profiles.

4.7 PREVENTION OF CONTAMINATIONS

4.7.1 By Using Inoculums from Recognized Source

To minimize the risk, it is advisable to obtain cells from a recognized source such as a culture collection that have confirmed identity of the cells as part of the banking process. Tests used to authenticate the cell cultures include isoenzyme analysis, karyotyping/ cytogenetic analysis and more recently molecular techniques of DNA profiling. While most of the techniques above are generalized tests and are applicable to all the cell lines but additional specific tests may also be required to confirm the presence of specific contaminations.

4.7.2 By Keeping Materials and Environment Aseptic

During the cell culture there is common problem of maintaining an aseptic condition. Cells in vitro get easily infected by the bacteria, viruses, mycoplasma, yeasts, and the fungal spores. Infection or contamination may occur at the small or large scale. A small infection in the culture or culture vessel may contaminate the whole experiment. The contamination in the cell culture can be seen by the microscopic examination. We can see the infection threads if the bacterial infection has been occurred. In case of mycoplasma it is difficult to observe the infection under the microscope. To prevent contamination, manipulations in the technique can be used such as using non-sterile surfaces, checking any spillage, and touching of pipettes to objects, splashing dust settling into containers, Operator contaminants like dust from skin, hair, clothing and aerosols from talking or coughing.

Table 4.1 Recommended antibiotic concentration (the concentrations given are for culture media containing serum; while serum-free media generally require lower concentrations of antibiotics)

Antibiotic	*Recommended concentration*	*Antibiotic spectrum*	*Stability in Media at 37°C; days*
Anti-PPLO Agent	10 – 100 µg/ml	Mycoplasma and gram positive bacteria	3
Kanamycin Sulfate	100 µg/ml	Gram positive and gram negative bacteria and mycoplasma	5
Polymixin B Sulfate	100 U/ml	Gram negative bacteria	5
Tylosin FUNGIZONE® (Amphotericin B)	0.25 – 2.5 µg/ml	Fungi and yeasts	3
Gentamicin Sulfate 5 days and mycoplasma	5 – 50 µg/ml	Gram positive and gram negative bacteria	
Neomycin Sulfate	50 µg/ml	Gram positive and gram negative bacteria	5
Nystatin	100 U/ml	Fungi and yeasts	3
Penicillin G	50 – 100 U/ml	Gram positive bacteria	3
Streptomycin Sulfate	50 – 100 µg/ml	Gram positive and gram negative bacteria	3

BRAIN QUEST

1. Classify the type of contaminations that usually occur during the culture.
2. How will you prevent the mycoplasma infection?
3. What are the different methods for identifying mycoplasma infections in culture?
4. What precaution will you take for media so that less chance of contamination may occur?
5. What are the modern techniques for identifying the contaminations in culture?
6. How will you determine the MIC for any antibiotics?
7. Why antibiotics are not a method of preventing complete contamination from the culture?
8. Write name of some compound that is antifungal?

STERILIZATION TECHNIQUES

5

In this Chapter we will learn about different techniques of sterilization involved in the animal cell culture

5.1 STERILIZATION

Over the years, heat has proved to be the most popular method of sterilization. Therefore, it is the technique by which unwanted microorganism are killed. It is the most economical, safe and reliable method. Heat is believed to kill microorganisms by denaturation and coagulation of their vital protein systems. Oxidation and other chemical reactions are also greatly accelerated as the temperature is increased (roughly doubling for every rise of 10°C). There are principally two methods of thermal sterilization: moist heat and dry heat.

There are other methods that are used to minimize the microorganism load quickly. These are disinfection and pasteurization.

5.1.1 Disinfection

A simple boiling water bath can quickly decontaminate items in the clinic and at home and the method is called as Disinfection not the sterilization. For disinfection, if a substance is exposed to 100°C for about 30 minutes and then sudden cooling. This method is called as pasteurization. All non-spore forming pathogen may get inactivated including *bacillus tubercle* and *staphylococci* by this technique.

5.1.2 Pasteurization

It is the technique of maintaining the fix number of organism by first heating at 90°C and then sudden cooling for few minutes. This method is often employed before the packing of milk which increases its life.

5.2 STERILIZATION METHODS

5.2.1 Dry Heat Sterilization

Dry heat sterilization can be done by using Bunsen burner or spirit lamp because of high temperature range (from 160°C to several thousand degrees). Dry heat has several effects on microbes; it can oxidize cells and at extremely high temperature reduces them to ashes. It may dehydrate cell components, there by concentrating the protoplasm. In mean time, it also denatures proteins and DNA, but because proteins are more stable in dry heat, higher temperatures are required to inactivate them.

5.2.2 Chemical Sterilizations

The first chemical used for the control of microorganism was chloride of lime and the iodine solution. These chemical were used to control the infection of the wound of the military people and for hand washing. At present nearly 10,000 antimicrobial chemical agents have been manufactured. The lists of such chemicals are given in Table 5.1. Antimicrobial chemical occurs in the liquid or in the gaseous or even solid-state form. They vary from disinfectants and antiseptics to sterilants and preservative. Liquid include mostly alcohol, water or mixture of two into which various solutes are dissolved. Solution containing pure water is called as aqueous; where those with pure alcohol or alcohol water mixture called as tinctures.

Table 5.1 List of chemicals used in sterilization

1. Chlorine (sporicidal)	2. Iodine
3. Phenolic	4. Alcohol
5. H_2O_2 hydrogen peroxide	6. Soaps
7. Mercurial	8. Silver nitrate
9. Glutaraldehyde	10. Formaldehyde
11. Ethylene oxide	12. Chlorhexidine

Principle of Chemical Sterilization

These chemical sterilant react with the membrane and penetrate the membrane of the microbes. The ratio of dilution is very important factor in dissolving and penetrating the membrane. Most halogens exert antimicrobial effect due to their nonionic state e.g. Chlorine and Iodine are most routinely used for the germicidal purpose. In the market most of the ingredients of antiseptic have halides. The major forms of the halide as chlorine used are the hypochlorite (OCl) and chloramines (NH_2Cl). They denature the enzymes that mostly interfere with the disulfide bridge. The resulting denaturation is permanent and suspends metabolic reactions. Chlorine kills not only bacterial cells and spore but also the fungi and viruses, but note that chlorine killings are ineffective at the alkaline pH. Excessive organic matter can reduce their activity; they are relatively unstable if exposed to light.

5.2.2.1 Hypochlorite

Hypochlorite is a good general purpose disinfectant and is also active against viruses. They are corrosive against metals and therefore should not be used on metal surfaces e.g. in centrifuges. They have one disadvantage that they are readily inactivated by organic matter and therefore should be made fresh daily and should be used at 1000 ppm for general use surface disinfection, 2500 ppm for washing pipettes, and 10,000 ppm for tissue culture waste and spillage.

5.2.2.2 Iodine and Phenols

Iodine is a pungent black chemical that forms brown coloured solution when dissolved in water or alcohol. Tincture Iodine is 2% solution of iodine in sodium iodide and diluted alcohol. Phenolic chemicals are called as carbolic acid formed from the distillation of the coal tar. This was first adopted by JOSHEPH LISTER as a surgical germicide.

5.3 IONIZING RADIATION

This method of sterilizing culture by the exposure of radiations called as cold sterilization. Device that emit ionizing radiation include gamma ray (cobalt 60) is X-ray. Gamma rays are most penetrating while X-ray intermediate and cathode-ray is least penetrating. They affect mostly DNA by breaking the bonds in between strand (H-bond by the gamma and X-rays). Most of the vegetative cells like bacteria, protozoa, worms, and insects are most sensitive to radiation and thus are killed completely.

5.4 NON-IONIZING RADIATION

Other than ionising radiations UV rays are non-ionising radiation. Ultraviolet rays are released naturally from the sun-rays, and special fluorescent tubes (UV lamp). Their wavelength range exits from 100 to 400 nm. It is most lethal between 240-280 nm. UV rays are not as penetrating as ionizing radiations. They are absorbed by the DNA and this brings replication of DNA to halt. Specific damage occurs to the pyrimidine bases because they form pyrimidine dimers due to which DNA polymerase cannot move further.

5.5 STERILIZATION BY FILTRATION

Sterilization of culture medium is done by filtration using porous filters having ranges of pore size from 0.45 to 0.025 μm. The advantage of filtration is that all microorganisms including virus can be filtered off without degrading medium component. Membrane filters are made up of porous sheet of cellulose acetate, nylon or Polyvinyl chloride. They have thickness of 150 μm and have low liquid retention capacity. Culture medium is filtered by porous filters having ranges of pore size from 0.45 to 0.025 μm. Other than this membrane filters are inert, non pyrogenic, non toxic and 100% efficient in filtering microorganism.

Their pore size is carefully standardized. Largest size is 8 micrometer, and smallest size is 0.02 micrometer. The largest size of virus is 300 nm and smallest bacteria have size 10 nm. The filter that commonly employed is of size 0.02 micrometer. Using 0.1 mm filter can also check the mycoplasma infection.

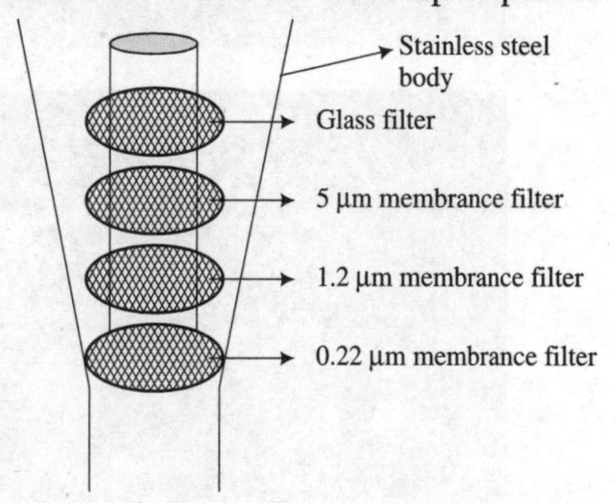

Fig. 5.1 Stacking filter assembled in glass fiber prefilter

5.5.1 Method of Filter Sterilization

Generally media is sterilized by filteration using nominal 0.2 micron sterilizing grade membrane filters. For media containing animal sera, a final sterilizing grade 0.1-micron-rated filter may be used for added protection against mycoplasma. Since natural media is heat labile due to presence of various protein and organic compounds and get denatured when they come in contact with high heat and pressure. Therefore, best option is to sterilize them using filter.

5.5.2 Filter Selection

Selection of filter size is a very important criterion in sterilization of media component. There are various filter available having different pore size. Since the aim of filtration is to remove pathogens or contaminant of every size. This is not possible by passing the media through single filter of smallest pore size. Therefore,, more recent practice is to pass the media slowly through a stacking filter having three to four type of filter in decreasing order (Fig. 5.1). **Stacking filter** can be purchased or alternatively media can be passed separately through filter of decreasing pore size from 5 micrometer to smallest pore size 0.22 micrometer. Gradually this is necessary because of high viscosity of serum media and due to presence of high particulate content. Note that before and after filtration filter should be washed with sterilized distilled water or with autoclaved buffer or with BSS. It prevents unwanted contamination during the filtration of media.

Fig. 5.2 Different type of filter apparatus used in the laboratory for filtration of heat labile media

One of the major problems arises even after the filtration is of mycoplasma infection. Mycoplasma can pass even through 0.22, micrometer, pore size filter and can destroy whole serum. They are difficult to detect. One most employed method is to filter the media again with the 0.1 micrometer filter which may prevent mycoplasma during the filtration.

Sterilizing cell culture media expeditiously is critical in controlling bio-burden and endotoxin levels. Therefore, filter selected must be able to provide high flow rates at typical operating pressures and consistently high throughput. The combination of flow rate (flux) and throughput (volume processed per unit area) determines the filter area required for a given batch process time. Potential interactions between a filter and culture media should be assessed carefully to ensure no inhibition of cell growth or protein expression.

5.6 HEAVY METALS AS MICROBICIDAL AGENTS

Some heavy metallic elements like Hg, Ag, Au Cu, As, and Zn have been applied in microbial control since these metals are toxic at higher concentration and breaks proteins SH bonds. The mercury and silver are common germicidal. Some metal are required in very small amount [in Pico mole part per million (ppm)]. Metal exerts microbicidal effects by forming complex with several functional group including sulfhydryls, hydroxyls, amines, and phosphate that results in fast inactivation of the of the protein. This is why they are used against the bacteria, fungus, algae, protozoa, and viruses. They have several drawbacks-the metals are very toxic, they commonly causes allergic reactions, and microbe develops resistance against the bacteria.

Table 5.2 Biological test organism and methods of validating sterilization processes

Process	*Physical methosd*	*Chemical methods*	*Biological test organism*
Dry heat	Temperature recording charts	Colour change indicator	*B. subtilis niger*
Moist heat	Temperature recording charts	Colour change indicator	*B. stearothemophilus*

Step 1 Assembly of filter sterilization & placing of filter over filtration apparatus with a sterile forceps

Step 2 Filtration of media and washing with buffer, after applying vacuum to filtration apparatus

Step 3 Removal of filter with sterile forceps after sterilization

Fig. 5.3 Detail view of method of membrane filtration for media

Table 5.3 Lab-wares and their sterilization method

Autoclaving	*Hot air oven*	*Ethylene oxide*	*Gutturaldehyde*	*Filtration*
Animal cages	Glass ware	Fabric	3 in 1 syringe tips	Antibiotics
Sugar tubes	Beakers	Bedding	Cystoscopes	Sera
Lab. coats	Flasks	Blanket	Endoscopes	Vaccines
Cotton	Petri dishes	Clothing		
Filters	Pipette	Mattresses		
Instruments	Slides	Pillows		
Culture media	Syringes	**DISPOSABLES**		
RUBBER	Test tubes	Blades		
gloves	Glycerin	Knives		
stopper	Needles	Scalpels		
tubing	Oils	Scissors		
GLASS		**PAPER**		
slides		cups		
syringe and needles		plates		
test tubes		Plastics		
Enamel metal trays		flasks		
Wire baskets		petridish		
WOOD		tubes		
tongue depressors		Rubber tubings		
applicator		catheters		
		drains		

BRAIN QUEST

1. Explain mechanism of UV sterilization.
2. Explain the main cases of failure of animal cell culture.
3. Explain the mechanism of working of HEPA filter.
4. Explain the process of medium filtration.
5. Describe the basic steps to be followed during animal cell culture.
6. How you can detect the mycoplasma culture after infection?
7. Write the name of different chemical used for the sterilization.

8. Why ethyl alcohol is better to clean the surface of safety cabinet at 70% & not 90%?
9. How antibacterial checks the bacterial growth?
10. Write the composition of tincture.
11. The first antibacterial discovered was?
12. Why heavy metals are toxic at higher concentration for the cells and protein?
13. Name the chemical most commonly used as the germicidal? Why halogens are best antiseptic?

CULTURE ENVIRONMENT

6

Here we will learn about the effect of culture environment on animal cell culture. Substrate is the most important factors after the media since it helps not only in attachment and growth but also in scale up of the culture of cells.

6.1 CULTURE ENVIRONMENT

An animal cell culture is often induced by various factors like substrate (to which any cell attaches and proliferate) and its surrounding physiological conditions (like gas constituents and incubation temperature maintained during the culture). Choice of culture-ware decides cell yield which is proportional to the available surface. The most common culture environment is: 1. Substrate; 2. Gas phase; 3. Media supplements; and 4. Conditioned medium e.g. feeder layer.

6.2 SUBSTRATE

Substrate provides the surface on which culture has to grow. Substrate is a very important criterion for the successful animal cell culture since substrate provides signals for normal cell attachment and for further growth and development.

The substrate may be solid (as in case of adherent cells) or semisolid (agar or collagen) or may be liquid (often called as suspension where transformed cells grow well). Most of the vertebrate cells covers the whole surface, forming monolayer of cells. Only few normal cells does not make any monolayer like hematopoietic cells. Monolayer culture attains confluency when they covers the whole solid surface. The cell that needs substrate to grow is called as **anchorage dependent**, while cell lines and the transformed cells are unable grow on solid substrate, so they are anchorage independent. This is the unique basis of identification of the normal cell with the tumor cells.

6.2.1 Glass

Glass as a substrate has several benefits. Glass can be easily washed, can be autoclaved, and can be reused. Also glass is transparent carrying correct charge for cell growth and therefore culture can be observed without opening the lid, and without infecting the culture. If substrate is layered by albumin, fibronectin and other glycoprotein in the media it improves the cell attachment. Washing with the magnesium acetate buffer also makes the substrate more favorable for the animal cell culture.

6.2.2 Plastics

Plastic substrate is a cheaper option since they can be singly used, but the disadvantages is they cannot be autoclaved and hydrophobic. Now a day Teflon made petri-dish are available which can be autoclaved and reused. Plastic petri dishe is often covered with collagen and poly-L-lysine to develop the correct charge on substrate surface before culturing. Better monolayer can be obtained on commercially available plastic culture-ware such as Roux bottle, roller bottle, or hollow fiber cartridge (Fig. 6.1 and 6.2). Gamma-irradiation or some chemical treatment can produce correct charge over the surface. Teflon plastics have the correct charge. It has been reported that addition of collagen improves attachment and cell proliferation uncharged PTFE-Teflon is used for growth for macrophages and some transformed all line, while charged Teflon is used for monolayer cells and organotypic culture.

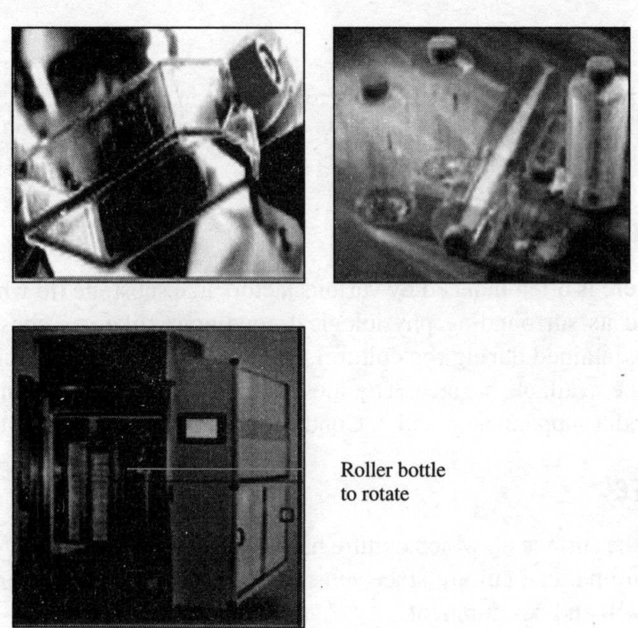

Fig. 6.1 (a) Roux bottle (b) Roller bottle culture (c) Roller bottle attached with motor to remote

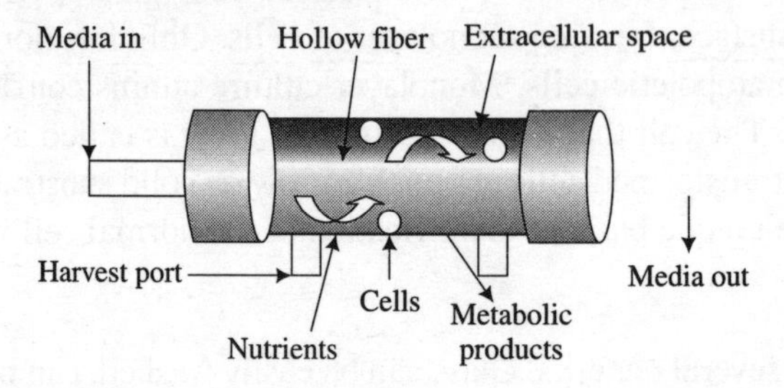

Fig. 6.2 Hollow fibre catridge for large scale culture of cells

6.3 MECHANISM OF CELL ADHESION

Growth of normal cell is substrate dependent but the proliferation of normal cells requires adhesion molecules. Adhesion of cells in culture is a multistep process which involves adsorption of the attachment factors (Cold insoluble globulin CIG) or with the help of attachment glycoprotein (that increases contact between the cell and the substrate surface and which enhances the attachment of the cells with the help of multivalent heparan sulfate synthesized by the cell itself) Fig. 6.3. Note that for propose differentiation, high density of cells are required beside this cell to cell interaction, cell to matrix interaction and differentiation factors required. A high cell density inhibits proliferation this monolayer show contact inhibition. Different cell secretes different extracellular matrix like fibronectin, laminin collagen, proteoglycan (more details are given in Chapter 18). This is the reason, we coat collagen or matrigel on solid substrate for normal cells.

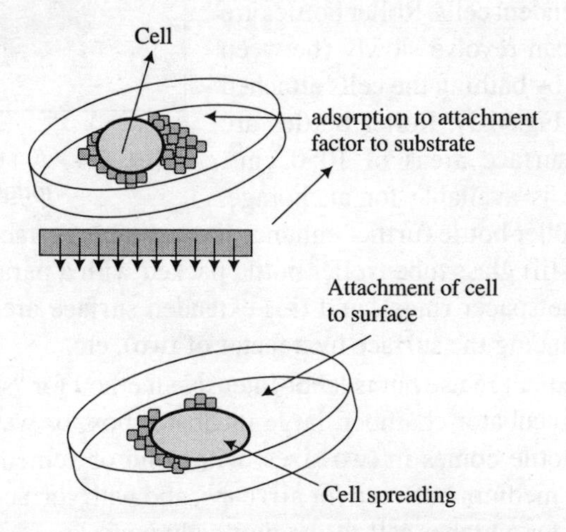

Fig. 6.3 Behaviour of cell on substrate. Cell grows and forms monolayer after attaching to substrate

6.4 SCALE UP OF ANIMAL CELL CULTURE

Scaling up is the technical term used often in Biochemical engineering for increasing the yield of the product. There are various commercially available vessel that has enhanced surface area to grow large number of cells. Most tissue culture is performed on a small scale where relatively small numbers of cells are required for experiments but for growing large number of cells usually requires special designed wares like T flasks (ranging from 25 cm^2 to 175 cm^2) and Roller bottle culture. There are different strategies adapted for scaling up of normal cell and suspension cells. At small scale shake flask, T-flask and roller bottle are used while at large scale stirred tank column reactors, membrane reactor and other hybrid designs are used. For suspension culture which require low density and high volume are grown in airlift reactor, hollow fiber reactor, perfusion reactor in which micro-carrier are used (e.g. hybridoma, CHO culture). First product produced was tPA (tissue plasminogen activator in CHO cell lines. Sometime spinner flask is also used for cell line culture at large scale. Now a days wave bioreactor is used for many suspension culture which moves and produce waves at rocker plateform mixing and airsupply in controlled environment is main feature of such bioreactor.

6.4.1 T-Flask

A standard T-flask has been shown in Fig. 6.4. Typical cell yields in a T-175 flask ranges from 1×10^7 for an attached line to 1×10^8 for a suspension line, however exact yields may vary, depending on the type of cell line. It is not practicable to produce much larger quantities of cells using standard T flasks, due to time compulsory for repeated passaging of the cells, and due to demand of space in incubator and the cost of equipment.

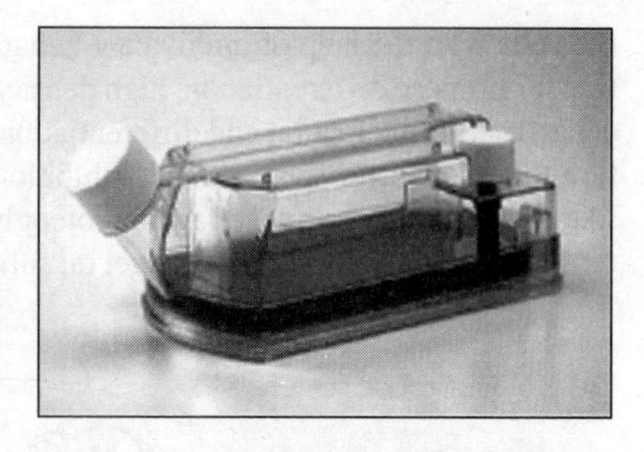

Fig. 6.4 A standard T-Flask (neck is upright to a ground)

6.4.2 Roller Bottle

This culture vessel is most commonly used for initial scaleup for anchorage dependent cells. Roller bottles are cylindrical in shape that can revolve slowly (between 5-60 revolutions per hour) by bathing the cells attached to the inner surface (see Fig. 6.5). Roller bottles are available typically with surface areas of 1050 cm^2. Its entire internal surface is available for anchorage. Several modifications of roller bottle further enhance the available surface, e.g., (i) Spira-Cell (spiral polystyrene cartridge), (ii) glass tube (roller bottle packed with a parallel cluster of small glass tubes separated by silicone spacer rings), and (iii) extended surface area roller bottle (the bottle surface is corrugated enhancing the surface by a factor of two), etc.

Roller bottles are convenient in use but is labor-intensive method for “scaling up.” A small roller rack can be placed in an incubator chamber, large incubator box, or warm room set to 37 °C. A sterile, disposable roller bottle comes in two sizes, with solid or vented (must be used in a CO_2 incubator chamber or with medium buffered for air) caps, and with surface area enlarged by “pleating” the surface, allowing for a higher cell-to-medium-volume ratio.

Cells should be plated in the roller bottles at a relatively high density (two- to fourfold more cells/cm^2 of than that used in plates. This is to allow for the reduced initial growth rate experienced by many cells in roller bottles due to the decreased control over pH if the bottles are not grown in a CO_2 incubator. Plate in the same medium as used in tissue culture plates. It is best to have the medium prewarmed and equilibrated to the correct pH before putting it in the roller bottles. If the cells are sensitive to change in pH, add an organic buffer and/or flush the airspace in the bottle with a 5% CO_2 and 95% air gas mix before sealing the cap. After placing the medium and cells in the roller bottle, seal the cap and place the roller bottle on the roller rack. Thus, for promoting cell attachment initially slow speed is required while later on speed may be increased.

The size of some of the roller bottles presents problems since they are difficult to handle in the confined space of a microbiological safety cabinet. Recently roller bottles with expanded surfaces are available which has made handling more manageable, but repeated manipulations and subculturing should be avoided if possible. A further problem with roller bottles is with the attachment of cells, since some cells do not attach evenly. This is a particular problem with epithelial cells. This may be overcome a little by optimizing the speed of rotation, during the period of attachment for cells with low attachment efficiency. See Fig. 6.5

6.4.3 Spinner Flask

This is the method of choice for culture of many anchorage independent cell lines (e.g. hybridoma) and for attached cell lines that have been adapted to grow in suspension (e.g. HeLa S3). Spinner flasks are either made up of plastic or glass bottles with a central magnetic stirrer shaft and side arms for the addition and removal of cells and medium, and for gassing with CO_2 enriched air (see Fig. 6.5). Spinner flask systems are designed to handle culture volumes of 1-12 liters and are

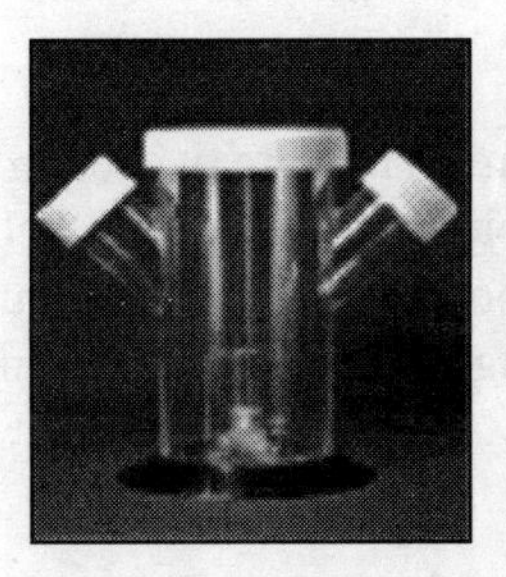

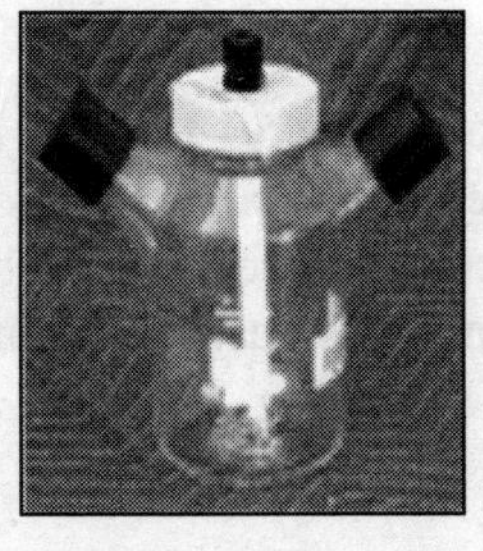

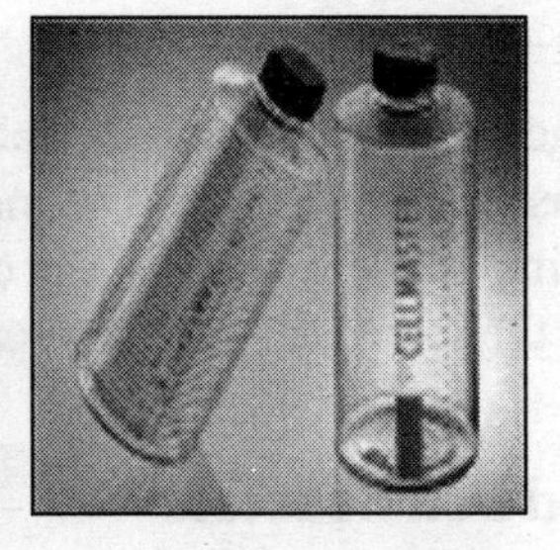

Fig. 6.5 Spinner flask of two different type (500 ml) and roller bottle

commercially available from Techne, Sigma, and Bellco company. Once cells starts growing, they can be transferred to stirred tank bioreactor for scale up.

6.4.4 Multitray Unit

A standard unit made up of 10 chambers stacked on each other which have interconnecting channels. This enables the various operations to be carried out in one go for all the chambers. Each chamber has a surface area of 600 cm^2 and the total volume of the unit is 12.5 liter. This polystyrene unit is disposable and gives good results similar to plastic flasks. See Fig. 6.6. This is used for cell factories or for production of interferous.

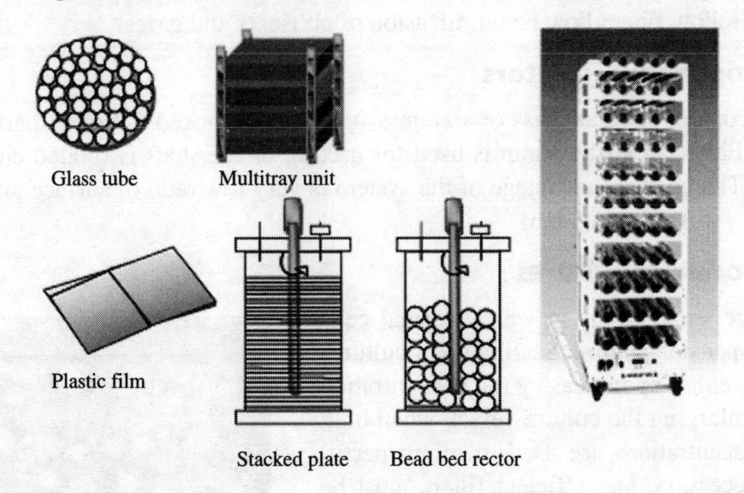

Fig. 6.6 Scaling up of culture for anchorage dependent cell, a multitray unit

6.4.5 Plastic Film

Teflon (fluoro ethylene propylenes copolymer) is biologically inert and highly permeable to gas and Teflon bags (5 × 30 cm) filled with cells and medium (2-10 mm deep) serve as good culture vessels. Cells are often attached to the inside surface of bags. Alternatively, teflon tubes are wrapped round a reel with a spacer and the medium is pumped through the tube; cells grow on the inside surface of tube (a culture vessel called **stericell** is available).

6.4.6 Heli-cell Vessels

The vessels packed with polystyrene ribbons ($3 \times 5 - 10 \times 100$ mm) that are twisted in helical shape were named as Heli cell vessel. The medium is pumped through the vessel, the helical shape of ribbons ensuring good circulation; the cells adhere to the ribbon surfaces. All the culture vessels, in addition to the increased surface area, allow further scaling up by the use of multiple units of the vessels.

6.4.7 Bead Bed Reactors

These reactors are packed with 3-5 mm glass beads (which provide the surface for cell attachment) and the medium is pumped either up or down the bead column. Use of 5 mm beads gives better cell yields than that of 3 mm beads. (See Fig. 6.6)

6.4.8 Hollow Fiber Reactor

Hollow fiber systems are designed to provide a very high cell-number-to-medium-volume ratio. The cells reside inside the apparatus and cannot be observed visually or sampled easily (see Fig. 6.2). These systems are designed primarily for the continuous production of harvest fluid. They are used often for the production of monoclonal antibody. This method also works well for proteins that are degraded or unstable in cell culture medium at 37 °C, since the medium can be harvested on an ongoing basis, thus minimizing the amount of time the protein is exposed to the culture milieu. Hollow fiber allow better diffusion of nutrients and gases.

6.4.9 Heterogeneous Reactors

These reactors contain circular glass or stainless steel plates stacked 5-7 mm apart and fitted to a central shaft. Either an airlift pump is used for mixing or the shaft is rotated either vertically or horizontally. The chief disadvantage of the system is very low ratio of surface area to medium volume (1-2 cm^2 / ml). (See Fig. 6.6)

6.4.10 Microcarrier Cultures

Microcarriers are small beads in which normal cells are grown in suspension culture. Scaling up of cultures can also be done either by increasing the concentration of beads or by enlarging the culture vessel, when high microcarrier concentrations are used, medium perfusion becomes necessary, and efficient filters must be used to allow medium withdrawal without cells and microcarriers. Generally 90-300 mm diameter beads used for cell attachment (see Fig. 6.7). Initially, Dextran beads (Sephadex A-50) were used by Van Wezelin 1967; that were not entirely satisfactory due to the unsuitable charge of beads and possibly due to toxic effects but later on microcarrier made up of other material like polystyrene, polyacrolein, glass, polyacrylamide, silica, DEAE Sephadex, cellulose, gelatin to collagen; and the specific gravity ranges from 1.02 to 1.05. Use of microcarriers permits the handling of monolayer systems as suspension cultures. Harvesting of cells from microcarrier beads requires treatment with trypsin or some other

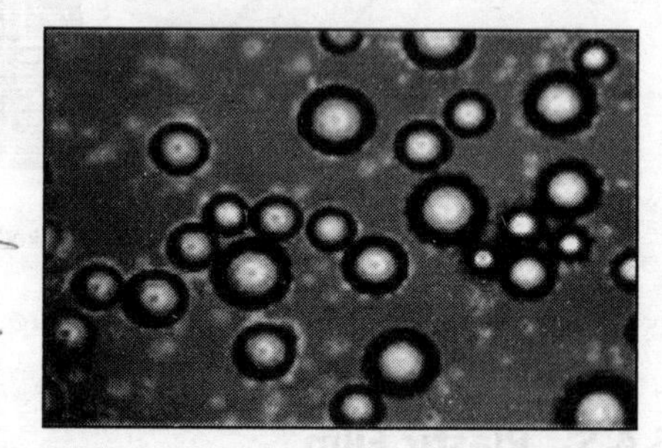

Fig. 6.7 Microcarrier beads structure in confocal microscope (Immunology lab)

suitable enzyme. Alternatively, the beads may be dissolved where possible, e.g., gelatin beads are dissolved by trypsin, collagen-coated beads are treated with collagenase and dextranase is used for dextran beads; these treatments leave the cells free which can be collected.

6.5 SUSPENSION CULTURES

When referring to mammalian cells, suspension culture is used for the maintenance of cell types, which do not adhere, including some types of blood cells, or in order to have cells express characteristics, which are not seen in the adherent form. Sometimes it is necessary to prevent adhesion by choosing a hydrophobic surface, which does not encourage cell adhesion. The absence of serum components from the medium will also help to prevent adhesion. Cell suspensions are clonally maintained by the routine transfer (subculture) of cells in the early stationary phase to a fresh medium. The following methods are available for growing cells as a suspension in Bioreactor.

6.5.1 Batch Culture

Until a few years ago, most biologicals were produced as a single batch. Growth is determined by the rate of nutrient depletion and toxic metabolite accumulation. Vaccines and interferons are still produced in this mode. The yields are low compared to other modes. In this type of culture, cells are inoculated into a fixed volume of medium. They consume nutrients as they grow and release toxic metabolites into the surrounding medium. The environment changes continuously. Eventually, cell multiplication ceases due to accumulation of toxic metabolites and depletion of nutrients (Griffiths 1992).

6.5.2 Fed-Batch Culture

The culture is fed intermittently either with the limiting nutrient or with the whole growth medium. Cell growth is limited not by nutrients but by accumulation of waste products. In this mode, a higher cell concentration is achieved and cell viability is maintained for larger periods compared to the batch mode (Griffiths 1992).

6.5.3 Continuous Culture

This mode is based on the principle established by Monod (1942). Homeostatic culture conditions are maintained in chemostats with no fluctuations of nutrients, metabolites or cell numbers. Once established they can be maintained at high cell density and high product yield for a long period (Griffiths 1992).

6.5.4 Perfusion Culture

Here medium is withdrawn continuously from the culture system and an equal volume of fresh medium is added which passes through barrier. Cell concentration increases constantly until it becomes limited at high cell population densities (Mizrahi 1989). Perfusion is used with glass bead and, particularly with microcarrier systems. Desirable, O_2 supply in the culture vessel can be enhanced from the normal 21% to a higher value and the air pressure can be increased by 1 atmosphere. This increases the O_2 solubility and diffusion rates in the medium, but there is a risk of O_2 toxicity.

6.6 FERMENTATION TECHNOLOGY

Fermentors have been used for the cultivation of bacteria and yeasts for a long time. It is natural that the same principles be applied for the mass cultivation of animal and plant cells. However, adaptation of these processes is required. Cultivation of animal cell is difficult, mainly due to the slow metabolism of these cells which is reflected by slow cell growth. Animal cells have complex nutritional requirements as compared to bacteria and yeasts.

Thus, the media are complex and costly and prone to contamination. Animal cells lack the cell wall present in bacterial cells making them fragile and shear sensitive. Therefore, the agitation and aeration systems have to be designed differently. Cell densities achieved are low resulting in low concentration of products. Downstream processing procedures required to concentrate and purify the product lead to increased costs (Mizrahi 1989). Inspite of these drawbacks, fermentors have been used for growing animal cells for the last few decades (Mc Liman et al. 1957; Telling and Elsworth 1965; Klein et al. 1971; Tovey and Brouty-Boye 1976). Different cell lines like **BHK-21** (Berg 1985; Altshuler et al. 1986; Telling and Elsworth 1965), LS, **Namalwa cells** etc have been grown in fermentors as submerged cultures (Tovey et al. 1973) for the production of viral vaccines and other products. The shear sensitivity and fragility of animal cells are overcome by the introduction of novel impellers, which are paddle shaped. Direct gas sparging produces bubbles capable of rupturing cells and hence supply of gases needs to be done via diffusion through a silicone tubing. The medium contains plenty of serum proteins capable of generating froth. Therefore, agitation has to be slow and gentle. For high-density cultivation, oxygen supply becomes critical (Balin et al. 1976).

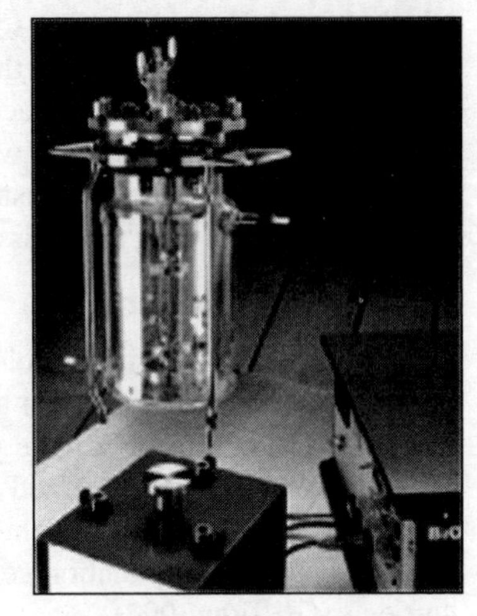

Fig. 6.8 Air lift Bioreactor

The silicone tubing method of aeration has many advantages. No bubbles are formed and the oxygen transfer rate has been found to be satisfactory. However, the winding of the fragile tubing is difficult and is limited to small scale reactors used in laboratories (Beyeler et al. 1989). Thus suitably modified fermentors can be used for the mass culture of animal cells growing as suspension cultures. If an anchorage-dependent cell line needs to be grown, it becomes necessary to use a carrier system such as microcarriers. The reactors used for large scale suspension cultures are of 3 main types: (1) Stirred bioreactor, (2) continuous flow reactor, and (3) airlift fermenter.

6.6.1 Stirred Bioreactors

Stirred bioreactors are closed systems and are usually agitated with motor-driven stirrers, with water jacket in, temperature control system, curved bottom for better mixing at low speeds. Many heteroploid cell lines can be grown in such vessels. In contrast to continuous culture where steady state is maintained by constant dilution of the culture, in perfusion cultures, the cells are physically retained in the vessel. These vessels have capacity 1-1000 liter or even 8000 liter (for interferon; but in practice their maximum size is 20 liter, since larger vessels are difficult to handle, autoclave and to agitate the culture effectively). (See Fig. 6.9) The needs for research biochemical's from cells are met from 2-50 liter reactors, while large scale reactors are mainly used for growing hybridoma cells for the production of monoclonal antibodies although their yields from cultured cells is only 1-2% of those obtained by passaging the cells through peritoneal cavity of mice.

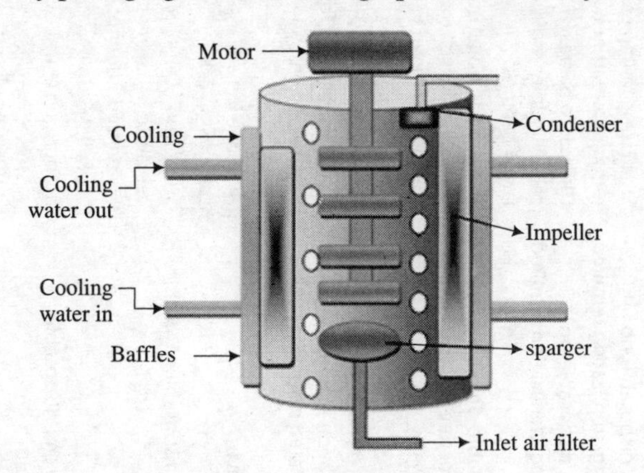

Fig. 6.9 Stirred tank bioreactor (a) external part (b) internal parts

6.6.2 Continuous-Flow Cultures

The continuous-flow cultures provide a continuous source of cells, and are suitable for product generation, e.g., for the production **of viruses and interferon's.** These culture systems are either of chemostat or turbidostat type. In both the types, cultures begin as a batch culture. In a chemostat type, inoculated cells grow to the maximum density. Fresh medium is added after 24-48 hours of growth, at a constant rate (usually lower than the maximum growth rate of culture) and at an equal rate the culture is withdrawn. When the rate of growth equals the rate of cell with drawn, the cultures are in a 'steady state', and both the cell density and medium composition remain constant. However, chemostat is the least efficient or controllable at the cell's maximum growth rate; hence the steady-state growth rates in them are much lower than the maximum. In contrast, in turbidostat cells grow to achieve a predecided density (measured as turbidity using a photoelectric cell). At this point, a fixed volume of culture is withdrawn and the same volume of fresh normal (not having a growth-limiting factor) medium is added; this lowers the cell density or turbidity of the culture. Cells keep growing, and once the culture reaches the preset density the fixed volume of culture is replaced by fresh medium. This system works really well when the growth rate of the culture is close to the maximum for the cell line.

6.6.3 Airlift Fermenter

The design of airlift fermenter is given in Fig. 6.8. It has same basic design as in other fermenter except one more unit for supply of air. Cultures in such vessels are both aerated and agitated by air bubbles introduced at the bottom of vessels. The vessel has an inner draft tube through which the air bubbles and the aerated medium rises since aerated medium is lighter than non-aerated one; this results in mixing of the culture as well as aeration. The air bubbles lift to the top of the medium and the air passes out through an outlet. The cells and the medium that lift out of the draft tube move down outside the tube and are recirculated. O_2 supply is quite efficient but scaling up presents certain problems. Fermenter of capacity 2-90 liter are commercially available, but 2000 Liter fermenter are being used for the **production of monoclonal antibodies.**

Table 6.1 Comparative property of substrate for attached and suspension culture

Technology	*Suspension*	*Attached*	*Max Vol (ml)*	*Max S/A (cm^2)*	*Max cells (susp)*	*Max cells (att)*	*Advantages*	*Disadvantages*
T Flask	Yes	Yes	150	225	1.5×10^8	$\sim 10^7$	Cheap, disposable, no cleaning/sterilization required	Small scale Multiples required for larger batches.
Triple Flask	(Yes)	Yes	150	525	1.5×10^8	3×10^7	Cheap, disposable, no cleaning/sterilization required	Difficult to harvest attached cells. Multiples required for larger batches.
Cell Factory	N/A	Yes	8000	25280	N/A	1.5 × 1010	Disposable–Single batch manufacture	Require additional equipment (vessels etc which may require cleaning). Difficult to harvest cells.
Roller Bottles	Yes	Yes	1000	1700	1 × 109	1 × 108	Cheap, disposable, no cleaning/ sterilization required. Versatile. Automated systems available.	Require "decks" to turn. Multiples required for larger batches. Automation very costly.
Expanded Roller Bottles	N/A	Yes	(1000)	3400	N/A	2×10^8	As above	As above (no advantage for suspension cells).
Shake Flasks	Yes	N/A	1000	N/A	1×10^9	N/A	Some disposables available.	Suspension Only*. Glass vessels to be cleaned & sterilized. Requires Shaker incubator.
Spinner Flasks	Yes	N/A	1000	N/A	1×10^9	N/A	Some semi-disposables available	Suspension Only*. Glass vessels to be cleaned & sterilized. Requires Stirrer-base + incubator.

Table 6.2 Different internal parts of a bioreactor and their functions

Part of Bioreactor	*Function in Bioreactor*
Motor	Provide energy to the impeller in order to generate it.
Impeller	Mix the media by stirring.
Sparger	Introduces air in the form of bubbles
Baffles	Prevent vortexing of culture.
Inlet Air Filter	Remove contaminants; adjust flow rate
Exhaust Air Filter	Filter used air moving out to the environment
Rotameter	Measures the flow rate of the air.
Pressure Gauge	Measures the pressure.
Temperature Gauge	Measures temperature, giving the culture broth an appropriate warmness to maintain in.
Cooling Jacket	Control Temperature
pH probe	Measures pH
Dissolved Oxygen Probe	Measures the amount of dissolved oxygen
Foam Probe	Detect the presence of foam
Acid	Added when pH too high (Alkaline)
Base	Added when pH too low (Acidic)
Antifoam	Add anti-foam agent when foam is present in the fermenter with a peristaltic pump.
Sampling Tube	Inoculation, addition of acid or base, and sample removal.
Control Panel	Control the two fermenter that is working
Level Probe	Measures the level of probe

6.7 IMMOBILIZED CELL CULTURE

There are two basic approaches to cell immobilization: (1) immurement and (2) entrapment.

6.7.1 Immurement Cultures

In such cultures, cells are confined within a medium permeable barrier. Hollow fibers packed in a cartridge are one such system (see Fig. 6.2). The medium is circulated through the fiber while cells in suspension are present in the cartridge outside the fiber. This is extremely effective for scales up to 1 liter and gives cell densities of 1-2 × 10^8 cells/ml; sophisticated units can yield up to 40 g monoclonal antibodies/month. Membranes permitting medium and gas diffusion are also used to develop Bioreactor of this type; both small scale and large scale versions of membrane Bioreactor are available commercially. The cells may be encapsulated in a polymeric matrix by adsorption, covalent bonding, cross-linking or entrapment. The materials used as matrix are gelatin, polylysine, alginate and agarose. This approach (1) effectively protects cells from mechanical damage in large fermenter and (2) allows production of hormones, antibodies, immunochemicals and enzymes over much longer periods than is possible in suspension cultures. (3) The medium diffuses freely into the matrix and into the cells, while cell products move out into the medium. For production of larger molecules like monoclonal antibodies, agarose in a suspension of paraffin oil is preferable to alginate since the latter does not allow diffusion of such products out of the alginate beads.

6.7.2 Entrapment Cultures

In this approach cells are held within an open matrix through which the medium flows freely. An example is the Opticell in which the cells are entrapped within the porous ceramic walls of the unit. Optic cell of surface area up to 210 m^2 are available, which yields up to 50 g monoclonal antibodies per day. The cells can also be enmeshed in cellulose fibers, e.g., DEAE, TLC, QAE, and TEAE. These fibers can be autoclaved.

6.7.3 Porous Micro Carriers

Porous micro carriers are small (170 mm 6000 mm) beads made up of gelatin, collagen, glass or cellulose which have a network of interconnecting pores. These provide a tremendous enhancement in surface area/volume ratio, permit efficient diffusion of medium and product, are suitable for scaling up, and are equally useful for suspension and monolayer cultures. These can be arranged as fixed bed or fluidized bed reactors or used in stirred Bioreactor. It is expected that future developments will make the immobilized cell systems the most dominant production system. Cultures based on immobilized cells offer several advantages:

(1) Higher cell densities (50-200 × 10^6 cells/ml),
(2) Stability and longevity of cultures,
(3) It is applicable to both suspension and monolayer cultures,
(4) Many systems protect the cells from shear forces due to medium flow, and
(5) Less dependency of cells at higher densities on external supply of growth factors which saves culture cost.

Table 6.3 *Products obtained from cells growing on microcarriers*

Products (Cells/Cellular products)	*/ Reference*
Interferon	Giard et al. 1979
Vascular endothelial cells	Davies 1981 Bing et al. 1991
Primary and established cell lines	Reuveny et al. 1982
Pancreatic islet cells	Bone et al. 1982
Proteolytic enzymes	Varani et al. 1986
Arachidonic acid	Varani et al. 1986
Tissue plasminogen activator	Nilsson et al. 1988
Growth inhibitor	Spier & Fowler 1985
Nerve growth promotor	Norrgren et al. 1983
Kallikrien	Kumar et al. 1999

BRAIN QUEST

1. Describe the factors that most commonly affect the culture of cells.
2. Describe the role of substrate in growth and differentiation of cells.
3. Describe the strategies followed for large scale culture of animal cells.
4. Name various substrate and reactors used for large scale culture of cells.
5. Describe different advantages of Immobilization.
6. Describe different part and their function in Stirred tank Bioreactor.

CULTURE MEDIA

7

Here we will learn about role of different media components and their formulation strategies. Animal cell culture is difficult since they need various complex nutrients and growth factors along with proper pH, temperature osmolality, viscosity and presence of oxygen. Serum is not only costly but also they contain various toxic materials. Now a day's various artificial serum substitutes is available for animal cell culture.

7.1 HISTORY

The earliest cell cultures involved clot of plasma as a media to grow explant (fragments of tissue) embedded in glass but that was far from experimental analysis. In the late 1940s, a major advancement was the establishment of cell lines that grow well in undefined media consisting of varying combinations of serum and embryo extracts. For example, a widely used human cancer cell line (called HeLa cells) was grown in a medium consisting of chicken plasma, bovine embryo extract, and human placental cord serum. Therefore, use of such complex and undefined culture media made impossible, the analysis of specific growth requirements of animal cells.

Harry Eagle was the first to solve this problem, by carrying out a systematic analysis of the nutrients needed to support the growth of animal cells in culture. Animal cells are more difficult to culture than microorganisms because they require many more nutrients and typically grow only when attached to specially coated surfaces. Despite these difficulties, various types of animal cells, including both undifferentiated and differentiated ones, were cultured successfully. Nutrient rich media of specific composition are required for specific culture of animal cells mimicing the natural environment.

7.2 PHYSIOCHEMICAL PARAMETERS

7.2.1 Temperature

Temperature should be maintained within the limits of natural habitat of animals because cold and warm blooded animals have different body temperature and some organs requires temperature below body temperature (e.g. testis, skin). Therefore adverse temperature can't support their proper growth and development.

7.2.2 pH

Optimal pH range for proper animal cell culture is narrow and usually from 7.2 -7.6. Any pH change can be detected if some pH indicators are present like phenyl red, orange dye (Yellow, Orange and Pink). Small, cell-to-cell type variations are also dependent on the pH. Optimal pH for cell growth varies relatively little among different cell strains.

7.2.3 Oxygen Tension

Cultured cells often rely on glycolysis to meet the energy requirements. Minimal amount of O_2 is needed for cell growth (between 1-10% of O_2) at which cell show better growth but O_2 beyond this concentration, leads to formation of toxic peroxide. For large tissue, high O_2 concentration is required that results in more formation of peroxide. To prevent inhibitory effect of O_2, adequate amount of selenium 'Se' must be supplemented.

7.2.4 CO_2 Tension

Carbon dioxide get dissolved in the medium (after respiration) and remains in equilibrium with HCO_3^- and lowers the pH. Since HCO_3^- has a fairly low dissociation constant, it tends to reassociate, leaving the medium acidic. The net result of increasing atmospheric CO_2 is to depress the pH. Therefore, increasing the bicarbonate concentration neutralizes the effect of elevated CO_2 tension. The increased HCO_3^- concentration pushes equilibrium to the opposite direction until the equilibrium is reached at a required pH. Addition of amino acid or pyruvate in the medium enables cells to increase their endogenous production of CO_2 as well as HCO_3^{-2} for e.g. **Leibovitz L15 medium** contains sodium pyruvate (550 mg/l) its buffering is achieved by high concentrations of amino acid. Some medium are not depend on CO_2 supply, so they are mostly supplemented with sodium β-glycerophosphate as buffer.

7.2.5 Buffering

Regulation of pH is particularly important just before cell seeding and it is usually achieved by; (i) a "Natural" buffering system where gaseous CO_2 balances with the CO_2/HCO_3 content of the culture medium and (ii) "Chemical buffering" using zwitterions like HEPES. Most cells require pH in the range 7.2 - 7.4. Control of pH is essential for optimum cell culture. There are major variations to this optimum e.g. Fibroblasts prefer a higher pH (7.4-7.7) whereas, continuous transformed cell lines require more acid conditions pH (7.0 -7.4). Cultures using natural bicarbonate/CO_2 buffering systems need to be maintained in an atmosphere of 5-10% CO_2 that can be easily supplied in a CO_2 incubator.

$$H_2O + CO_2 \rightarrow H^+ + HCO_3$$

HCO_3^{-2}/CO_2 system have poor buffering capacity, still it is used more frequently than any other buffer, because of its low toxicity. Although other organic buffers have more buffering capacity than bicarbonate/CO_2 buffer such as β-glycerophosphate, TES [N-tris (hydroxy methyl)-2-amino ethane sulfonic acid], and BES [N, N′-tris (2-hydroxy ethyl)-2-amino ethane sulphonic acid and HEPES buffer. HEPES buffer (N-2-hydroxyl ethyl piperazine-N′-2-ethane sulphonic acid) is a much stronger buffer in the pH range 7.2-7.6 and is frequently used at the concentration of 10 or 20 mM, but major disadvantage of it is relatively high cost and also it can be toxic to some cell types at higher concentrations. It has a advantage that it don't require, a controlled gaseous atmosphere. Most commercial culture media had pre added phenol red as a pH indicator so that the pH status can be checked of the medium is constantly indicated by the change of color. Phenol red gives different colour at different pH; Red at 7.4; Orange at 7.0; Yellow at 6.5; Purple at 7.8. Usually the culture medium should be changed/replenished if the color turns yellow (acid) or purple (alkali) the culture.

7.2.6 Osmolality

Most cells have a fairly wide tolerance for osmotic pressure. Osmolality varies with the cell type and average requirement is 300 m Osm/Kg. Osmolality for human plasma is about 290 m Osm/kg. Osmolality should be kept consistent at ± 10 m Osm/kg. Slightly hypotonic medium may be better for petri dish or open plate culture to compensate for evaporation during incubation.

7.2.7 Viscosity

The viscosity of a culture medium is influenced mainly by the serum content. Low viscosity can be damaging to the cells if the suspension is agitated or when cells are dissociated after trypsinization. Low viscosity also hampers cell-cell contact and but it can be increased by addition of serum, carboxy methyl cellulose (CMC) or polyvinyl pyrolidone (PVP) or agar.

7.2.8 Surface tension and foaming

Protein denaturation causes foam formation and thus surface tension decreases and if the cell gets entraps in the bubble then cell dies due to shear stress as happens in lager fermenter which result in huge cell death. The addition of a silicone antifoam (Dow Chemical) or Pluronic F68 (Sigma), 0.01- 0.1 % helps in preventing foaming in this situation by reducing the surface tension and may also protect cell against shear stress from bubbles.

7.3 MEDIA

Media provides essential amino acids, fatty acids, trace elements, salts, vitamins and cofactors and maintains proper chemical environment like - pH and osmolarity mostly through ions and bicarbonates. It also provides required energy (carbon source) derived after metabolism of substrate. Therefore, choice of culture medium and supplements have a major impact on growth, function and phenotype of cells *in vitro.*

7.3.1 Basic Components of Media

Basic Components of medium includes energy sources, nitrogen sources, vitamins, bulk ions, lipids and phospholipids precursors, Non-Nutrient substances like antibiotics, water, low molecular weight nutrients, reducing agents, and buffers (along with indicator dye), protective agents, and attachment factors. Animal cells that have to be grown in culture must be supplied with the nine essential amino acids, namely, histidine, isoleucine, leucine, lysine, methionine, phenylalanine, threonine, tryptophan, and valine. In addition, most cultured cells require cysteine, glutamine, and tyrosine. In the intact animal, these three amino acids are synthesized by the specialized cells; for example, liver cells make tyrosine from phenylalanine, and both liver and kidney cells can make glutamine.

Animal cells both within the organism and in culture can synthesize the eight remaining amino acids; thus these amino acids need not to be present in the diet or culture medium. The other essential components of a medium for culturing animal cells are vitamins, which the cells cannot make at all or in inadequate amounts.

7.3.2 Type of Media

Media are of two types: 1. Natural media (undefined media) 2. Artificial media (defined media).

7.4 NATURAL MEDIA

7.4.1 Serum Media

The serum is obtained primarily from the blood after clotting. It is obtained from variety of sources such as bovine calf, fetal calf, horse etc. The serum contains following constituents- Proteins, polypeptide, hormones, growth factors, nutrients, amino acids, glucose, and inhibitors. The human serum sometimes is used by the scientist for the growth of specific cell lines.

Serum is a complex mixture of albumins, growth factors and growth inhibitors and is probably one of the most important components of cell culture medium. The most commonly used serum is fetal bovine serum. The quality, type and concentration of serum, all affects the growth of cells. Therefore, it is important to screen batches of serum for their ability to support the growth of cells. In addition there are many tests that may be used to select the batch of serum (such as cloning efficiency, plating efficiency and the preservation of cell characteristics). Serum is also able to increase the buffering capacity of cultures that can be important for slow growing cells or where the seeding density is low (e.g. cell cloning experiments). It also protect cells against mechanical damage in stirred Bioreactor or while using a cell scraper.

A further advantage of serum is they support wide range of cells, and is able to bind and neutralize some toxins. However, serum composition is subject to batch-to batch variation that makes standardization of production protocols difficult. There is also a risk of cross contamination associated with the use of serum. These risks can be minimized by obtaining serum from a reputable source since suppliers of large quantities of serum perform a battery of quality control tests and supply a certificate of analysis with the serum. In particular serum is screened for the presence of bovine viral diarrhea virus (BVDV) and mycoplasma. Heat inactivation of serum (incubation at 56ºC for 30 minutes) can help in reducing the risk of contamination since some viruses are inactivated by this process. However the routine use of heat inactivated serum is not an absolute requirement for cell culture. The use of serum also has a cost implication not only in terms of medium formulation but also during downstream processing. A 10% FBS supplement contributes 4.8 mg of protein per milliliter of culture fluid, which complicates downstream processing. Fetal Bovine Serum (FBS) is the most widely used type serum media but it is most expensive. It has high levels of growth stimulatory factors and lower levels of growth inhibitory factors.

7.4.2 Serum Collection

Serum is the defibrinated plasma. Serum provides a number of growth promoting factors in a balanced amounts. Serum can be collected by cardiac puncture and umbilical exsanguination

7.5 SERUM MEDIA COMPOSITION

7.5.1 Proteins

Serum is very rich in proteins. Various proteins present in the serum are albumin, globulins, transferrin, protease inhibitor, and cell attachment factors etc. Functions of some of these proteins are unknown.

A. Functions of Serum Albumin

(i) It helps in transport of naturally bound ligand for e.g. billirubin, fatty acid, hormones, vitamins, amino acids, peptides, globulins, metal ions etc.
(ii) Help in detoxification of free fatty acids, heavy metal ions, endotoxins etc.
(iii) It has buffering action.
(iv) It provides mechanical protection against shear forces.

B. Functions of Transferrin Protein are

Major role is in transport of ions out of the cell by using apotransferrin and siderotransferrin (for iron) and also in transfer of Vanadium and other metal ion detoxification.

7.5.2 Serum Inhibitors

Serum contains certain inhibitors that inhibits the growth of the cell and their further proliferation e.g. presence of TGF- β or γ-globulin (antibodies that can cross react with the culture). Heat treatment may inactivate them and can reduce their cytotoxic action (of the immunoglobulins) without damaging the polypeptide growth factors. TGF (Tissue growth factor) inhibitor are antimitotic to some cells; while presence of **Complement proteins**- activate a series of protein and thus causes cell lysis in culture.

7.5.3 Hormones

Many of the hormones secreted naturally from the cells has important role in cell growth and proliferations 1. Insulin, 2. Hydrocortisone (for cell attachment); 3. Growth hormones (it provides mitogens for providing signal to divide to cell in culture); 4. Triiodothyroxin (I).

Table 7.1 *Procedure for erum extraction from Blood*

Procedure for serum extraction is as follows:	→ Let it stored at room temperature for 1 ho ur (clotting will take place)	→ Leave it at 4°C	→ Centrifuge and discard the pellet	→ Yellowish supernatant obtained is serum → Serum is filtered through 1 microne membrane and sterilized by is gamma irradiation.

7.5.4 Nutrients and Metabolites

There are various dissolved nutrients and metabolites present in the serum which has major role in cell growth and proliferations such as 1. Glucose; 2. Keto acids; 3. Intermediary metabolites; 4. Calcium chloride; 5. Fe, K^+, PO_4^{3-}, Se, Na & Zn; 6. Polyamines-Putresine and Spermidine; 7. Pyruvic acid, Lactic acid, Hexamins 8. All amino acids ; 9. Lipids 10. Cholesterol 11. Fatty acids; 12. Phospholipids 13. Linoleic acid; 14. Vitamins A and B complex

7.5.5 Growth Factors

There are various growth factor present in serum that acts as cell signalling molecule such as [1] Plasma growth factor; [2] Fibroblast growth factor (It can lead to fibroblast cell formation); [3] Vascular endothelial growth factor; [4] Insulin like growth factor; [5] Epidermal growth factor; [6] Endothelial cell growth factor.

7.6 ADVANTAGES OF USING SERUM

1. It is composed of various factors that is required for the growth and maintenance of cells
2. It has universal growth supplement, and thus all the cell types can be grown in a medium containing serum.
3. Serum provides buffers to the cultured cells.
4. It prevents toxic effects due to active oxygen species and pH, heavy metals ions, endotoxins etc.

7.7 DISADVANTAGES OF SERUM MEDIA

1. Its batch to batch variations leads to lack of reproducibility of results of cell culture.
2. It has several ill defined components that may affect cell functions in unknown way.
3. Serum can be infected by various organisms.
4. It is much costlier.
5. It shows growth inhibitory activity due to presence of various proteins like complement, antibodies, TGF β, and endotoxins.
6. Fibroblast growth promoting factor can cause high fibroblast cell growth if kept for longer duration.
7. Downstream processing of products (desired purification) is very complicated as it gets mixes with other components.
8. Security of serum supplements and its regular supply is not guaranteed since cattle population is directly proportional to the serum cost.

7.7.1 Special Treatment can Deactivate the Inhibitors Present in the Serum

1. By γ irradiation;
2. By heat inactivation (De-complementation of serum) at 50°C for 30 minutes (inactivate complement proteins.)
3. By separation of proteins by ethanol fractionation and ion exchange chromatography.
4. By dialysis of serum-dialysate against 0.15 M NaCl, until glucose level goes to 5% w/v;
5. Therefore by applying any of these process, low molecular weight components can be removed.

7.8 DEFINED MEDIA

7.8.1 Medium Formulation

One of the major considerations in using animal cell lines is the medium formulation.The medium formulation is important because of high cost of serum and its effect on growth and productivity of the cell line. The basic cell culture medium consists of 30 to 50 ingredients. If media composition can be defined quantitatively it is called as defined media. e.g. RPMI 1640, had wide range of serum free formulation. RPMI 1640 is used for lymphoblastoid cell lines culture Ham's F-12 with higher content of vitamin and amino acids is used for various cell lines. Various supplements are usually added to the basic medium. These include glucose, serum (added to 5 to 10%) and antibiotics to complete the growth medium.

7.8.2 Balance Salt Solution (BSS)

In 1880, **Ringer** used a solution to store the blood cells and in 1910, **Tyrode** coined the term Balanced Salt Solution and also defined its properties. It is a mixture of salt with glucose (some time) along with the inorganic salts and the bicarbonates. Balanced salt solutions contains essential inorganic ions in optimal concentration, correct pH (7.2), pH indicator, correct osmolality (~300 m Osm/Kg) and a source of energy (optional). Addition of $NaHCO_3$, become helpful, when it is required for the short term cell culture, or if CO_2 incubator is not there. Phenol red dye is often added to check the pH (Acidic- yellowish, Neutral-orange red, Alkaline-purple to pink). Therefore, balanced salt solution contains various salts in a way as to take care of osmotic pressure, concentration, inorganic ions, carbohydrate and gases. Sometime calcium, magnesium is used in BSS, in order to reduce cell aggregation and attachment as they help in cell agglutination. BSS can be used for variety of purposes;

7.8.2.1 Functions

1. Provides base (isotonicity) for culture medium.
2. For washing tissue/cells, dissection of tissue, for diluting media.
3. Provides water and inorganic ions to the cells, while maintaining its pH and the osmotic pressure. Buffering is required to maintain the pH.
4. Provides a short term source of energy.
5. However Balanced Salt Solution cannot support full potential and functions of cells.

7.9 FUNCTION OF MEDIA COMPONENTS

A mixture of inorganic salts and other nutrients capable of sustaining cell survival in vitro is called as medium. Basically there are two type of media 1. **Growth medium** is the medium which is used in routine culture such that the cell number increases with time. 2. **Maintenance medium** is the medium which will retain cell survival without cell proliferation (e.g. a low serum-free medium used with serum dependent cells). [1] Carbohydrates [2] Proteins and peptides [3] Amino Acids [4] Vitamins [5] growth factors and Hormones [6] Inorganic salts [7] gases [8] buffer.

7.9.1 Carbohydrates

The main source of energy is derived from six-carbon sugar e.g. Glucose, galactose but mixture of pyruvate, maltose or fructose and ribose may also be used. The concentration of sugar varies from (basal media containing) 1 g/l to 4.5 g/l (in some more complex media). Media containing the higher concentration of sugars are able to support the growth of a wider range of cell types. Most of the eukaryotic cells are aerobic, since they require oxygen to meet their necessary energy requirements and for operating the citric acid cycle. Glucose is metabolized via Glycolysis to form pyruvate which later on metabolized by the citric acid cycle in the presence of the oxygen into ATP, while transformed cell, and other tumor cell can't metabolized the pyruvic acid and the lactic acid. There occurs an accumulation of lactic acid and thus form the basis of identification of these cells. It has been reported that glutamate is the essential requirements than the glucose.

7.9.2 Proteins

Proteins are major component of serum, but mainly they acts as carrier of the minerals, fatty acids and hormones. The most common proteins and peptides include albumin, transferrin, fibronectin and fetuin which can be used to replace serum component. There may be other proteins as yet uncharacterized, essential for cell attachment and growth.

7.9.3 Amino Acids

Amino acids are required for the formation of the peptide and the proteins by the translation. Proteins take parts in the structural framing or in the metabolic pathway. Usually 20 amino acids are involved in protein synthesis. These are included depending on the individual needs and aims to reduce the biosynthetic load. Essential amino acids includes Arginine, Lysine, Histidine, Isoleucine, Leucine, Cysteine, Methionine, Phenylalanine, Threonine, Tryptophan, Tyrosine Valine and Glutamine; nonessential amino acids like Alanine Asparagine, Asparatic acid, Glutamic acid, Glycine, Proline, and Serine. Differentiation of cells may lead to reduction in cell capacity to synthesize amino acids since biochemical specialization lost *in vitro* and failure to synthesize a particular amino acid *in vitro* may represent the lack of correct precursor or cofactor in artificial environment. Ex: Monkey kidney cells have reduced ability to convert folic to folinic acid. The later is required as a cofactor for the conversion of Serine to Glycine. Adherent cells divide at a very active pace, thus insufficient amounts of amino acid are synthesized. Glutamine supports cell growth and amino acid uptake glucose utilization, and support protein turnover. They get decompose in the medium in a time and temperature at 37°C. Their half life is 8 days, at 4°C on decomposition ammonia is produced which is potentially toxic.

> ***The Experiment of Eagle***
>
> Eagle studied the growth of two established cell lines: HeLa cells and a mouse fibroblast line called **L cells**. He was able to grow these cells in a medium consisting of a mixture of salts, carbohydrates, amino acids, and vitamins, supplemented with serum protein. By systematically varying the components of this medium, Eagle was able to determine the specific nutrients required for cell growth. In addition to salts and glucose, these nutrients included 13 amino acids and several vitamins. A small amount of serum protein was also required. The basal medium developed by Eagle is described further from his 1955 paper. The medium developed by **Eagle** is still the basic medium used for animal cell culture. Its use has enabled scientists to grow a wide variety of cells under defined experimental conditions, which has been critical to studies of animal cell growth and differentiation, including identification of the growth factors present in serum now known to include polypeptides that control the behavior of individual cells within intact animals.

7.9.4 Vitamins

Vitamins are precursors for numerous cofactors. Many vitamins especially of B groups are necessary for cell growth and proliferation. The vitamins commonly used in media include vitamins B-complex such as biotin, folic acid, niacinamide, pantothenic acid, pyridoxine, riboflavin, thiamine and Vit. B-12 etc. Both vitamin E and K is toxic when in excess. Lack of vitamins may hinder the survival and the cells size. Serum is an important source of vitamins in cell culture. However, many media are also enriched with vitamins making them consistently more suitable for a wider range of cell lines.

7.9.5 Salts

The inclusion of inorganic salts in media performs several functions. Primarily they helps in the osmotic balance of the cells that regulats membrane potential.

Bulk ions has many functions for e.g. sodium maintains osmotic pressure, potassium maintains osmotic pressure in the cell, calcium and magnesium is essential for intracellular enzymes and participate in cell attachment and spreading. Calcium is essential for cytoplasmic progress. Iron is needed for respiratory pigments-cytochromes. Major parts of the metals are used in maintaining the osmolality of the cell, maintaining the membrane potential and in the transport of the necessary component.

7.9.6 Hormones and Natural Growth Factor

There are various polypeptides (secreted by the cells itself) or growth factor e.g. PDGF (platelet derived growth factor) secreted from the blood clot. Which stimulates the growth (in the fibroblast and glia (in the neurons)). Growth factors act synergistically with each other. Many growth factors are required in the culture like interferon, interleukin, and heparin-binding growth factor for e.g. IGF-1 and IGF-2 (interleukin growth factor) which binds to insulin receptor and stimulate glucose and amino acids uptake; Hydrocortisone promotes cell attachment. Several other hormones like somatotropin, insulin, improves the plating efficiency, hydrocortisone improves cloning efficiency of the glia and fibroblasts (has been found necessary for maintenance for epidermal keratinocyte). 5-tri-iodo-tyrosin (T3) is a necessary supplement for epithelial cell, MDCK (dog kidney cell) and various combinations of estrogen, androgen, progesterone with hydrocortisone and prolactin is necessary for the maintenance of mammary epithelium cell.

7.9.7 Other Organic Nutrients

Choline and Inositol acts as a substrate in the biosynthesis reactions. Purine can be derived from adenine or hypoxanthine, thymidine, polyunsaturated fatty acid, cholesterol, polyamines, citric acid cycle intermediates, and pyruvate.

7.9.8 Trace Elements

These are also equally required by cultured cells but their functions are unknown. E.g. Fe, Zn, Se, Cu, Mn, Mo and V.

7.10 SUPPLIMENT OF THE SERUM FREE MEDIA

The serum free media must be supplemented with the following components:

1. It must have Fibronectin and the laminin that will help in growth and attachment of cell with the surface.
2. It must have the protease inhibitor like soya bean trypsin inhibitor.
3. It must have various mixture of hormones to increases the plating efficiency, and to improve cloning efficiency.

4. It must be supplied with the growth factor like EGF, IFN-γ1, IFN-γ2, TNF-β, platelet derive growth factor nerve growth factor NGF-α, NGF-β, interleukin to provide the mitogens, and cellular differentiation. Mixture of growth factor sometime is necessary because it provides the synergistic effect on growth.
5. It must be supplemented with selenium (required for lipids or lipid precursor) Transferring is the carrier of iron and is mitogenic so it must also be supplied.

7.11 ADVANTAGE OF SERUM FREE MEDIUM

Periodical supply of serum is not possible due to various reasons. Supply may be limited due to cultural, political, or economical reason. Serum free media does not have chances of early degradation and also is fixed in composition. Using serum free media have another advantages during downstream processing; it does not creates problem during the product recovery while serum media posses various problems because of interference by the presence of serum.

Serum is prone to infection with the mycoplasma and the viruses and is difficult to eliminate while SFM have less such chance and easy to eliminate. The most important advantage of the serum free media is its easy replacement and low cost against serum media. Serum action is unpredictable in term of either it may restricts growth or it may enhance the growth of particular cell type. It's very difficult job to standardize the experimental protocol regarding the use of serum, while serum free media is defined, so not difficult to standardize.Serum free media had known composition and can be used according to the requirements of the cell type. We can add extra biochemicals like mitogens and the cytokines that may have different effect on the cell. For mitogenic activity we can add PDGF platelet derived growth actor for development of the fibroblast cell, and TGF–α for cytostatic growth for some epithelial cell.

7.12 DISADVANTAGES OF SERUM FREE MEDIUM

1. Cells become more fastidious (uncontrolled) in absence of serum.
2. Culture conditions can be very critical in serum free medium, so better control of gases and pH is required.
3. Serum free medium have reduced capacity to activate the toxic components. Animal antibodies are more toxic to the medium.
4. Mechanical protection of cells from shear is reduced.
5. Adaptation period is very long as a result it becomes very labour intensive.
6. Multiplicity of the media presents a problem for laboratories maintaining cell lines of different origin as each cell type require a different set of medium.
7. Defined medium tends to select some specific cell types and leads their overgrowth.
8. Reagent added should be of absolute purity otherwise they will become a limiting factor.
9. Growth is often slower in serum free media, and fewer generations are achieved with finite cell lines.
10. Availability of properly controlled serum free medium is quite limited and the products are often more costly than conventional media.

Table 7.2 List of defined media their name and uses

Media type	*Examples*	*Uses*
Balanced salt solutions	PBS, Hanks BSS, Earles salts DPBS HBSS EBSS	Form the basis of many complex media
Basal media	MEM	Primary and diploid cultures.
	DMEM	Modification of MEM containing increased level of amino acids and vitamins. Supports a wide range of cell types including hybridoma.
	GMEM	Glasgows modified MEM was defined for BHK-21 cells
Complex media	RPMI 1 640	Originally derived for human leukaemic cells. It supports a wide range of mammalian cells including hybridoma
	Iscoves DMEM	Further enriched modification of DMEM which supports high density growth
	Leibovitz L-15 liquid)	Designed for CO_2 free environmen ts
	TC 100 Grace's Insect Medium, Schneider's Insect Medium	Designed for culturing insect cells
Serum Free Media	CHO Ham F10 and derivatives Ham F12 DMEM/F12	For use in serum free applications. Note: These media must be supplemented with other factors such as insulin, transferrin and epidermal growth factor. These media are usually HEPES buffered
Insect cells	Sf-900 II SFM, SF Insect-Medium-2	Specifically designed for use with Sf 9 insect cells

Table 7.3 Artifical media and their references

Artificial media	*References*
Medium 199	Morgan et al. 1950
MEM	Eagle 1959
CMRL 1066	Parker et al. 1957
DMEM	Dulbecco 1959
Ham's F-12	Ham 1965
RPMI 1640	Moore et al. 1967

Common media composition	
Amino acids	0.1-0.2 mM
Vitamins	1 μM
Salts	
NaCl	150 mM
KCl	4-6 mM
CaCl	1 mM
Glucose	5-10 mM

7.13 SOME COMMERCIAL SERUM FREE MEDIA

SFM formulation is done in the same way as the serum components. There are various type of SFM which has been named after the discoverer.

1. Eagle's Basal Medium
2. Eagle's Minimal Essential Medium (Membrane)
3. Dulbecco's modification of Membrane (DMEM)

Eagle's medium and their derivatives

1. Eagle's medium
2. Minimal essential medium with Earle's salt (EMEM)

3. Minimal essential medium with a-modification (AMEM or)
4. Dulbecco's modified Eagle's medium (DMEM)
5. Glasgow modification of Eagle's medium (GMEM)
6. Minimal essential medium with Joklik's modification (JMEM). Mouse cells are cultured in a-MEM, and Murine cells are cultured in DMEM and a-MEM.

7.14 ARTIFICIAL SERUM

Totally protein free media is most desirable but is hard to produce. It must contain some serum substitutes like HL1, CPSR1-5, Nutridoma+, Nuserum. It also contains certain defined proteins and additives of Basal Salt Mixtures, a mixture of calcium chloride, magnesium chloride, potassium chloride. Therefore, serum may be partially or completely removed. Other example include Serxtend, (NEN), Ventrex, Nutricyte and CLEX.

7.14.1 Commercial

1. Supplementation of serum free formulations along with cytokines, nutrient, transport factors and adhesion factors required for targeted cell functions.
2. Adapting genetically modifying parental cells to reduce or eliminate their serum factors e.g. transformation of cells as they have reduced serum requirement.
3. Replacing whole serum with serum component after removing undesired constituent of serum e.g. removal of albumin, Antibodies, Microbes.

7.14.2 In Laboratory

1. Take a known residue of a completely defined medium and alter the constituent until the medium is optimized for given targeted cells. It was designed by Ham and is very time consuming.
2. Use a completely defined medium and restrict the manipulation of ingredients to a shorter list of substances e.g. selenium, transferrin, albumin, insulin, hydrocortisone, estrogen, tri-iodothyronine, ethanolamine, phosphoethanolamine, growth factors, prostaglandins, and other which may have relevance for given cell types. To gradually reduce serum and r.place serum with defined components.

Table 7.4 Serum-containing medium (eagle's medium)

Medium	*$NaHCO_3$*	*% CO_2*
Eagle's Membrane	(Ham's salts)	4
Grace's medium	(Ham's salts)	4
IPL-41	(Ham's salts)	4
TC100	(Ham's salts)	4
Schneider's medium	(Ham's salts)	4
IMDM	36	5
TC199	26	5
DMEM Ham's F12	29	5
RPMI 1640	24	5
GMEM	42	10

BRAIN QUEST

1. Name some substances used as feeder layer.
2. Write the importance of feeder layer during the cell culture.
3. What are the components cells obtained from the feeder layer?
4. Define confluency.
5. How confluency is harmful for monolayer culture?
6. Name some Defined media.
7. Write the advantage and disadvantage of serum free media.
8. Write the name of inhibitors present in the serum.
9. Write the component of serum that supports the animal cell culture.
10. Write the use of Balance salt solution.
11. Describe how CO_2 acts as buffer naturally.
12. Describe the importance of serum in animal cell culture?
13. Describe the protocol how will you obtain the serum from the blood?
14. Name some protein that are important and present in the blood?
15. Give the advantages and disadvantages of using serum in animal cell culture?
16. Describe the way in which pH of the system is controlled during the culture in serum?
17. Describe the constituents of a defined media for animal cell culture.
18. Describe the method of sterilization of serum media.
19. Name some properties of the substrate used in animal cell culture? How cells attached to the substrate during the cell culture.
20. Describe the alternative way to grow animal cell at large scale culture.
21. Name some growth factor that is important for attachment of cell in culture.
22. Describe and differentiate the properties of adherent cell and non-adherent cell.
23. How media is conditioned by feeder layer?

Fill in the blanks

1. Serum protein that inactivates toxin is ________.
2. Hormone that helps in attachment is ________.

EXPLANT ISOLATION

8

Here we will learn the different aspects of explant isolation from already differentiated cells or embryonic cells. Isolation of explants with minimum damage is the important aspects of the animal cell culture and often two method employed; enzymatic and mechanical method for explant isolation.

8.1 EXPLANT

Explant is an excised cell obtained after tissue disaggregation. Explant is used for the purpose of seeding to get a monolayer culture mostly from young embryonic cells and sometime from mature cells. Cells that are cultured first time are known as **primary cells**. With the exception of some tumors, most primary cell cultures have limited life-span. After certain number of population doublings, cells undergo the process of senescence and apoptosis. (for more detail please refer to Chapter 11.2)

8.2 METHODS OF EXPLANT ISOLATION

The methods of Explant isolation depends on the origin and source of the tissue. Mostly enzymatic desegregation method is employed, but some time mechanical desegregation method are also employed. The protease enzymes may be used alone or in combinations to increase the efficiency and viability of cells. Various proteases like Trypsin (for embryonic tissue), collagenase (for normal and malignant tissue), elastase, hyaluronidase, DNAse, pronase, and dispase have been used. Trypsin and the pronase are harmful, since they have strong protease activity and digest cells badly, yet they gives complete disaggregation. DNAse is used to separate the DNA attached with lysed cell protein. The purpose of using the enzyme is the digestion of various cell adhesion molecules (CAMs) present between the cells. When the cells become semi-confluent then cells are disaggregated.

8.3 MECHANICAL ISOLATION

This method is quick and easy but disrupt cells which results in significant cell death. This method is best when target is to harvest many different samples of cells, i.e. when viability is not important. The steps involved in physical disaggregation have been depicted in Fig. 8.1. A rubber spatula can be used to physically remove the cells from the well grown culture surface. The tissue is chopped aseptically and isolated in buffer (kept first in a sieve of 100 μm sieve and then passed through sieves of decreasing pore size 50 μm and 20 μm mesh). The disaggregated cells can be counted using a hemocytometer. In other method cells are isolated after slicing with knives and by pressing

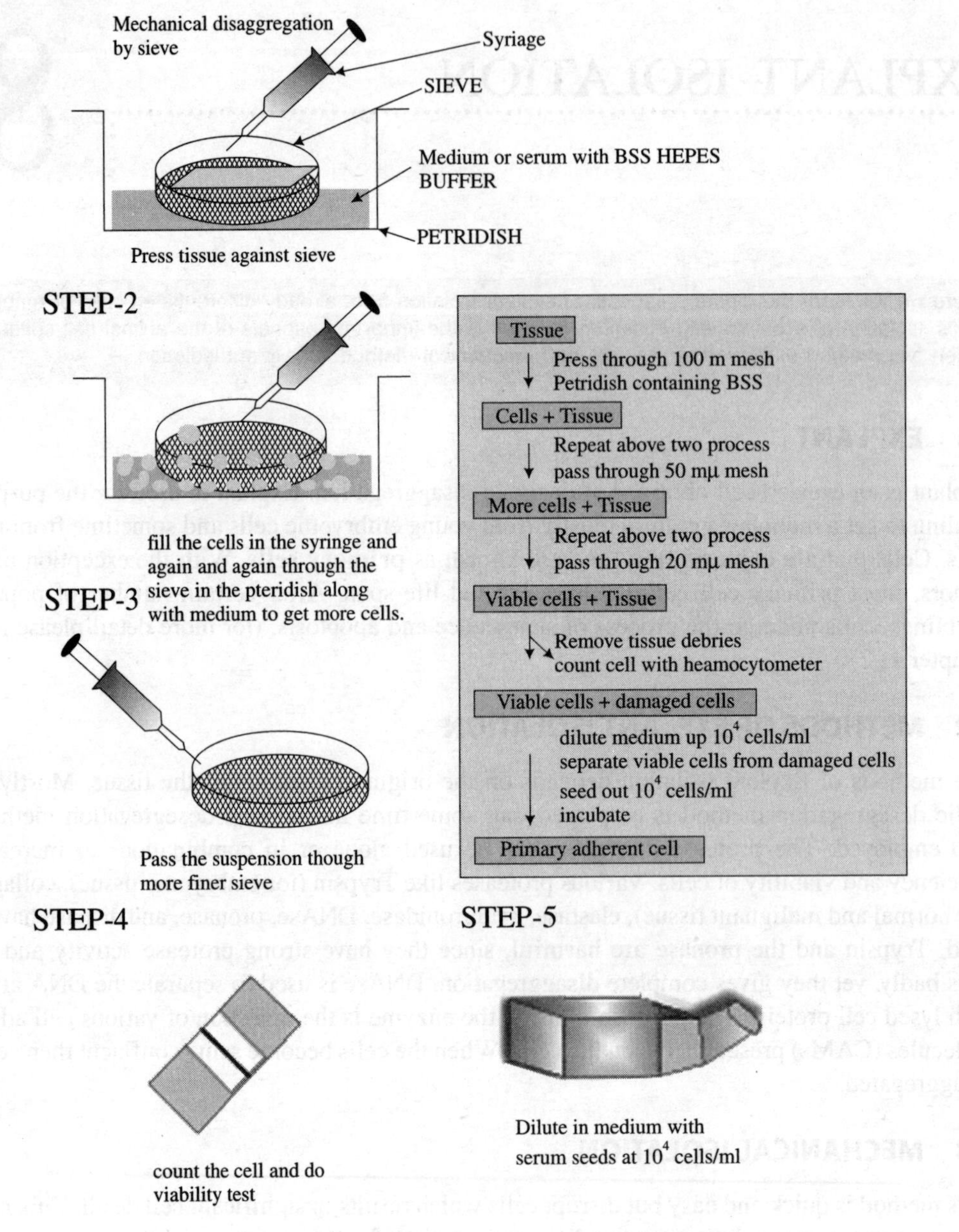
STEP-1
Mechanical disaggregation
by sieve
Syriage
SIEVE
Medium or serum with BSS HEPES
BUFFER
PETRIDISH
Press tissue against sieve
STEP-2
Tissue
Press through 100 m mesh
Petridish containing BSS
Cells + Tissue
Repeat above two process
pass through 50 mμ mesh
More cells + Tissue
Repeat above two process
pass through 20 mμ mesh
Viable cells + Tissue
Remove tissue debries
count cell with heamocytometer
Viable cells + damaged cells
dilute medium up 10⁴ cells/ml
separate viable cells from damaged cells
seed out 10⁴ cells/ml
incubate
Primary adherent cell
fill the cells in the syringe and
again and again through the
sieve in the pteridish along
with medium to get more cells.
STEP-3
Pass the suspension though
more finer sieve
STEP-4
STEP-5
Dilute in medium with
serum seeds at 10⁴ cells/ml
count the cell and do
viability test

Fig. 8.1 Mechanical disaggregation of explant by seive method

or by forcing with syringe against the mesh or by repeated pipetting. This gives more desaggregated cells than the enzymatic methods with minimum damage. When the availability of tissue is in plenty and efficiency of yield is not important, then this method is more applicable.

8.4 ENZYME DESEGREGATION

Enzymatic digestion is less labor intensive but yields less surviving cells if strong protease is used because of uncontrolled digestion. Therefore, it is an alternative method to isolate the explants. The aim is to break connection between the cells that helps in attachment of one cell to next cell. Always it should be kept in mind that cells are alive and therefore must be kept in buffer at low temperature otherwise maximum cell death may results. Other things that should be in mind is to complete all these operation inside the closed hood.

This method is fast and reliable but can damage the cell surface by digesting exposed cell surface proteins. Various proteolytic enzymes has been used such as Trypsin, Collagenase, or Pronase, usually in combination with EDTA to remove metal ions holding cells. The reaction can be quickly terminated by transferring the cells in complete medium containing serum. Since EDTA is a metals ion chelators and membrane attached with Ca^{+2}, and other divalent ions can be removed easily and thus, complete disaggregation of cells occur.

Many time combination of enzyme is used to desegregates various components of the extra cellular matrix (ECM) that adheres cells tightly for e.g. CAM (calcium dependent cadherins), fibronectin, laminin, and proteoglycan. Cadherins are calcium dependent ECM, therefore, EDTA is also used. Alternatively, other enzyme like Glyconase, as hyaluronidase, and heparinases can be in place of trypsin, since they are mild in action.

8.4.1 Collagenase

Collagenase is generally used for disaggregation of embryonic cell, normal cell as well as malignant tissues (to detach collagen). Following steps are involved in disaggregation of epithelial tissue.

1. First the biopsy of epithelial tissues are kept in medium (containing antibiotics).
2. Thereafter, the tissue is dissected into pieces in BSS.
3. The chopped tissue after washing is transferred to incomplete medium containing collagenase. After five days of treatment, the mixture is pipetted 2-5 times so that the medium may get dispersed and the epithelial cells are settled.
4. The settled cluster of the epithelial cells is washed thoroughly and the cell suspension is made free from the enzyme collagenase by centrifugation.
5. The suspension consists of enriched fibroblast fraction which is plated out on medium.

In some cases complete separation of cells from the substrate is not possible then combination of proteases are tried to increase the cell number. Embryonic tissue requires less use of enzyme treatment because newly born cells adhere less tightly and therefore, more dead cells results. Various factors that may results in more dead cells are: a) Impure enzyme, b) Longer digestion period, and c) Age of tissue. Toxicity of enzyme can be decreased by the following method by washing the tissue by the buffers and by use of pure enzyme.

8.4.2 Trypsinization

Trypsinization is the treatment of monolayer culture with trypsin enzyme. Depend on temperature provided. It is of two types 1. Cold and 2. Warm trypsinization. see Fig. 8.2

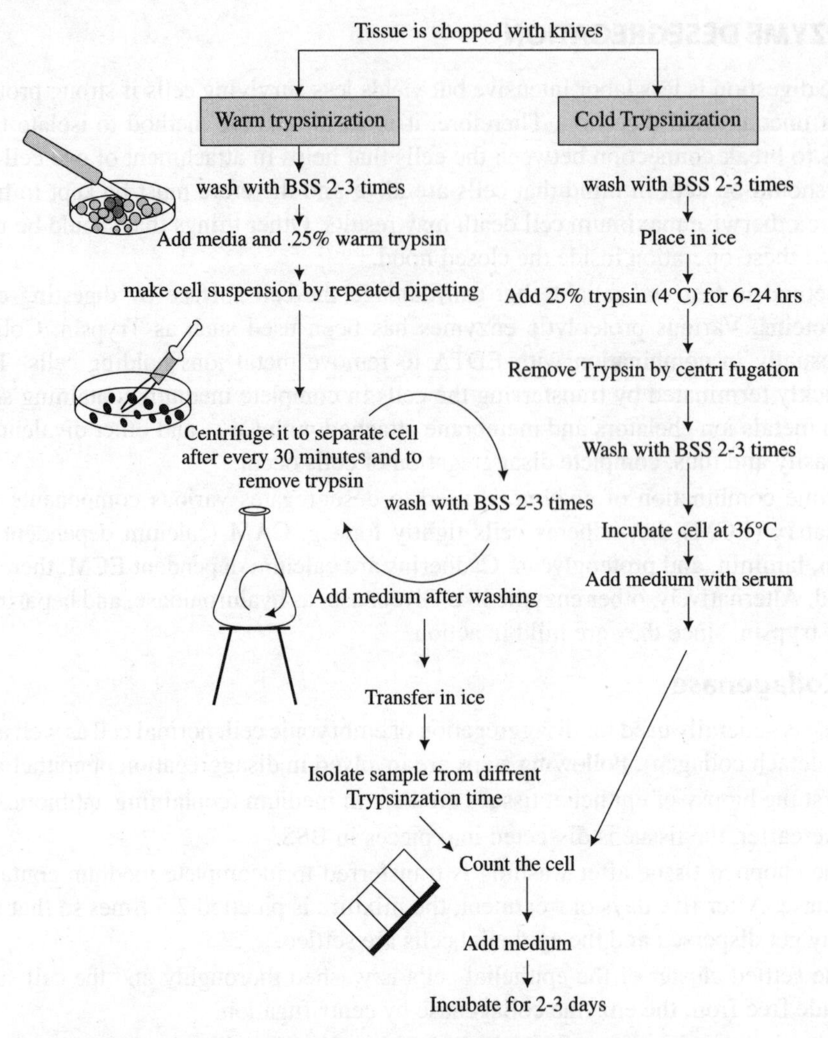
Tissue is chopped with knives
Warm trypsinization
Cold Trypsinization
wash with BSS 2-3 times
wash with BSS 2-3 times
Add media and .25% warm trypsin
Place in ice
make cell suspension by repeated pipetting
Add 25% trypsin (4°C) for 6-24 hrs
Remove Trypsin by centri fugation
Centrifuge it to separate cell after every 30 minutes and to remove trypsin
Wash with BSS 2-3 times
wash with BSS 2-3 times
Incubate cell at 36°C
Add medium with serum
Add medium after washing
Transfer in ice
Isolate sample from diffrent Trypsinization time
Count the cell
Add medium
Incubate for 2-3 days

Fig. 8.2 Steps in warm and cold trypsinization methods

8.4.2.1 *Cold Trypsinization*

If the tissue and trypsin enzyme are incubated first at 4°C for 6 to 8 hours and then completly at 37°C for 20 to 30 minutes then cell can be disaggregated without any harms. Generally 0.25% Trypsin is used for treatment in buffer or Hams F12 media.

8.4.2.2 *Warm Trypsinization*

This is a method to isolate the cells rapidly in very short period of time and employed generally for the free moving cells rather than the still cells. For this, cells are treated at 37°C and then centrifuged after 4 hours. Cells can be collected at every 30 minutes. The method is employed generally for the digestion of human tumors, mouse kidney, human adult fetal brain, lung and many tissue of epithelium. Many epithelial tissue like kidney (tubular and glomeruli cluster), carcinoma cells of breast, gastro intestinal tract and lung alveoli that are not disaggregated by collagenase treatment because connective tissue are completely disaggregated by the collagenase.

During dissociation of cells odds cells are died at large number. Therefore, their separation is necessary. The most commonly used method is the culture of cells in the media, where viable cell are attached to the bottom while non viable cell floats in the medium. Another popular method is centrifuging the cells along with the ficoll and hypaque (sodium metrizoate). This technique results in the separation of two layers and the viable cells are separated from the interface. The more in-depth discussion is available in next chapter.

BRAIN QUEST

1. Describe in detail the mechanical separation method of cells.
2. Describe the type and advantage of trypsinization.
3. Describe the advantage of using EDTA along with collagenase enzyme.

CELL COUNTING AND VIABILITY ASSAY

9

Here we will learn about the different cell assay techniques, since choice of cells decide total fate of the experiments. Cell counting method has taken a major breakthrough due to introduction of many new instruments for counting of cells at large scale within few minutes. Instruments like cell counter and flow cytometry has revolutionized the animal cell culture.

9.1 CELLS VIABILITY ASSAY

The first step in any cell culture experiment is the decision whether to test for cell proliferation or viability parameters, depending on the objective of a study. These parameters are measured by assaying for "vital functions" characteristic of healthy and growing cells. Cell viability measurements assess healthy and growing cells in a sample. This can be accomplished either by directly counting the number of healthy cells or by measuring an indicator in cell populations (e.g., in a microplate assay). Normally whether the cells are actively dividing or quiescent is difficult to distinguished. An increase in cell viability indicates cell growth, while a decrease in viability can be interpreted as the result of either toxic effects of compounds/agents or suboptimal culture conditions. In contrast to cell viability, **cell proliferation assessment** measures actively dividing cells in a sample. It can be expressed either as the actual number or proportion of proliferating cells in cell culture, tissues, or as relative values in assays for cell populations. Quiescent (non growing healthy) cells can not be detected by cell proliferation assays.

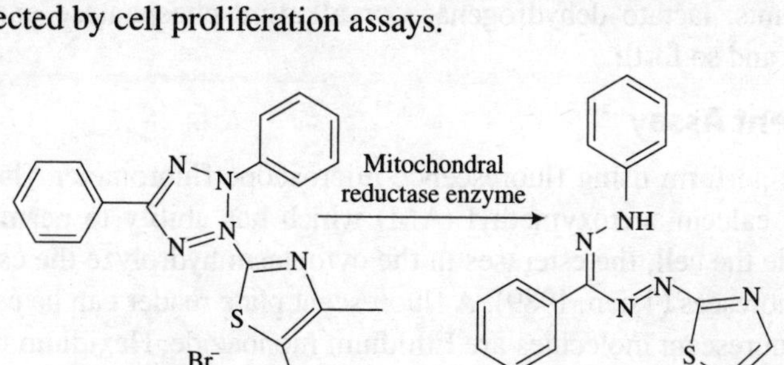

Mitochondrial reductase enzyme

MTT (Yellow) *FORMAZAN (blue/purple)*

Fig. 9.1 Reaction of color formation in MTT assay

Most viability assays are based on one of two characteristic parameters: metabolic activity (enzyme) or cell membrane integrity of healthy cells. Usually the metabolic activity is measured in cell populations via incubation with a tetrazolium salt (e.g., MTT, XTT*, WST-1**) that is cleaved into a colored formazan product by metabolically active cells. The ATP status of cells can also give and gives an indication for cellular energy capacity and thus viability. The other assay is based on dye-exclusion that takes advantage of the ability of healthy cells with uncompromised cytoplasmic membrane integrity to exclude dyes such as Trypan blue. Dead cells are stained and thus healthy cells can be counted directly. In cytotoxicity studies, dying cells with leaky cytoplasmic membranes releases lactate dehydrogenase or [^{51}Cr] release which can be measured.

9.2 CELL PROLIFERATION ASSAY

The most prominent parameter for analyzing cell proliferation is the measurement of DNA synthesis (a specific marker for replication). In this assay, labeled DNA precursor ([^{3}H]-Thymidine or 5-bromo-2, deoxyuridine [BrdU]) is added to cells (or animals) and their incorporation into genomic DNA (during the S phase of the cell cycle) is quantified. The amount of labeled precursors incorporated into the DNA is quantified either by measuring the total amount of labeled DNA in a population, or by microscopically detecting the labeled nuclei. Since incorporation of the labeled precursor into DNA is directly proportional to the rate of cell division. It is useful while studying non synchronized cell (cells in different phases of the cell cycle) from synchronised cell. Cell proliferation can also be measured using indirect parameters such as, molecules that regulate the cell cycle are quantified (by measuring their activity) e.g., cyclin-dependent kinase assays or by direct quantification (e.g., Western blots, ELISA, or immunohistochemistry).

9.3 ASSAY TECHNIQUES

There are three different type of assay (cell proliferation) used for cell proliferation such as fluorescence-based assay, colorimetric assay, and radiometric assay using [3H] thymidine incorporation. Each requires fairly expensive equipments, availability of which then determines the assay of choice. There are many different alternative assays that include total protein/culture, total DNA/ culture, nuclear counts, lactate dehydrogenase or alkaline phosphatase enzyme measurement, packed cell volume, and so forth.

9.3.1 Fluorescent Assay

Fluorescent assay is perform using fluorescence microscope fluorometer, fluorescent microplate reader eg. A probe, calcein-acetoxymethyl (AM) which has ability to permeates the intact cell membrane and inside the cell, the esterases in the cytoplasm hydrolyze the esterified fluorophore and the molecule fluoresces (Tsien, 1989). A fluorescent plate reader can be used to read the plate. Other example of fluorescent molecules are Ethidium monoazide, Hexidium iodide, etc.

9.3.2 MTT Assay

MTT [3-(4,5-dimethylthiazolyl)-2-5-diphenyltetrazolium bromide] colorimetric assays is based on the reduction of MTT by living cells to an insoluble (in aqueous solutions), dark purple formazan precipitate (utilized for reading by a standard plate reader). This is a frequently used in high-throughput assay method for cytotoxicity by utilizing [3H] Thymidine incorporation method into

*(2, 3-bis(2-methoxy-4-nitro-5-sulphophenyl)-2H-tetrazolium-5 carboxanilide

**Water soluble tetrazolium salts

DNA during nuclear replication (or repair) and incorporated DNA can be precipitated by trichloroacetic acid (TCA) and measured in a scintillation counter.

9.3.2.1 MTT Assay for Cell Proliferation

The reduction of tetrazolium salts is now widely accepted as a reliable way to examine cell proliferation. The yellow tetrazolium MTT (3-(4, 5-dimethylthiazolyl-2)-2, 5-diphenyltetrazolium bromide) is reduced by metabolically active cells, in part by the action of dehydrogenase enzymes, to generate reducing equivalents such as NADH and NADPH.

The resulting intracellular purple formazan can be solubilized and quantified by spectrophotometric means. The MTT Cell proliferation assay measures the cell proliferation rate and conversely, when metabolic events lead to apoptosis or necrosis, the reduction in cell viability. The MTT Reagent is ready to use and stable at 4°C in the dark for up to eighteen months, provided there is no contamination. (Care should be taken not to contaminate the MTT Reagent with cell culture medium during Pippetting).

9.3.2.2 Advantages of MTT Assay

1. This is the most prevalent *in vitro* assay and is supposed as most rapid, versatile, quantitative and highly reproducible technique and can be adaptable to large-scale screening of viable cells.
2. MTT reduction correlates to indices of cellular protein and earlier cell number and is very sensitive and more earlier predictor of toxicity than classical LDH or neutral red measurements

9.3.2.3 Disadvantage of MTT Assay

1. Production of the formazan is dependent on the MTT concentration in the medium. The kinetics and degree of saturation are dependent on the cell type.
2. Assay is less effective in the absence of cell proliferation.
3. MTT cannot distinguish between cytostatic and cytocidal effect.
4. Individual cell numbers can't be quantitated and results are expressed as a percentage of control absorbance.
5. Test is less effective if cells have been cultured in the same media that has supported growth for a few days, which leads to under estimation of control and untreated samples.

9.3.3 Acid Phosphatase Assay

The action of Acid phosphatase in many of tissue is to cleave a waste product called pyrophosphate and effectively convert it to a usable phosphate Pi. P-nitrophenyl phosphate is used as substrate which gives nitrophenol as product. Nitrophenol is colorless in acidic pH but gives yellow color in alkaline pH. See Fig. 9.2 a.

9.4 VIABILITY STAINS

These stains are used to determine the proportion of living cells in a population. A living cell is metabolically active and selectively uptakes the materials into the cytoplasm, while dead cells loose this selective capacity as a result everything pass into the cell. Certain dyes qualify for selective exclusion by a live cell while a dead cell are rapidly stained. The dyes used for this test are *Eosin Y, erythrosine B, and trypan blue*, the latter being the most common. The Sigma catalog has a detailed procedure for the use of the trypan blue stain in the tissue culture section. Living cells will appear refractile and colorless, while dead cells will stain blue. (See Fig 9.2 b)

9.4.1 Dye Exclusion Test (Trypan Blue)

Trypan blue dye exclusion is a rapid reliable and recommended method that relies on the membrane integrity. Trypan blue is one of several stains recommended for use in dye exclusion procedures for viable cell counting. This method is based on the principle that live cells do not take up certain dyes, whereas dead cells do. *Following steps are adapted for viable cell count.*

(a)
O PO2 — OH → OH + Pi
O NO — O NO
PNPP — P-nitrophenol
OH ↓
O NO
P-nitrophenolate

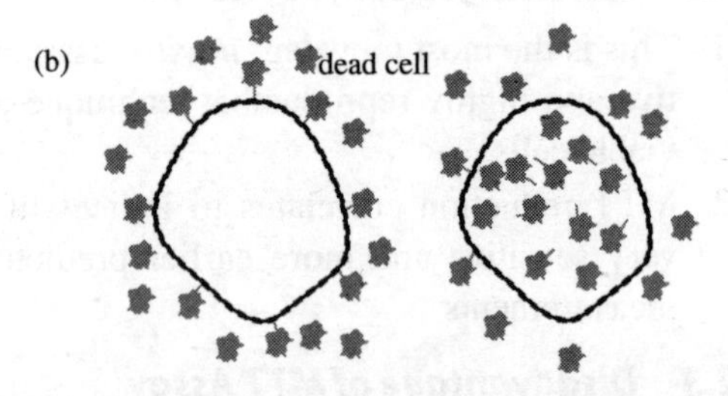

Fig. 9.2 (a) Complete reaction of substrate P-nitrophenyl phosphate and acid phosphatase which lead to formation of Nitrophenolate (colour). [P-nitrophenyl phosphate 'Acid phosphatase' Nitrophenol (Colourless) "Nitrophenolate (colour)] (b) Dye exclusion test: Alive cell don't allow dye to go inside while in dead cell dye enters inside the cytoplasm

1. Prepare a cell suspension after trypsinization, either directly from a cell culture or from a concentrated or diluted suspension (depending on the cell density) and combine 20 μl of cells with 20 μl of Trypan blue suspension (0.4%). Mix it thoroughly for 5-15 minutes.
2. Add a small amount of Trypan blue-cell suspension to both chambers of the hemocytometer by carefully touching the edge of the cover slip with the pipette tip and allowing each chamber to fill by capillary action. Do not overfill or underfill the chambers.
3. Count all the cells in each square (1 × 1) for viable and nonviable cells separately.
4. If there are too many or too few cells to count, repeat the procedure either concentrating or diluting the original suspension as appropriate.
5. Do not count cells touching middle line at bottom and right. Count all 4 corners of squares and middle square and calculate the average in each square of the hemocytometer. Note: each square represents a total volume of 0.1 mm^3 or 10^{-4} cm^3. Since 1 cm^3 is equivalent to approximately 1 ml, the total number of cells per ml can be determined using the following calculations:

$$\text{Cells per ml} = \text{average cell count per square} \times \text{dilution factor} \times 10^4$$

Precautions

1. Right coverslip should be used otherwise it will disturb the volume.
2. Coverslip should be properly placed on the platform.
3. The aliquot must be suspended before counting.
4. Adherent cells must be counted immediately after trypsinization.
5. Not less than 200 cells should be counted.

9.5 CELL COUNTING (SORTING) TOOLS

9.5.1 Hemacytometer

Hemacytometer is used to count the cell number within a defined area of known depth (0.1 mm)and the cell concentration is derived from the count.

Figure 9.3 show grid and chamber of hemocytometer. The chambers are covered with a glass coverslip that rests exactly 0.1 mm above the chamber floor.

Thus, volume of neubauer's chambers is 0.1 mm^3 and the number of cells in 1 mm^2 (cells/ml) can be calculated by the formula:

$$\% \text{ Cell viability} = \frac{\text{total viable cells (unstained)}}{\text{total cells} \times 100 \text{ Cells/ml}}$$

$$= n \times N \times 10^4 \times \text{dilution factor}$$

where n = no. of chamber in 1 mm^2 area,

N = Average no. of cells in 0.1 mm^2

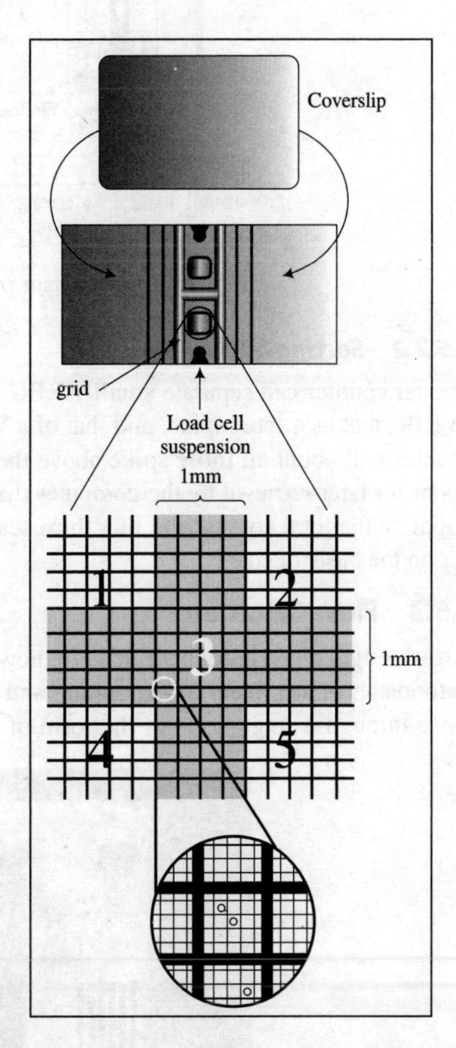

Fig. 9.3 Hemacytometer slide (Improved Neubawer) and coverslip. hemacytometer (numbers added show cell counting chamber)

9.5.2 Coulter Counter

The Coulter Particle Counter has been depicted in Fig. 9.4. It helps in the counting the number of cells per ml and also the number of the cells of different sizes. It is manufactured by Coulter electronics limited and was designed by Wallace H. Counter. Particles under pressure are drawn through the orifice, changing the electric resistance in the current flowing through the orifice resulting in generation of electric pulses that can be counted. Orifice of different sizes can be obtained. Sizes of the pulse are proportional to the volume of the cells so the size of pulse determines the density and size of the cells. It helps in complete blood count in hospitals an also in characterizing nano-scale size particles.

9.5.2.1 Counting

The basis of counting depends on the fact that whenever a particle goes through the hole, the electronic system detects a sudden and momentary increase in resistance (a partial interruption of current flow) and a green vertical line appears on the screen, as shown in Fig. 9.4

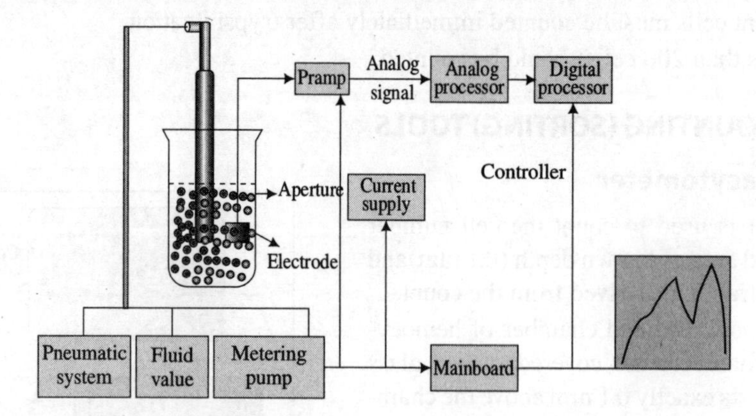

Fig. 9.4 Mechanism of cell counting by coultour counter

9.5.2.2 Sorting by Size

Coulter counter can separate smaller RBC from larger white blood cells (WBC). The passage of an RBC makes a short spike, and that of a WBC makes a tall spike on the oscilloscope. Thus, the counter will count all those spike above the imaginary line and put that number into one storage scalar for later retrieval by the computer that is usually connected to the device. In storage, scalar might be the total counts, and in a third scalar all the "short" counts. Thus, blood can be shorted out on the basis of size.

9.5.3 Flow Cytometer

It was designed by Heisenberg in 1976, now it is manufactured by the Coulter electronics limited, Orthodiagnostics system incorporation and Beckten and Dickinson, UK. In flow cytometer, cells move through a single tube in the form of droplets. The cells of different sizes and charged are

Fig. 9.5 Coulter counter

first stained by fluorescent dye (propidium iodide or chlorogenic A3 for DNA) and then are passed through jet former. Cells coming out of the jet former, when intercepts a laser beam then light scatters and fluorescence is produced. Photomultiplier detects the fluorescence emitted and sends data to the computer. Computer analyses the cells passing the laser by the wavelength of fluorescence produced, optical density, spectrum produced and intensity.

9.5.3.1 Advantages

Cells of different types can be counted separately and simultaneously. It has high efficiency and can count 2000-5000 cells per ml. Cells can be separated on the basis of parameters like DNA content, receptor and cell cycle etc.

9.5.3.2 Disadvantages

(a) Instrument are very expensive and need trained operators
(b) Rotor gets infected often due to presence of cells.

9.5.3.3 Applications

1. To initiate cultures from a primary sources when cell suspension is a mixture of different cell types.
2. When individual cells have to clone to develop cell strains.
3. To separate Melanoma and Hybridoma.

9.5.4 High-throughput Assays

Sometimes it is preferable to use a more rapid readout method for assaying the proliferative effect of a factor. This format becomes convenient when the factor to be assayed is in short supply. Alternatively, obtaining a sufficient number of cells may be a rate-limiting step. Often, the factor to be assayed needs to be titrated over a wide range or a large number of samples are to be tested. In such cases, a rapid readout format can be quite useful. This type of assay is also easier to adapt to high-throughput, including automated (i.e., robot), formats. It is difficult to develop a high-throughput assay for actually counting cell number. Therefore, many investigators uses a secondary endpoint, which usually correlates well with increased or decreased cell number, as a screen. The thymidine incorporation assay described above is one such secondary endpoint. Others are based on colorimetric readouts such as crystal violet, a protein stain, or fluorometric assay, such as Calcein-AM, whose level in cells is proportional to esterase activity in the cells. Many of these assays are extremely sensitive and therefore can be automatized with cells in 96 well plates.

The results can be automatically read on a plate reader, analyzed by an attached computer, and printed in a form suitable for publication.

9.6 DIRECT METHODS OF CELL COUNTING

9.6.1 Acoustic Resonance Densitometry (ARD)

This method has been successfully used for measuring the concentration of hybridoma and human lymphoma cell culture. The method utilizes use of the linear relationship between density of the fluid and the resonance frequency of the cell culture sample. Cell culture sample is enclosed in a test chamber and is excited electromagnetically to vibrate at its natural frequency. Fluid density is measured from the square of the oscillation period of the sample. A hollow fiber device filters out the supernatant and its density is also determined. This is then subtracted from the density of the cell culture and gives the density of the cells.

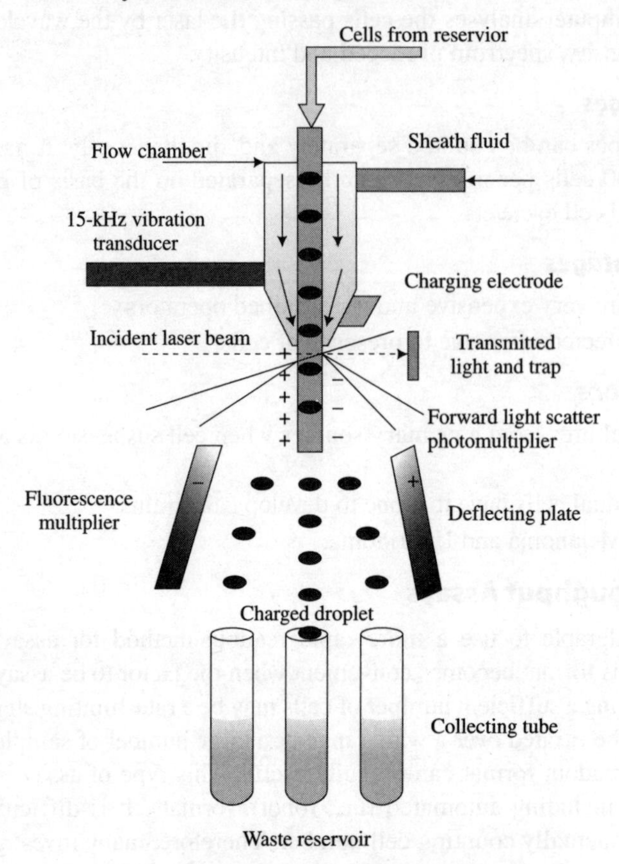

Fig. 9.6 Flow cytometer working flow chart

9.6.2 NMR Spectroscopy

It is based on the measurement of total Na^+ per unit volume of cell culture. The culture is placed in a magnetic field and exposed to electromagnetic radiation and the reaction of the Na^+ nuclei is measured. Since intracellular Na^+ concentration is low relative to extracellular Na^+ concentration, with increase in cell concentration, total Na^+ will decrease and will give an estimate of the cell concentration. NMR continuously monitors the high density of viable cells in hollow-fiber Bioreactor. Other methods fail at high concentrations.

9.6.3 Methods Based on Electrical Properties

Electrical properties of a culture solution are exclusively dependent on the cell concentration, and so this property can be used for measuring cell concentrations. The electrochemical method (fuel cell) measures the intensity of electric current between two electrodes immersed in a cell culture.

Since the current results from direct oxidation of cells on the surface of the anode, it gives a direct measure of cell concentration. **A conductivity meter** can measure electrical conductivity. This is directly proportional to the cell concentration and can be easily measured in a hollow fiber reactor. Another electrical property made use of for measuring cell concentration is capacitance (C). Each viable cell works as a capacitor.

9.6.4 Optical Techniques

These are the most popular techniques, but cannot distinguish between viable and nonviable cells. Fluorescence Cells are irradiated with UV (340 nm), and fluorescence occur due to NADPH. Fluorescence is proportional to cell concentration, but at high concentration it is less sensitive. Improvements are being made to overcome this problem. Light absorbance scattering is also used as a measurement of cell concentration. Most of the commercial instruments are designed to measure high cell density and are less sensitive at low concentration. An image analysis system with a CCD camera may lead to a breakthrough in bioprocess monitoring. The cells can be viewed on the screen and cell concentration can be monitored.

9.6.5 Laser-beam-based Commercial Equipment

These are being used in microbial bioprocess monitoring and may be used for cell culture bioprocess monitoring in future. They measures real-time particle size or cell concentration.

9.6.6 Magnetic Activated Cell Sorting (MACS)

Magnetic activated cell shorting is based on cell separation after tagging of magnetizable ferritin beads conjugated with specific antibodies which are available in the market in the name of Dynabeads from Dynal. These magnetized beads attracts conjugated antibodies which are directed to surface markers of the cells to be separated. This technique has been utilized in the separation of populations of Lymphocyte and tumor cells. See Fig. 9.7.

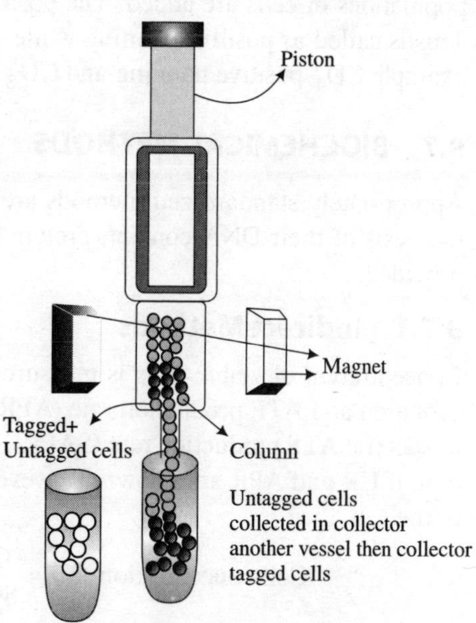

Fig. 9.7 Magnetic activated cell shorter

Table 9.1 Comparative cell selection assay

Category of viability assay	*Assays*	*Principles*
Membrane integrity assay	– Exclusion dyes – Fluorescent dyes – LDH leakage	The determination of membrane integrity via dye exclusion from live cells
Functional assay	– MTT, XTT assay – Crystal violet/Acid phosphatase (AP) assay	Examining metabolic components that are necessary for cell growth

Contd..

	- Alamar Blue oxidation reduction assay - Neutral red assay - [3H]-thymidin/BrdU incorporation	
DNA labeling assay	- Fluorescent conjugates	Simultaneous cell selection and viability assay
Morphological assay	- Microscopic observation	Determination of morphological change
Reproductive assay	- Colony formation assay	Determination growth rate

9.6.7 Immune Panning

This technique has been utilized in separation of population of lymphocyte or to surface specific cell expressing cell surface markers from the mixed populations for example to separate CD_{34}^{+} cells from bone marrow mononuclear cells or to separate T cells from peripheral mononuclear cells. The specific markers are attached and incubated overnight to the bottom of the petri dish in which populations of cells are added. The positive cells are attached while negative cells are washed of. This is called as positive panning while separation of other cell is called as negative panning (for example CD_{4}^{+} positive panning and CD_{8}^{+} negative panning).

9.7 BIOCHEMICAL METHODS

Appropriately standardized methods are used to count the cell number. Cells can be counted on the basis of their DNA content, protein content etc. but true reflection of the cell number is not provided.

9.7.1 Indirect Methods

Concentration of viable cells is measured in terms of oxygen uptake rate (OUR), carbon dioxide evolution and ATP production rate (APR). A constant called specific oxygen uptake rate (SOUR) or specific ATP production rate (SAPR), which is determined experimentally using cell concentration, if UR and APR are known. However SOUR and SAPR fluctuate with the physiological state of the cell.

$$\text{Cell concentration (X)} = \frac{\text{OUR}}{\text{SOUR}}$$

or

$$= \frac{\text{X.APR}}{\text{SAPR}}$$

9.8 SEPARATION OF CELLS

9.8.1 Velocity Sedimentation (Celsap and Staput Apparatus)

In this method cells are layered on to a fluid column containing various concentrations of 0-2 % BSA, 0-30 % serum in PBS, or 2-4 % Ficoult (density between 1.0074-1.0113 gm/ml). Cells are mixed with them and when the cells are allowed to fall down under gravity for several hours. Cells get separated depends on their size and density. This can be done by the equipment named as Celsap and Staput apparatus. See Fig. 9.8.

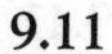

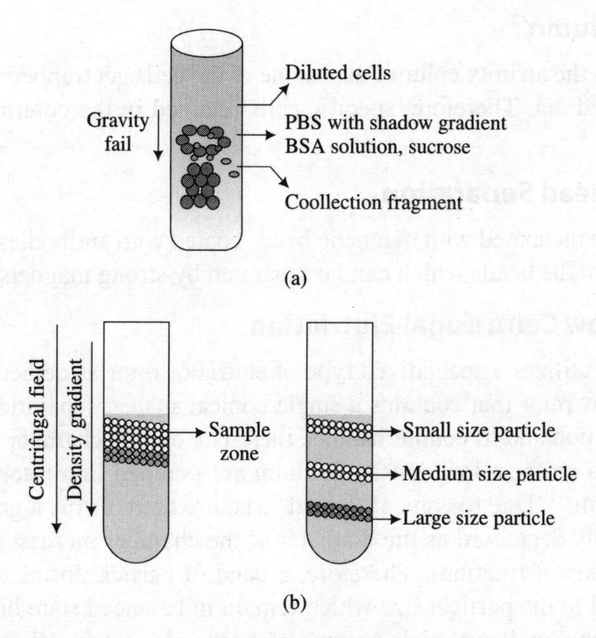

Fig. 9.8 Velocity sedimentation and density gradient formation due to medium and centrifugal field

9.8.1.1 *Advantages*

(a) Separates cells from a mixed population, into several populations in bulk, depending on their size and density.
(b) It helps to separate cells in aseptic conditions.
(c) It is very useful in separation of bone marrow stem cells, Chondrocytes etc.

9.8.1.2 *Disadvantages*

(a) Most of mammalian cells have overlapping density.
(b) Very small number can be processed.
(c) It is expensive equipment.

9.8.2 Density Gradient Centrifugation (Buoyant Density Centrifugation)

In this method cells are layered on a density gradient made up of mixture of Ficoll and Hypaque and centrifuged with sufficient force for sufficient time so that they can end up in layer which is equal to their own density. Ficoult (Synthetic sucrose, MW 77,000, Conc. 5.6 % v/v) Hypaque (Sodium Metrizoate, 9.6% v/v) is used to maintain the density gradient. Sodium Metrizoate is 3, 5-bis (acetlyamino)-2, 4, 6 triiodobenzoic acid sodium, and is named as **Isoprake** by some companies. 1.077 gm/ml is the desired density for the separation of Lymphocytes.

9.9 CELL SEPARATION BASED ON CHARGE AND AFFINITY

9.9.1 Electrophoresis

Type and amount of charge on cells helps in cell separation by this method. (The cells are separated by flow cytometer utilizing this property).

9.9.2 Affinity Column

Cells if passed through the affinity column then some of the cells get trapped in the column matrix while other cells passed out. Therefore, specific cells retained in the column can be eluted later on.

9.9.3 Magnetic Bead Separation

In this process cells are incubated with magnetic beads coated with antibodies, hence cells specific to antibodies adhered to the beads which can be separated by strong magnets. See Fig. 9.7

9.9.4 Counter-flow Centrifugal Elutriation

Centrifugal elutriation utilizes a specialized type of elutriator rotor as depicted in Fig. 9.8. This a type of continuous flow rotor that contains a single conical shaped separation chamber, which is away from axis of rotation and to counterbalance there is a bypass chamber on the opposite side. Particle suspended in a uniform low density medium are pumped into rotor spinning at a speed of 1000 to 3000 Rev min^{-1}. Due to conical shaped structure there forms a gradient of liquid flow velocity which gradually decreased as the diameter of the chamber increases towards its centripetal end (toward the axis of rotation). Therefore, a band of particle forms due to sedimentation velocity is proportional to the particle size which remain in balanced state due to liquid flow rate in the opposite direction. Smaller particle accumulates towards centripetal end and larger particle towards centrifugal ends. Various type of cells has been separated by this technique such as monocyte from lymphocyte, bulk separation of rat brain cells and different type of mammalian testis cells. Thus in brief following advantages are there by this techniques. See Fig. 9.9.

(a) Cells are separated on the basis of their size and density
(b) It is a rapid technique
(c) Large numbers of cells (10^8-10^9) are separated simultaneously
(d) High yield and viability of cells is obtained
(e) Centrifugal forces provides in the medium hence helps in maintaining stability and viability

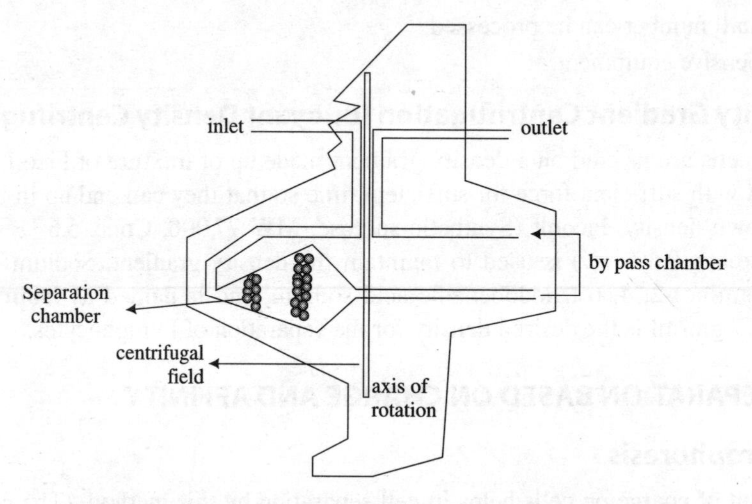

Fig. 9.9 Separation of particle by centrifugal elutriation

9.9.5 Cells Selection on the Basis of Selective Killing

Selective killing can be done by using antigen- antibody or by using complement protein (by activating complement system). Dead cells can be separated by using antibody against CD_4^{++} rabbit/guinea pig and by using serum as a complement source. Cell adhesion helps in macrophage separation; Dispase II, while selectively dislodges epithelial cells leaving fibroblasts.

9.9.6 Cytotoxicity

Cell cultures can be used for checking the toxicity of variety of pharmacological agents and are also used for toxicological investigations.

9.10 METABOLISM OF TOXICITY

Toxic substance are neutralize after being metabolized in the liver. Toxicity cannot be checked unless and until the agent is pretreated with enzyme extracted from the liver or it is genetically transformed cell line which can metabolize the substance in similar manner as liver does.

9.11 ASSAY SYSTEM FOR CYTOTOXICITY

9.11.1 Thymidine Release Assay

^3H is a γ-emitter and scintillations are counted to measure the cytotoxicity. General protocol followed is: Dividing cells is incubate in a medium containing $^3H_1$ thymidine for incorporation in the dividing cells after washing unincorporated Thymidine by PBS buffer; labeled cells are treated with toxic agent; if lysis occurs the $^3H_1$ will be released and released amount will be proportional to the amount of cells lysed by the agent measure the radioactivity in the culture supernatant.

9.11.2 Cytotoxicity Assay by Utilizing MTT

MTT assay can also be used to determine cytotoxicity of potential medicinal agents or any other toxic materials, since those agents would result in cell toxicity and therefore mitochondrial dysfunction and therefore decreases performance in the assay. Yellow MTT [3-(4,5-Dimethylthiazol-2-yl)-2,5-diphenyltetrazolium bromide, a tetrazole] is reduced to purple formazan with mitochondrial reductase enzymes (can be used as a measure of viable cells. However, it is important to keep in mind that other viability tests sometimes give completely different results, as many different conditions can increase or decrease metabolic activity. Changes in metabolic activity can give large changes in MTT results while the number of viable cells is constant. When the amount of purple formazan produced by cells treated with an agent is compared with the amount of formazan produced by untreated control cells, the effectiveness of the agent causing death, or changing metabolism of cells, can be deduced through the production of a dose-response curve.

9.12 PROLIFERATION ASSAY

9.12.1 Thymidine Uptake Method

This proliferation assay is used in determination of the number of cells that are growing in the absence or presence of certain proliferation affecting agents, e.g., TNF-alpha or anti-Fas antibody

(IPO^{-4}). Thymidine uptake is measured in 96 well plates coated with scintillant or using filter binding methods. Plates containing scintillant detect the thymidine taken up by the cells by using 14°C or ^{3}H.

9.12.1.1 Principle

A certain number of cells seeded in the wells of a 96 well plate is added with a proliferation affecting agent. The cells are incubated for a certain time (e.g. 24 hrs) at 37°C in presence of ^{3}H-Thymidine. Cells that are incubated without any growth inhibiting agent will grow during each cell division and the cells will incorporate ^{3}H-Thymidine into its DNA. The more will be the cell divisions (or the higher the proliferation rate) the more radioactivity will be incorporated into DNA. Thus cells that are incubated in the presence of the growth inhibiting agent (e.g. TNF-alpha) will incorporate less radioactivity. After incubation the cells are harvested and during harvesting, the cells are washed out of the wells of the 96 well plate with water. The cells and organelles burst and the cell's DNA is set free. The cell fragments and DNA are passed through a filter membrane (glass fiber) but only particles of smaller than 1.5 μm can pass the filter. So, intact DNA (with a fragment length in the range of millimeter or even centimeters) will not be able to pass the filter but be collected on the filter membrane.

The higher the proliferation rate of the cells and the more radioactive DNA will be collected on the filter. The filter membrane is dried and the amount of radioactivity (what corresponds to the number of cells in the well or the number of cell divisions during incubation) is counted in a scintillation counter.

To calculate the inhibition of proliferation in presence of a growth inhibitor compare the counted radioactivity (counts per minute = cpm) of cells that were not treated, cpm (untreated), with the cpm in cells that were treated with agent, cpm (treated):

$$\text{Inhibition of Proliferation [\%]} = \frac{\text{cpm [(untreated) – (treated)]}}{\text{cpm (untreated)}}$$

$$\%\ \text{Cytotoxicity} = \frac{\begin{pmatrix}\text{Radioactivity released (cpm) in}\\ \text{target cells incubated with toxic agent}\end{pmatrix} - \begin{pmatrix}\text{Radioactivity released (cpm)}\\ \text{without toxic agent}\end{pmatrix}}{\text{Total radioactivity incorporated in target cells}}$$

9.12.2 WST Method

This type of a colorimetric assay is used for the quantification of cell proliferation and cell viability, based on the cleavage of the tetrazolium salt WST-1 [2-(4-iodophenyl)-3-(4-nitrophenyl)-5-(2,4-dis-. ulfophenyl)-2H tetrazolium, monosodium salt] by mitochondrial dehydrogenases in viable cells (a soluble version of MTT or MTX). Tetrazolium salt (WST-1) is cleaved to soluble formazan dye by the succinate-tetrazolium reductase (EC 1.3.99.1), which exists in mitochondrial respiratory chain and is active only in viable cells. Total activity of this mitochondrial dehydrogenase in a sample rises with the increase of viable cells. As the increase of enzyme activity leads to an increase of the production of formazan dye, the quantity of formazan dye is related directly with the number of metabolically active cells in the medium. The formazan dye formed by metabolically active cell can be quantitated by measuring its absorbance by ELISA reader. The absorbance is directly proportional to the number of viable cells. This product is also designed for nonradioactive and spectrophotometric quantification of cell growth and viability in proliferation and chemosensitivity assay.

9.12.2.1 Application

1. The analysis of cytotoxic and cytostatic compounds such as anticancer drugs, etc. can be done.
2. The physiological mediator and antibodies which inhibits the cell growth can be assayed.

9.12.3 Alamar Blue (Colorimetric Assay)

Analysis of cell proliferation and cytotoxicity is a vital step in evaluating cellular health and in the drug discovery process. Alamar Blue is a proven cell viability indicator that uses the natural reducing power of living cells to convert Resazurin to the fluorescent molecule, Resorufin. The active ingredient of Alamar Blue (Resazurin) is a nontoxic, cell permeable compound that is blue in color and virtually nonfluorescent. Upon entering cells, resazurin is reduced to resorufin, which produces very bright red fluorescence. Viable cells continuously convert resazurin to resorufin, thereby generating a quantitative measure of viability-and cytotoxicity.

9.13 APOPTOSIS ASSAY

To study Apoptosis is the process of programmed cell death which leads to various changes such as cell shrinkage, nuclear fragmentation and chromosomal changes. Necrosis and apoptosis, cytotoxicity assays can be used. These assays are principally of two types. Radioactive and nonradioactive assays that measures increase in membrane permeability, (since dying cells become leaky). Nonradioactive assays includes colorimetric assay that measure reduction in the metabolic activity of mitochondria. Mitochondria in dead cells cannot metabolize dyes, while mitochondria in live cells can that has been described earlier.

Following apoptotic parameters can be used for the assay:

1. **Fragmentation of DNA** occur in populations of cells or in individual cells, in which apoptotic DNA breaks into different length pieces.
2. **Alterations in membrane asymmetry.** Phosphatidylserine translocates from the cytoplasmic to the extracellular side of the cell membrane.
3. **Activation of apoptotic caspases.** This family of proteases sets off a cascade of events that disable a multitude of cell functions.
4. **Release of cytochrome C into cytoplasm by mitochondria.**

9.13.1 General Annexin V Staining Procedure

Phospholipid phosphatidylserine (PS) is located in the cytosolic leaflet of the plasma membrane lipid bilayer. PS redistribution from the inner to the outer leaflet is an early and widespread event during apoptosis. However, in necrosis, PS becomes accessible due to the disruption of membrane integrity. Apart from necrosis and apoptosis, PS also becomes accessible in activated platelets, in certain cell anomalies such as sickle cell anemia, in erythrocyte senescence, upon degranulation of mast cells, and in certain stages of B cell differentiation. PS exposure also serves as a trigger for the recognition and removal of apoptotic cells by staining with propidium iodide (PI), can be directly detected through their staining with fluorochrome-conjugated Annexin V. Dead cells are stained with both Annexin V and PI, whereas viable cells cannot be stained with either. Macrophages, Annexin V is a 36 K Dalton phospholipid-binding protein has a high affinity to PS in the presence of physiological concentrations of calcium (Ca^{2+}). Apoptotic cells, which are otherwise undetectable by staining with propidium iodide (PI), can be directly detected through their staining with fluorochrome-conjugated Annexin V. Dead cells are stained with both Annexin V and PI, whereas viable cells cannot be stained with either.

9.13.2 Radioactive DNA Fragmentation Assay

The DNA fragmentation assay determines the amount of DNA degraded when treated with certain toxic agents, e.g. with TNF-alpha or anti-Fas antibody (IPO^{-4}).

9.13.3 DNA Laddering Assay (for Treated Cells)

DNA laddering can be detected from samples containing upto 8% apoptotic cells. Alternatively, the cells can be stained with DAPI and analyzed by flow cytometry.

1. First prepare lysis buffer (1 ml) : 10% NP-40 100 µl + 200 mM EDTA 100 µl + 0.2 M Tris-HCl (pH 7.5) 250 µl + D.W. 550 µl
2. Add 1 ml of trypsin to monolayer culture and, harvest the cells by centrifugation (3,500 rpm, 5 min), and wash cell pellets with 1X PBS
3. Now add 100 µl of lysis buffer for 10 sec. centrifuge it and obtain supernatant. Add 10 µl of 10% SDS solution to pooled supernatant, treat with 10 µl of 50 mg/ml RNase A (final 5 µg/µl) and incubate for 2 h at 56 C.
4. Add 10 µl of 25 mg/ml Proteinase K (final 2.5 µg/µl) to digest protein and incubate for 2 h at 37°C.
5. Add 1/2 vol. (65 µl) of 10 M ammonium acetate. Add 2.5 vol. (500 µl) of ice-cold ethanol and mix thoroughly. Stand for 1 h in - 80°C freezer ("ethanol precipitation" to precipitate DNA).
6. Centrifuge it for 20 min at 12,000 rpm, wash the white pellet with 200 µl 80% ice-cold ethanol and air-dry for 10 min at room temperature. Dissolve the pellet with 50 µl of TE buffer Determination of DNA concentration can be done at Abs. 260 and 2% agarose gel electrophoresis of the same concentration of DNA (about 4 µg).

9.13.4 Cytochrome 'C' Release Assay

Apoptotic cell death is a fundamental feature of virtually all the cells. It is an indispensable process during normal development, tissue homeostasis, and development of the nervous system and the regulation of the immune system. Insufficient or excessive cell death can contribute to human disease, including cancer or degenerative disorders. Like most mitochondrial proteins cytochrome 'C' is encoded by a nuclear gene and synthesized as a cytoplasmic precursor molecule, apocytochrome 'C', which becomes selectively imported into the mitochondrial intermembrane space. The mitochondria turned out to participate in the central control or executioner phase of the cell death cascade.

Cytochrome 'C' was identified as a component required for the crucial steps in apoptosis, during caspase-3 activation and during DNA fragmentation. Cytochrome 'C' was shown to redistribute from mitochondria to cytosol during apoptosis in intact cells. Mitochrondrial cytochrome 'C' is a water - soluble protein of 15 k Da with a net positive charge, residing loosely attached in the mitochrondrial intermembrane space.

Cytochrome 'C' functions in the respiratory chain by interaction with redox partners and is highly conserved. The molecular mechanisms responsible for the translocation of cytochrome 'C' from mitochondria to cytosol during apoptosis are unknown. Measurement of cytochrome 'C' release from the mitochondria is a tool to detect the first early steps for initiating apoptosis in cells. Cytochrome 'C' release in the cytosol occurs prior to the activation of Caspases and DNA fragmentation which is considered the hallmark of apoptosis. Detection of cytochrome 'C' released from the mitochondria to the cytoplasm can be achieved by a selective lysis of the cell membrane.

9.13.5 Caspase-3 Assay

Cells suspected to undergo apoptosis are first lysed to collect their intracellular contents. The cell lysate can then be tested for protease activity by the addition of a Caspase-specific peptide conjugated to the fluorescent reporter molecule (7-amino-4-trifluoromethyl coumarin (AFC)). The cleavage of the peptide by the caspase releases the fluorochrome which get excited in light (at 400 nm wavelength), and emits fluorescence at 505 nm. The level of caspase activity in the cell lysate is directly proportional to the fluorescence signal detected with a fluorimeter or a fluorescent microplate reader. Caspase-3 cleaves a variety of cellular molecules that contain the amino acid motif DEVD such as poly ADP-ribose polymerase (PARP), the 70 kD protein of the U1-ribonucleoprotein and a subunit of the DNA dependent protein kinase. The presence of caspase-3 in cells of different lineages suggests that caspase-3 is a key enzyme required for the execution of apoptosis.

According to sigma protocol, the Caspase 3 Fluorometric Assay Kit is based on the hydrolysis of acetyl Asp-Glu-Val-Asp 7-amido-4-methylcoumarin (Ac-DEVD-AMC) by caspase 3, resulting in the release of the fluorescent 7-amino-4-methylcoumarin (AMC). The excitation and emission wavelengths of AMC are 360 nm and 460 nm respectively. The concentration of the AMC released can be calculated from a calibration curve prepared with AMC standards (AMC standard included with the kit).

9.14 ACTIVATION ASSAY

9.14.1 Measurements of Intracellular Second Messengers

Activation of metabolic pathways: measurement of phosphorylation events intracellular or intranuclear components measured by using radioactive ligands

9.14.2 Morphological Assay

Large-scale, morphological changes that occur at the cell surface, or in the cytoskeleton, such as nuclear swelling, chromatin flocculation, loss of nuclear basophilia (during necrosis) or cell shrinkage, nuclear condensation, and nuclear fragmentation (Apoptotic cells). Damage can be identified by large decrease in volume secondary to losses in protein and intracellular ions of due to altered permeability to sodium or potassium.

9.15 CELL SYNCHRONIZATION

When cells are cultured (10 and 20 cultures) due to different phases of cell cycle, (hence the cell population is said to asynchronous) heterogenous cell population is produced. Cell synchronization is the process by which cells are made to start at the same phase their cell cycle simultaneously. There are various methods to make a cell in synchronization for producing homogenous cells. Sometime treatment of some chemicals makes the entire cell in synchronization. These chemicals make the process possible because of interference in the cell cycle at some stage and therefore all the cells remain in the same phase. (See Fig. 9.10)

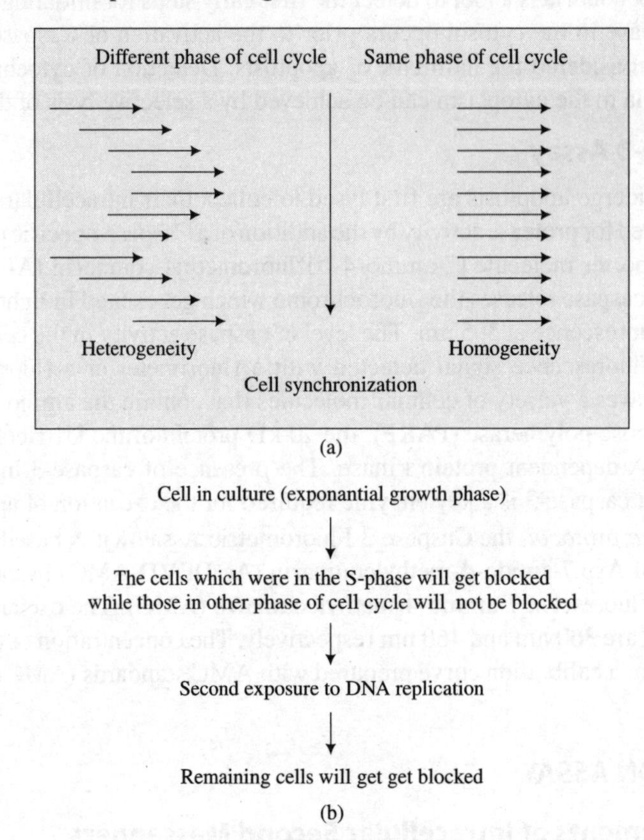

Fig. 9.10 (a) Cell synchronization by nutrition deprivation method (b) Graphical representation and flow chart of cell synchronization by using S-phase blocker

9.16 METHOD OF CELL SYNCHRONIZATION

9.16.1 Nutrition Deprivation Method

By making cells arrested in Go phase and then allowing them to re-enter cell cycle together. It is done by nutrition deprivation i.e. withdrawl of some growth factors helps in production of homogenous populations.

9.16.2 Mitotic Block Method (M- phase arrest)

Mitotic blocking agent such as Colchicine, Demecolcine (colcemid, (100 ng/ml)), Vinblastine, Nocadazole (50 ng/ml) etc. blocks the microtubule assembly which prevents the mitotic spindle formation and hence the mitosis is blocked. Nitrous oxide can also be used for mitotic block at high pressure (80 psi) for 6-8 hours. In long duration treatment with blocking agents, cells can escape or irreversibly arrested.

9.16.3 Arresting the Cells in S- Phase of Cell Cycle

It can be done by using following chemicals:

9.16.3.1 Hydroxyurea (2mM)

It inhibits ribonucleotide reductase by destroying tyrosine free radical on this enzyme. DNA replication gets affected as dCTP formation is inhibited.

9.16.3.2 High Quantity of Thymidine (2mM)

Excess of thymidine will lead to over accumulation of thymidine phosphate and this will allostearically inhibit cytidine diphosphate by ribonucleotide reductase as a result dCTP formation is inhibited.

9.16.3.3 Aphidicolin (3mM)

It binds to DNA polymerase and inhibits DNA replication. Doses can very from cell to cell and species to species and is used for He La cells and Chinese Hamster Ovary cells. Prolonged exposure to S- phase blocking agents is toxic to the cells.

9.16.4 Cell Synchronization by Mitotic Shake Off

Mitotic cells tend to round up and loosely attach to the substratum. These cells can be easily harvested by gentle shaking. However this technique is applicable only to anchorage dependent adherent cells. It is advantageous as there is no need to arrest the cell in phase of cell cycle but there is very low yield (only approximate 2 % of population is in mitotic phase at a given time).

9.16.5 Cell Synchronization by Centrifugal Elutriation

In this method there is no use of drugs to arrest the cells. In G1 phase cell size is half as compared to the G2 or M phase therefore, cells can be centrifuge and seperated on basis of size by density layring. It is useful for both adherent and non adherent cells and yield is very high 10^{-7} cells in one elutriation.

9.16.5.1 Disadvantages of Synchronization

1. Absolute synchronization is never achieved.
2. Some degree of contamination of cells in other phase of cell cycle.
3. Cells in intermediate stages also disturb synchronization.

BRAIN QUEST

1. Thymidine is the ________ emitter.
2. MTT assay is done for ________.

3. What are different assay for cytotoxicity?
4. Describe the mechanical disaggregation of the cell?
5. Define trypsinization. Why it is better than mechanical assay.
6. How a cell explants is obtained? Describe any one method in detail.
7. Describe different method used for culturing of isolated explants.
8. T-cells can be separated by ________.
9. MTT is ________.
10. Least count of the Hemocytometer is ________.
11. Coulter counter is used for ________.
12. Define cell synchronizations. Why it can not be achieved.
13. Discuss the different method of getting cell synchronizations.
14. Discuss the advantages and disadvantages of cell synchronizations.

GROWTH PHASES OF CELL IN CULTURE

10

Here we will learn about the growth behavior and their different phases which decide the final fate of successful animal cell culture. Kinetic study helps in deciding various aspects like time of subculture, their nutrient requirements, and the aspects of increasing their life for cell line production.

10.1 INTRODUCTION

Cells at different phases behave differently with respect to proliferation, enzyme activity, glycolysis respiration and synthesis of specialized products, etc. Normal cells usually cease to proliferate in a cell-cycle-specific way. They arrest in G_1 or enter a state of quiescence (G_0) from G_1 after depletion of serum or growth factors (Temin 1971; Pardee 1974; Baserga 1976) or nutrients (Prescott 1976) or after cell crowding (Nielhausen and Green 1965; Zetterberg and Auer 1970). The ideal method for determining the correct seeding density for subculture of cell is to draw a growth curve at different seeding concentrations. This helps in determination of the minimum cell concentration with short lag period and early entry into logarithmic growth. Therefore, It is essential to become familiar with the growth cell cycle for each cell.

10.2 GROWTH PHASES

Cell follows a characteristic pattern of the growth such as lag, log (or exponential), stationary (plateu) and death phase. The log or exponential phase gives vital information about the cell characteristic such as maximum growth and the doubling time. It also helps in monitoring time for exact serial passage of the culture and in the calculation of the cell yield. There are following type of growth phases of the cell. generally cells follows four phases of the growth as described below.

1. *Lag phase*
2. *Log phase*
3. *Stationary phase*
4. *Decline or Death phase*

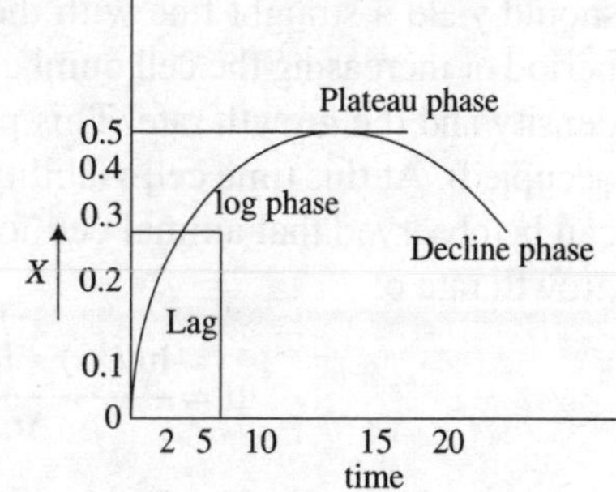

Fig. 10.1 Growth phases of cells

10.2.1 Lag Phase (Adapting Phase)

Initially, the cell remains in the adopting mode and prepares themselves for the cell attachment to the substrate and for cell-cell interaction. DNA polymerase increases due to increase in DNA contents.

10.2.2 Log Phase (Dividing Phase)

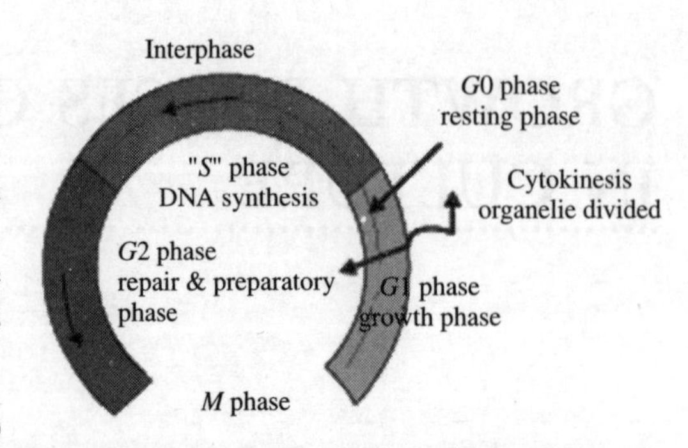

Fig. 10.2 Growth cycle phases

In the log phase the growth fraction remains high and the cell is in its most reproducible form. Mathematically, this exponential growth can be described in term of increase in biomass (X). Therefore, the rate of growth is dependent on the biomass concentration. This can be described as follows:

$$\text{Rate of change of biomass } \left(\frac{dx}{dt}\right) = X \qquad (10.1)$$

where X = concentration of biomass (g/L),
μ = specific growth rate (per hour) and
t = time (h).

When graph is plotted of cell biomass verses time, the product is a curve with a constantly increasing slope (see Fig. 10.1). Equation 10.1 can also be rearranged to estimate the **specific growth rate**

$$\mu = \frac{1}{X} \cdot \left(\frac{dx}{dt}\right) \qquad (10.2)$$

During any period of true exponential growth, equation 10.1 can be integrated to provide the following equation:

$$X_t = X \cdot e^{-\mu t} \qquad (10.3)$$

where X_t = biomass concentration after time' t,
X_o = biomass concentration at the start exponential growth, and
e = base of the natural logarithm.

$$\ln (X_t) = \ln (X_0) + t \qquad (10.4)$$

Taking natural logarithms, log (In), gives equation in the form $y = mx + c$ (intercept on y axis) where m = gradient, which is the general equation for a straight-line graph. For cells in exponential phase, a plot of natural log of biomass concentration against time, gives a semilog plot, which should yield a straight line with the slope (gradient) equal to Eq. 10.5 (Fig. 10.3). Log phase is a period of increasing the cell number exponentially. The length of the log phase may depend on cell density and the growth rate. This phase stops at once after the confluency (all the space has been occupied). At this time cell viability is high and is in the most uniform stage. From the Fig. 10.3 it can be observed that animal cell follows monod kinetics which gives equation in term of specific growth rate or

$$\mu = \frac{\ln (X_t) - \ln (X_0)}{\Delta t}$$

$$\mu = \left(\frac{\mu_{max} \cdot S}{k_S + S}\right) \qquad (10.5)$$

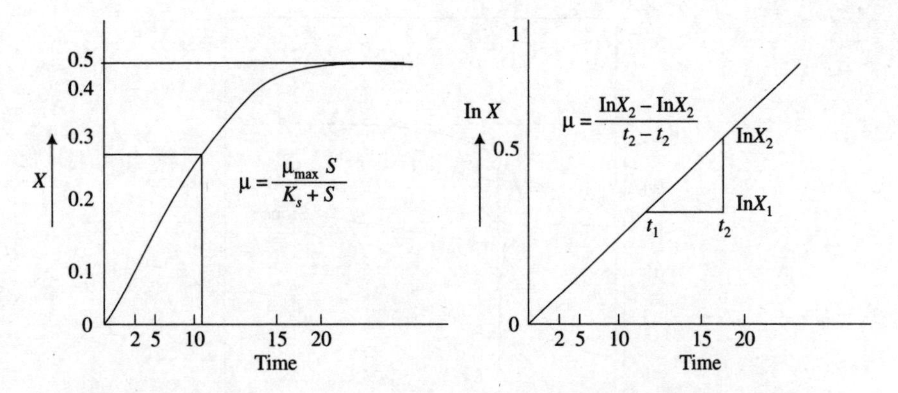

Fig. 10.3 Variation of cell mass with time and plot of log X with time

Doubling Time

The time taken by the culture to increase two folds is called as doubling time. During exponential growth of particular microorganism the cell dry weight get doubles at regular interval, so the number of doubling after time *t*.

$$N_0 = \left(\frac{N_t}{t_d}\right) \tag{10.6}$$

Where *n* is the number of doubling, t is the time period elapsed t_d is the doubling time of culture in h^{-1} hours.

If we consider a situation where at time zero, the cell biomass is X_0, then after a fixed period of time (*t*) of exponential growth, equivalent to one doubling time (t_d), the microbial biomass will double to $2X_0$, i.e. $X_t = 2X_0$, when $t = t_d$, Substituting these parameters into Eq. 10.3 gives

$$\text{if } X_t = 2X_0$$

$$2X_0 = X_0\, e^{-\mu}$$

taking log

$$t_d = \left(\frac{\ln 2}{\mu}\right) \tag{10.7}$$

Monod showed that growth rate is an approximate hyperbolic function of the concentration which is growth-limiting (*s*). This impact of essential nutrient depletion on- growth can be described mathematically by the Monod equation, in a form similar to that used in biochemistry, where Michaelis-Menten kinetics define the rate of an enzyme catalyzed reaction in relation to its substrate concentration.

$$\mu = \left(\frac{1}{2} \cdot \mu_{max}\right) \tag{10.8}$$

where μ_{max} = maximum specific growth (per hour) of the cells, i.e. when substrate concentration is not limiting; S = concentration of limiting substrate (g/L); K_s = saturation constant, concentration (g/L) of limiting nutrient enabling growth at half the maximum specific growth rate, and is a measure of the affinity of the cells for this nutrient.

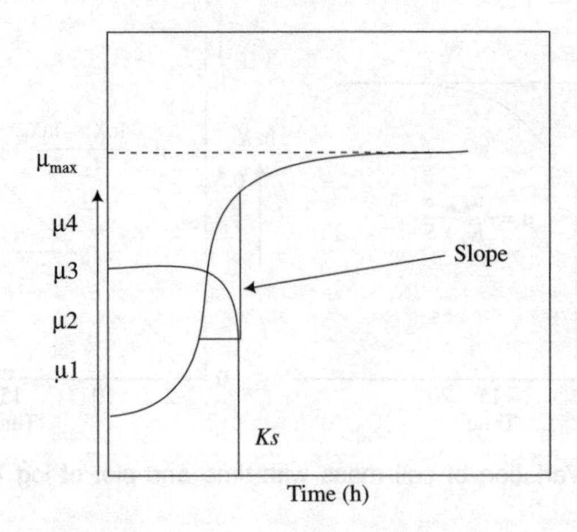

Fig. 10.4 Pattern of specific growth rate, for various medium component in different time interval that gives separate specific growth rate which can be used to plot a graph (between specific growth rate and time (h))

10.2.3 Stationary Phase

Stationary phase starts at the end of the log phase when confluency is acheived i.e. the entire substrate surface available for the growth is occupied and thus growth is reduced due to depletion of substrate. Further cell proliferation ceases, cell becomes less motile, membrane may become raffling, and contact inhibition started. Thus a monolayer is formed except cell line (except by those cell that are transformed by viruses, or chemical carcinogens) that acheives high cell density during this phase. Its may be due to loss of density inhibition which converts them into the cell line.

10.3 GROWTH RESPONSE CURVE

Growth response curve is different for each cell type (specific), therefore, a balance among interrelated components of culture system is required. Cells multiply only when all of their requirements have been satisfied. Ham & Mckeehan (1978) proposed that for every required component in a cell culture medium a growth response curve can be plotted between colony formations (CFU) and different nutrient concentrations.

Thus growth response curve can be divided into three phases (see Fig. 10.5)

(1) Growth stimulation phase
(2) Plateau & mid plateau of optimal growth
(3) Growth inhibition.

Growth of cells occurs until all the media get depleted and at this stage maximum cell mass can be obtained which indicates the plateau of growth. Growth inhibition may occur beyond this point due to one or several reasons, like depletion of nutrients, or due to accumulation of some toxic materials. The more detailed description has been given further.

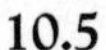

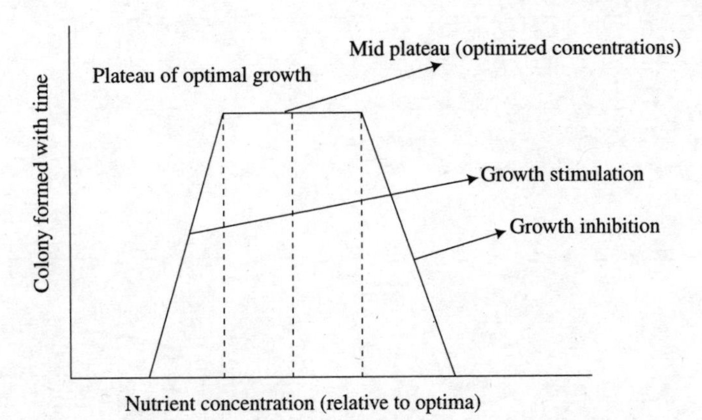

Fig. 10.5 Growth response curve plotted between nutrient concentration and colony formation

10.3.1 Stimulation of Growth

Any factor can acts as stimulant which can be identified by its treatment with cell for its effect on cell growth in different time intervals, since multiplication of cell is proportional to the amount of the nutrient component added.

10.3.2 Plateau of Growth

At saturation stage, growth is independent of the nutrients/component added, which can be obtained at the midpoint of the plot. This concentration is called as optimal "concentration".

10.3.3 Inhibition of Growth

Inhibition of growth occurs above a critical concentration of nutrients/component, after which the component added becomes inhibitory and growth response become negative. Thus growth response curve is positive till optimum concentration of nutrients is achieved.

BRAIN QUEST

1. Define doubling time of cells.
2. Describe growth response curve and its different phases.
3. Define Monod kinetics.
4. Describe different types of reactors used in animal cell culture to obtained the product.

CELL DIFFERENTIATION

11

In this Chapter we will study the different type of differentiations. Differentiation can be promoted by addition of several factors such as collagen and feeder layer. It helps in various type of tissue culture.

11.1 DIFFERENTIATION

In-Vivo Phenotypic expression is the result of differentiation. A well-differentiated cell is highly specialized, structurally highly organized and clearly recognizable. Differentiation involves changes in numerous aspects of cell physiology; size, shape, polarity, metabolic activity, responsiveness to signals, and gene expression profiles. Differentiation is often considered as a 'final step' of development in which cells take on their mature function. A cell that is able to differentiate into many cell types is known as **pluripotent**. A cell that is able to differentiate into all cell types is known as **totipotent**. In mammals, only the zygote and early embryonic cells are totipotent, such cells are called stem cells. Pluripotent stem cells undergo further specialization into multipotent progenitor cells that give rise to functional cells for example Hematopoietic stem cells (adult stem cells) from the bone marrow give rise to red blood cells, white blood cells, and platelets; Mesenchymal stem cells (adult stem cells) from the bone marrow give rise to stromal cells, fat cells, and types of bone cells; Epithelial stem cells (progenitor cells) give rise to the various types of skin cells and Muscle satellite cells (progenitor cells) that contribute to differentiated muscle tissue.

11.2 CLASSIFICATION OF DIFFERENTIATION

There are several type of differentiation invitro.

11.2.1 Adaptive Differentiation

Adaptive differentiation occur in the explant isolated from mature cells due to any factor provided from outside, therefore can be artificially be regulated for the expression.

11.2.2 Terminal Differentiation

Fully matured cell are said to be due to terminal differentiation.

11.2.3 Dedifferentiation

Dedifferentiation is usually as part of a regenerative process. Loss of basic properties to differentiate again, when a terminally differentiated cell is grown in culture is called as dedifferentiation. Certain drug such as Reversine, which is a purine analog, induces dedifferentiation in myotubes.

These dedifferentiated cells were then able to redifferentiate into other cells such as osteoblasts and adipocytes. Usually it is an adaptive process (may be recovered).

(i) tissues that replicate frequently (blood, skin)

Tissue that replicate frequently are Totipotent or Pluripotent stem cells. They proliferate and progress to terminal differentiation and lose capacity to divide again.

(ii) tissues that selectively replicate

Resting tissue has little chance of proliferation capabilities but in certain events (e.g. trauma) cell may lose specialization and they reenter again in division cycle and rapidly replicate when needed. In normal cell proliferation and differentiation cant go together except some Tumor cells are exceptions. They can undergo both proliferation and differentiation together.

11.3 PARAMETERS GOVERNING THE ENTRY OF CELLS INTO DIFFERENTIATION

If differentiation is required, then it is necessary to withdraw cells from the cell cycle. This can be achieved by removing, or changing, the growth factor supplementation; for example, the O_2, A common precursor of astrocytes and oligodendrocytes remains as a proliferating precursor cell in growth factors such as PDGF and FGF, whereas combining FGF with ciliary neurotropic factor (CNTF) results in differentiation into a type 2 astrocyte , and embryonal stem cells, which differentiate in the absence of leukemia inhibitory factor (LIF) and Phorbol myristate acetate (PMA, also known as TPA).

There are four main parameters governing the entry of cells into differentiation:

11.3.1 Growth Factors

Some growth factors like EGF, KGF, TGF-S and HGF, NGF), cytokines (IL-6, oncostatin-M, GM-CSF, interferon), vitamins (e.g., retinoid, vitamin D_3, and vitamin K) calcium [Table 16.1, Freshney, 2000], and planar polar compounds (e.g., DMSO and NaBt) [Tables 17.1,17.2, Freshney, 2005] increases cell differentiation in different cells depends on source.

11.3.2 Interaction with Matrix

Some cell adhesion molecule interacts with collagen IV, laminin, and proteoglycans thus increases interaction of cell. Some proteoglycans such as Heparan sulfate (HSPGs), in particular, have a significant role not only in binding to cell surface receptors but also in binding and translocating growth factors and cytokines to high-affinity cell surface receptors and thus helps in early establishment of the cell for further differentiations.

11.3.3 Enhanced Cell-cell Interaction

Certain cell to cell interaction will also promote differentiation. There are two type of interactions.

11.3.3.1 Homotypic Contact Interactions

This interaction (acts via gap junctions) tend to coordinate the response among many similar cells in a population by allowing free intercellular flow of second messenger molecules such as cyclic adenosine monophosphate (c-AMP) and via cell adhesion molecules such as E-cadherin or NCAM, which provides signal via anchorage to the cytoskeleton [Juliano, 2002].

11.3.3.2 Heterotypic Interactions

Heterotypic interaction (acts across a basal lamina) is likely to involve in direct cell-cell contact between dissimilar cells.

11.3.4 Effect of the Position, Shape, and Polarity of the Cells

The position, shape, and polarity of the cells may induce, or at least make the cells permissive for the induction of, differentiation for e.g. epidermal keratinocytes [Maas-Szabowski et al. 2002] and bronchial epithelial cells [Petra et al. 1993] require air/liquid interface, presumably to enhance oxygen availability. Establishment of polarity in the cells is promoted by the collagen. Similar conditions may be created in a perfuse capillary bed or scaffold or by adding serum.

11.4 FIBRONECTIN HELPS IN CELL ATTACHMENT

Glycoprotein such as cold insoluble Globulin (CIG) or Fibronectin (secreted from certain cells) promotes cell attachment and spreading under normal conditions. Many proteoglycans such as multivalent heparan sulfate mediate adhesion of cells to the culture surfaces by binding to glycoproteins present on the cell surface.

Many established and transformed cells secrete very small amounts of fibronectin. Certain cell such as diploid fibroblasts cells secretes significant quantities of fibronectin (therefore they do not require any exogenous glycoprotein for attachment to substrate). Cultures supplemented with 10% fetal calf serum contains - 2-3 μg fibronectin/ml, a large proportion of the fibronectin adsorbed by the cultures within a few minutes. A minimum of 15 nanogram of adsorbed fibronectin is required for spreading of BHK cell line. Thus a standard culture procedures usually insures that the culture surface is coated with adequate amounts of glycoproteins to promote cell attachment to substrate.

11.4.1 Feeder Layer

Feeder layer is provided by inactivating collagen with UV rays to promote cell adhesion of normal cells. Two such treatments generally given are lethal irradiation (Puck and Marcus, 1995) and mitomycin *C* treatment which releases large amount of important substance (feed) needed for the growth of the cell. The cells need to be cultured are then cocultured on the feeder layer and thus cells rapidly forms the monolayer and reaches the confluency (i.e. fill all the available space), e.g. culture of the neuron cell reached to confluency after they are grown on the glial cell. (See Fig. 11.1)

Various cell cultures show that survival of the animal cell can be improved by addition of inactivated collagen, fibronectin, proteoglycan and the growth factor such as heparin binding group, FGF, insulin growth factors (IGF-1 and 2), PDGF, which provides extra support to the substrate; such substances are called as **feeder layer**. They have property to secrete certain extra growth factor, cytokines, and metabolites; therefore used for **conditioning** of the medium.

Within the intact tissue, most of the cells attached tightly and interact specifically with other cells via various cellular junctions. The cell also interacts with the extracellular matrix, (a complex network of secreted proteins and carbohydrates that fills the spaces between cells). The matrix, whose constituents are secreted by cells themselves, helps in binding the cells in tissues together; it

also provides a lattice through which cells can move, particularly during the early stages of animal differentiation.

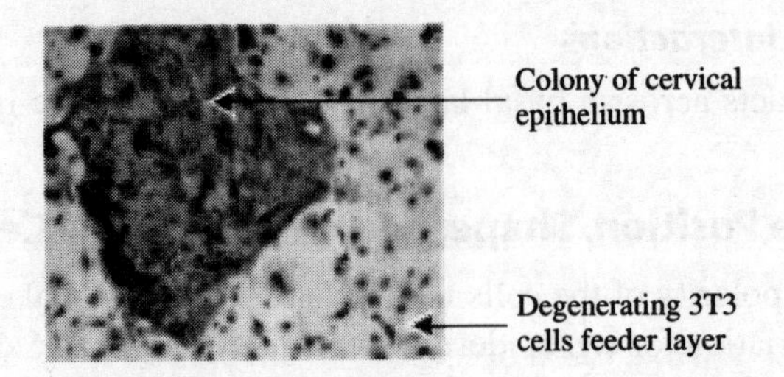

Fig. 11.1 Mechanism of cell attachment with and without glycoprotein

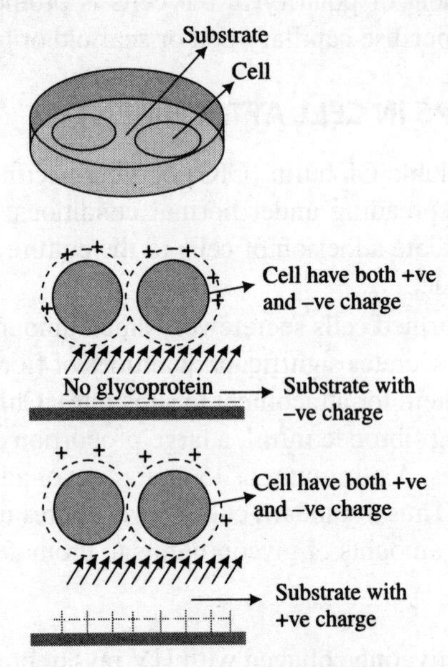

Fig. 11.2 Co-culture of cervical epithelium with irradiated 3T3 fibroblast leads to cell attachments without glycoproteins

BRAIN QUEST

1. Write down the different type of differentiations in various type of cells.
2. Write down the role of feeder layer in supporting the animal cell culture.
3. What are the different factor that effects the differentiation.
4. Write down the advantage and disadvantages of using feeder layer.

PRIMARY AND SECONDARY CELL CULTURE

12

Here we will study about the method of culture that increases the life span of the cells. Here we will learn about different method of sub culturing of animal, cell also called as passaging.

12.1 INTRODUCTION

First culture is called as Primary cell culture. Primary cells grow and devide to form monolayer of the cells. Basic steps are shown in Fig. 12.1. First a piece of tissue is disaggregated and is diluted (up to 10^6 cells/ml) to form suspension cell culture from which few cells are sucked to be used as explant. After 24-72 hours whole plate is filled with single layer of cells called as monolayer cells and stops growing. The primary culture is again cultured in fresh media to establish **Secondary Cultures**. The concept of maintaining cells as cell lines separate from their original tissue was discovered in the 19th century by English physiologist Sydney Ringer.

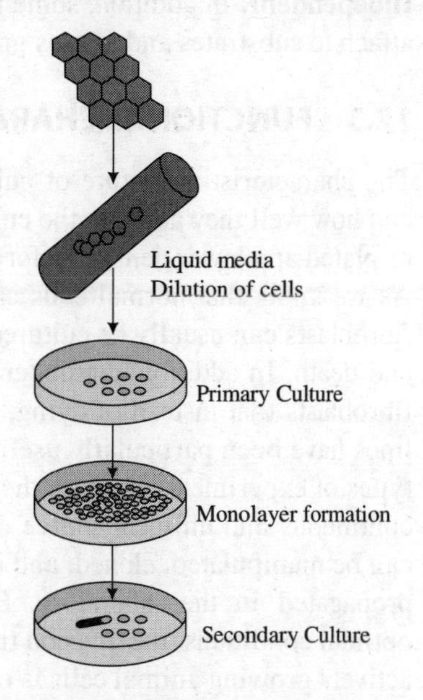

Fig. 12.1 Steps in primary and secondary cell culture.

Ross Granville Harrison established the methodology of tissue culture. The Salk polio vaccine was one of the first products produced using cell culture since then many products such as enzymes, hormones, immunobiologicals (monoclonal antibodies, interleukins, lymphokines), and anticancer agents were produced. Cell culture is a fundamental component of tissue culture and tissue engineering, as it establishes the basics of growing and maintaining cells ex vivo. Thus, the creation of cell lines that maintains some functional properties (of adrenal cells, pituitary cells (Bounassisi *et al.* 1962), neurons (Augusti-Tocco and Sato, 1969), myocytes (Yaffe, 1968), and hepatocytes (Thompson *et al.* 1966)) have been used for the study of growth, response to hormones and other environmental factors and in the production and secretion of hormones.

12.2 CELL CULTURE SYSTEMS

Two basic culture systems are used for growing the cells. These are based primarily upon the ability of the cells to either grow attached to a glass or treated plastic substrate (**Monolayer Culture Sytems**) or float freely in the culture medium (**Suspension Culture Systems**).

Monolayer cultures are usually developed in (tissue culture) treated dishes, T-flasks, roller bottles, or multiple well plates, the choice being based on the number of cells needed, the nature of the culture environment, cost and personal preference.

Suspension cultures are usually grown either:

1. In magnetically rotated spinner flasks or shaken Erlenmeyer flasks where the cells are kept actively suspended in the medium;
2. In stationary culture vessels such as T-flasks and bottles where, although the cells are not kept agitated, they are unable to attach firmly to the substrate.

Many cell lines, especially those derived from normal tissues, are considered to be **Anchorage-Dependent,** that is, they can only grow when attached to a suitable substrate. Some cell lines that are no longer considered normal (frequently designated as **Transformed Cells**) are frequently able to grow either attached to a substrate or floating free in suspension; they are **Anchorage-Independent.** In addition, some normal cells, such as those found in the blood, do not normally attach to substrates and always grow in suspension.

12.3 FUNCTIONAL CHARACTERISTICS

The characteristics feature of cultured cells is derived from both their origin (liver, heart, etc.) and how well they adapt to the culture conditions. Cells from primary cultures can be isolated and re-plated at a lower density to form secondary cultures. This process can be repeated many times. As we know that normal cells can be grown for a limited number of times e.g. normal human fibroblasts can usually be cultured for 50 to 100 populations, after which they undergo Apoptosis and death. In addition, a number of rodent cell lines have been isolated from cultures of normal fibroblasts that instead of dying, a few cells proliferate indefinitely, forming cell lines. Such cell lines have been particularly useful for many types of experiments because they provide a continuous and uniform source of cells that can be manipulated, cloned, and indefinitely propagated in the laboratory. Even under optimal conditions, the division time of most actively growing animal cells is on the order of 20 hours, ten times longer than the division time of yeasts.

Biochemical markers can be used to determine if cells are still carrying on specialized functions that they performed in vivo (e.g., liver cells secreting albumin). Morphological or ultrastructural markers can also be examined (e.g., beating heart cells). Frequently, these characteristics are either lost or changed as a result of an artificial environment. Some cell lines will eventually stop dividing and show signs of aging.

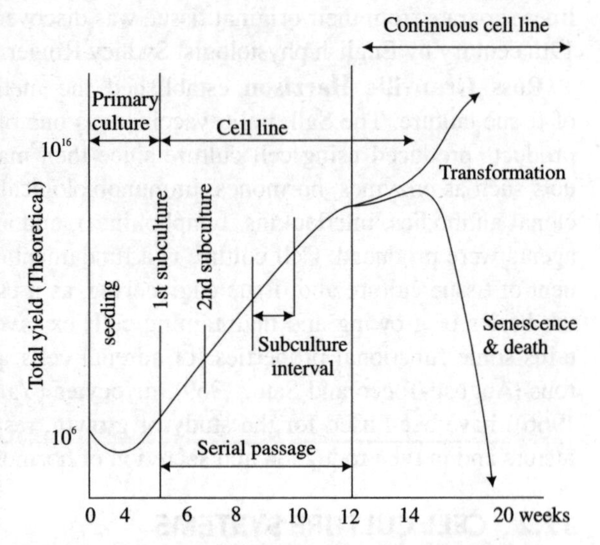

Fig. 12.2 Illustration of cell line formation after 12 weeks (forms continuous cell line while some undergo senescence and death).

These lines are called **Finite cell line**. Other lines, become immortal; and divide indefinitely and are called as **Continuous** cell lines (See Fig. 12.2). A finite cell line becomes immortal after it has undergone (a fundamental irreversible change) "transformation". This can occur spontaneously or be brought intentionally using drugs, radiation or viruses. **Transformed Cells** are usually easier and faster growing, may often have extra or abnormal chromosomes and frequently grow in suspension culture. Cells having the normal number of chromosomes are called **Diploid**; other than the abnormal number are called **Aneuploid**. If the cells form tumors when they are injected into animals, they are considered to be **Neoplastically Transformation**.

Senescence is determined by a number of intrinsic factors regulating cell cycle, such as Rb and p53 and results in shortening of the telomeres on the chromosomes [Wright and Shay, 2002]. Once the telomeres reach a critical minimum length, the cell can no longer divide. Telomere length is maintained continously by telomerase, which is downregulated in most normal cells except germ cells. The more detail are given in next chapter-14. (See Fig. 12.3)

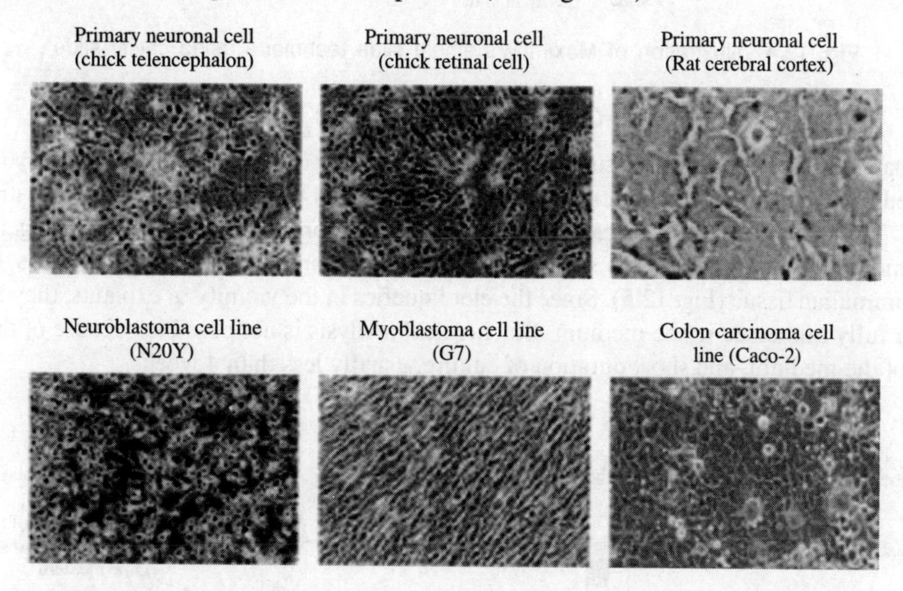

Fig. 12.3 Examples of primary cultures and continuous cell line cultures

12.4 PRIMARY EXPLANTATION TECHNIQUES

Harrison (1907) and **Carrel** (1912) developed primary explantation techniques for cultivation of fresh tissue derived from the organism. These were modified into various specialized techniques for embryo and organ culture. The different explantation techniques are.

1. *Slide or Coverslip Culture*
2. *Carrel Flask culture and*
3. *Roller Test Tube culture*

12.4.1 Slide/Coverslip/Culture

The oldest and most widely used method of tissue culture was the use of slide and coverslip. It is still used often for morphological studies and micrographic investigations. The fragment of tissue (explant) is placed on coverslip, which is subsequently inverted over the cavity of a depression slide containing plasma clot (see Fig. 12.4). Live cells are spread out in a manner suitable for microscopy and photography. Since cells grow directly on the coverslip, it can be fixed and stained to make permanent slides. This technique has also been utilized in organ culture. The technique is relatively simple and inexpensive. Slide culture has a disadvantage that only a very small amount of tissue can be cultured. Since supply of oxygen and nutrients are exhausted rapidly, the medium becomes acidic. So, rapidly growing tissue is transferred to new medium. Moreover sterility cannot be maintained for a long period in such type of culture.

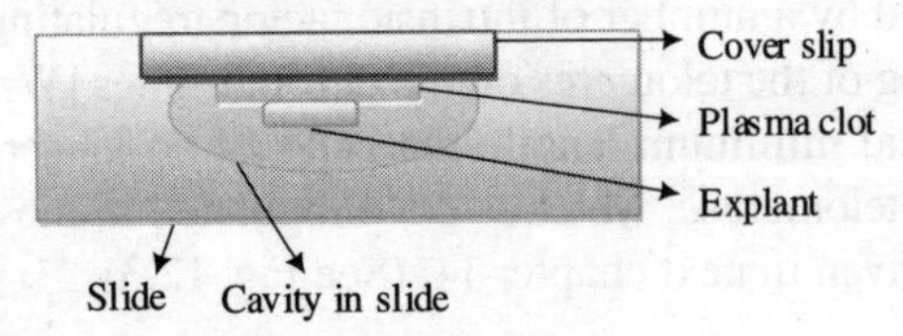

Fig. 12.4 Illustration of Maximow's single slide technique using cavity slide

12.4.2 Plasma Clot (Watch Glass Technique)

The explant culture is done on the surface of a clot (consisting of plasma and chick embryo extract contained in a watch glass), which may or may not be closed with a glass lid, sealed with paraffin wax. This has been the classical (standard) technique for studying tissue morphogenesis in embryonic organ rudiments and to study the action of hormones, vitamins, carcinogens, etc. On adult mammalian tissue (Fig. 12.5). Since the clot liquefies in the vicinity of explants, they become partly or fully immersed in the medium. Biochemical analysis is not possible because of the complexity of the medium, and short duration of culture, usually less than 4 weeks.

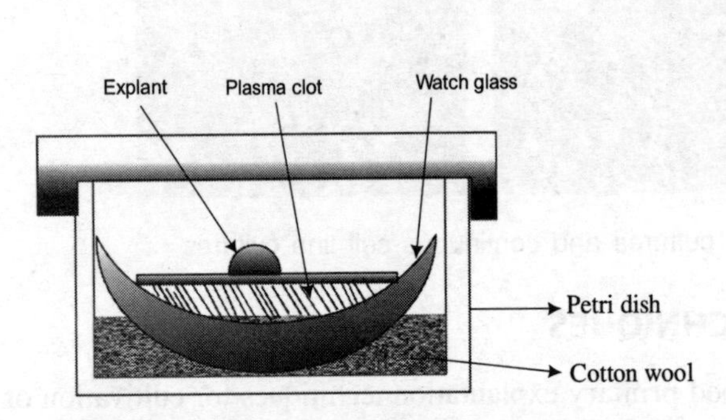

Fig. 12.5 Technique of organ culture, watch glass method of culture on a plasma clot

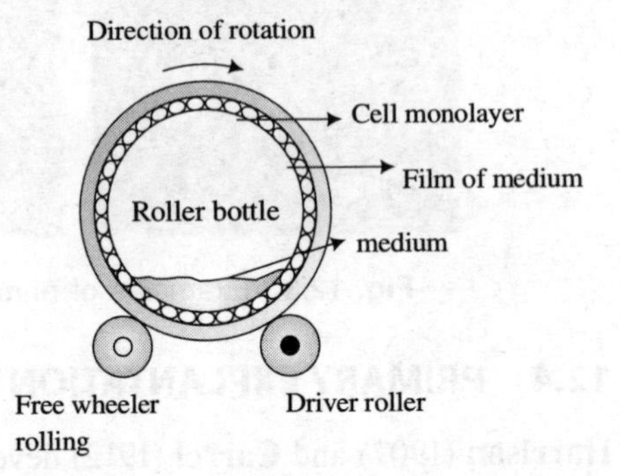

Fig. 12.6 Roller tube culture

12.4.3 Flask Cultures

The Carrel Flask Technique is used for the establishment of a strain from explants of a tissue, since flask has excellent optical properties for microscopic observation. There are two types of flask techniques. Thick dot culture, that allow rapid growth suitable for short term cultures, and thin dot culture which can be maintained for a considerable period. See Fig. 12.7

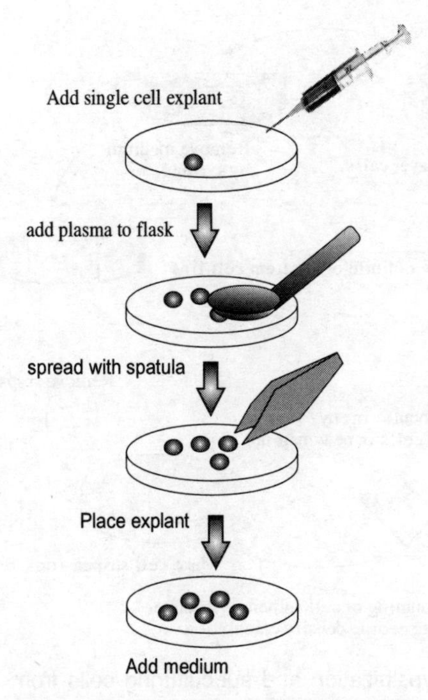

Fig. 12.7 Preparation of flask culture

12.4.4 Roller Tube Culture

See Section 6.4.2 (See Fig. 12.6)

12.5 SUBCULTURING

When disaggregated tissue pieces or explant are cultured, they attaches to the substrate, and forms a monolayer of cells. After a period of time, the available substrate are depleted and the cell stops growing and they are revived by seeding into a fresh vessel. The process of inoculating cultured cells into fresh culture vessels is termed as passaging or subculturing. This is usually done by removing them as gently as possible from the substrate using enzymes (See Fig. 12.8). Subculturing, or "splitting of cells," is required periodically for normally growing cells. The frequency of subculture and the split ratio, or density of cells plated, will depend on the characteristics of each cell. If cells are split too frequently or at too low a density, the line may be lost. If cells are not split frequently enough, the cells may exhaust the medium and die, or a different type of cell may be selected for in mixed cell cultures. In order to analyze the growth characteristics of a particular cell type or cell line, a growth curve can be established from which one can obtain a population doubling time, a lag time, and a saturation density. The culture becomes relatively more stable by the third subculture. This is the most striking that after the first subculture many cell types may be lost due to their inability to tolerate trypsinization and reseeding in fresh culture vessels.

Subculturing may be done every 7 days, but some fast growing lines require a change of medium after 3-4 days (feeding). There are two considerable factor involve before subculturing-is growth pattern of the cell and harvesting time.

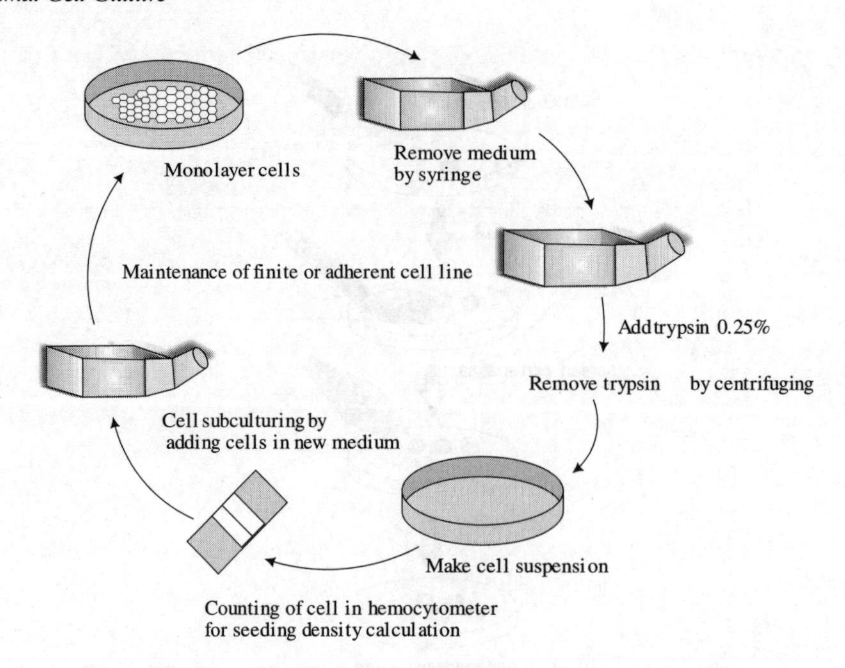
Monolayer cells
Remove medium
by syringe
Maintenance of finite or adherent cell line
Addtrypsin 0.25%
Remove trypsin by centrifuging
Cell subculturing by
adding cells in new medium
Make cell suspension
Counting of cell in hemocytometer
for seeding density calculation

Fig. 12.8 Trypsinization and subculturing cells from a monolayer

12.5.1 Growth Pattern

A cell generally follows lag, log or exponential, continuous phase and lastly death phase but it also depends on the cell type, the seeding density, the media components, and previous handling. The cells showing the highest metabolic activity in exponential confluent phase is achieved in stationary phase where the number of cells is constant. (for details see chapter 10).

12.5.2 Harvesting

Cells are harvested when the cell reached at confluent stage. Ideally, cells are harvested when they are in a semi-confluent state or still they are in log phase. At confluent stage, contact inhibition occur which results in apoptosis of cells.

12.6 PROCESS OF SUBCULTURE

Subculturing starts with process called as disaggregation of tissue which is done by the use of mild treatment of protease or mechanically at low temperature. The enzymatic treatment (like trypsin) are given in order to breaks the fibers that hold the cell. Such cells are always kept in the buffer. Enzyme can be removed by simple washing with BSS (balanced salt solution or by spinning them with medium many times.

We need to subculture the cells in following conditions.

1. When pH became acidic.
2. When cells are deprived of the nutrients in media.
3. When attached cells covered the whole surface available for growth.
4. Change in morphology is observed.

Sometimes change of medium is enough if they used nutrients but did not outgrow the surface.

12.7 SERIAL SUBCULTURE, SPLIT RATIO AND DOUBLING TIME

Whenever a culture undergoes subculture, the growth cycle is repeated. The number of doublings should be recorded with each subculture that can be simplified by reducing the cell concentration at the time of subculture by a power or two, the so-called split ratio (Fig. 12.9). A split ratio of two allows one doubling per passage; four, two doublings, and eight, three doublings, and so on. The number of elapsed doublings should be recorded so that the time to senescence can be predicted and new stock prepared. Recurring growth curves during serial subculture contains one or two detailed growth curve analyses and each cycle should be a replicate of the previous one, such that the same terminal cell density is achieved after subculture at the same seeding concentration. See in Fig. 12.9. The graph is plotted between the passage number, i.e., the number of times the culture has been subcultured and the generation number, i.e., the number of times the population has doubled. In this example, we can see that cell population get doubles three times between each subculture, suggesting that the culture should be split 1 : 8 to regain the same seeding concentration each time.

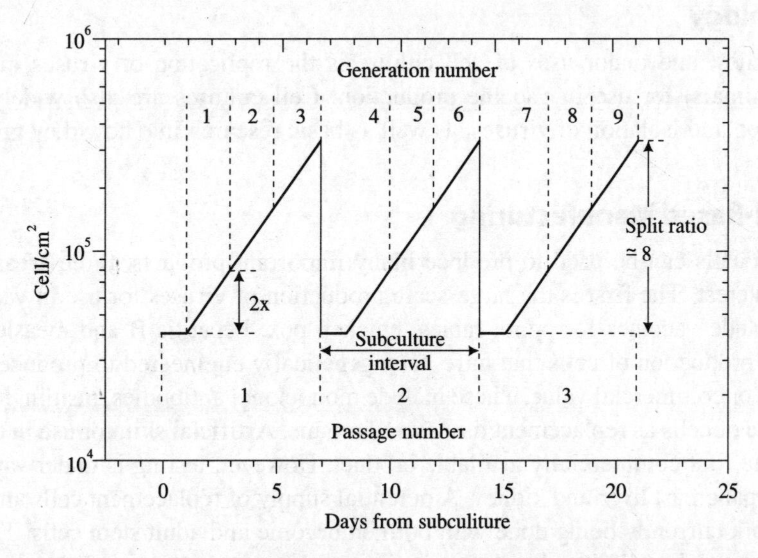

Fig. 12.9 Recurring growth curves during serial subculture to calculate split ratio. The lower number represents the passage number, i.e., the number of times the culture has been subcultured. The upper number numbers represent the generation number, i.e., the number of times the population has doubled. In this example, the cell populatin doubles three times between each subculture, suggesting that the culture should be split 1 : 8 to regain the same seeding concentration each time. [modified From Freshney, 2005.]

12.8 APPLICATIONS OF CELL CULTURE

Some of the important areas where cell culture is currently playing a major role are briefly described below:

12.8.1 Model Systems

Cell cultures provide a good model system for studying (1) basic cell biology and biochemistry, (2) the interactions between disease-causing agents and cells, (3) the effects of drugs on cells, (4) the process and triggers for aging, and (5) nutritional studies.

12.8.2 Toxicity Testing

Cultured cells are widely used alone or in conjunction with animal tests to study the effects of new drugs, cosmetics and chemicals on survival and growth in a wide variety of cell types. Especially important are liver- and kidney-derived cell cultures.

12.8.3 Cancer Research

Since both normal cells and cancer cells can be grown in culture, the basic differences between them can be closely studied. In addition, it is possible, by the use of chemicals, viruses and radiation, to convert normal cultured cells to cancer causing cells. Thus, the mechanisms that cause the change can be studied. Cultured to determine suitable drugs and methods for selectively destroying types of cancer.

12.8.4 Virology

One of the earliest and major uses of cell culture is the replication of viruses in cell cultures (in place of animals) for use in vaccine production. Cell cultures are also widely used in the clinical detection and isolation of viruses, as well as basic research into how they grow and infect organisms.

12.8.5 Cell-Based Manufacturing

While cultured cells can be used to produce many important products, three areas are generating the most interest. The first is the large-scale production of viruses for use in vaccine production. These include vaccines for polio, rabies, chicken pox, hepatitis B and measles. Second, is the large-scale production of cells that have been genetically engineered to produce proteins that have medicinal or commercial value. These include monoclonal antibodies, insulin, hormones, etc. Third, is the use of cells as replacement tissues and organs. Artificial skin for use in treating burns and ulcers is the first commercially available product. However, testing is underway on artificial organs such as pancreas, liver and kidney. A potential supply of replacement cells and tissues may come out of work currently being done with both embryonic and adult stem cells. These are cells that have the potential to differentiate into a variety of different cell types. It is hoped that learning how to control the development of these cells may offer new treatment approaches for a wide variety of medical conditions.

12.8.6 Genetic Counseling

Amniocentesis, a diagnostic technique that enables doctors to remove and culture fetal cells from pregnant women, has given doctors an important tool for the early diagnosis of fetal disorders. These cells can then be examined for abnormalities in their chromosomes and genes using karyotyping, chromosome painting and other molecular techniques.

12.8.7 Genetic Engineering

The ability to transfect or reprogram cultured cells with new genetic material (DNA and genes) has provided a major tool to molecular biologists wishing to study the cellular effects of the expression of theses genes (new proteins). These techniques can also be used to produce these new proteins in large quantity in cultured cells for further study. Insect cells are widely used as miniature cells factories to express substantial quantities of proteins that they manufacture after being infected with genetically engineered baculoviruses.

12.8.8 Gene Therapy

The ability to genetically engineer cells has also led to their use for gene therapy. Cells can be removed from a patient lacking a functional gene and the missing or damaged gene can then be replaced. The cells can be grown for a while in culture and then replaced into the patient. An alternative approach is to place the missing gene into a viral vector and then "infect" the patient with the virus in the hope that the missing gene will then be expressed in the patient's cells.

12.8.9 Drug Screening and Development

Cell-based assays have become increasingly important for the pharmaceutical industry, not just for cytotoxicity testing but also for high throughput screening of compounds that may have potential use as drugs. Originally, these cell culture tests were done in 96 well plates, but increasing use is now being made of 384 and 1536 well plates.

BRAIN QUEST

1. Describe the primary and secondary culture and their importance.
2. How one can determine the split ratio from primary culture.
3. Define the phases when cell strain can be obtained.
4. Describe detail procedure of serial subculture.
5. How will you determine suitable time of serial subculture.
6. Describe the technique of primary explantation.

TECHNIQUES IN CLONING OF THE CELLS

13

Here we will learn about the procedure of cloning of the cells. Success of cloning depends on various factors and the plating efficiency finally helps in quantification of the cells.

13.1 CLONING

Cloned cell population is often derived from single parental cell and they are genetically similar. Cloned cells are advantageous due to production of single cell pupulation with minimal genetic variability, since in cell culture most of the cells are heterogeneous in nature. Cloning of adherent cells is possible on all type of the substrates. Colony Stimulating factors play main role in colony formation and which may be provided by the feeder layer. for cloning of the cell. First single cells are obtained which are then seperated by various technique in new environment to grow in single population called as clone of the cells.

13.2 METHODS OF CLONING

13.2.1 Microdrop Technique

In this technique cells are seeded as microdrop under liquid paraffin which seperats clone from each other.

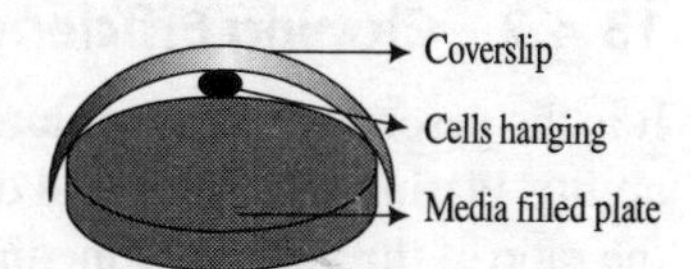

Fig. 13.1 Cloning by Hanging drop method [123]

13.2.2 Hanging-Drop Technique

Hanging-drop culture a culture in which the material to be cultivated is inoculated into a drop of fluid attached to a coverglass, which is inverted over a hollow slide.

13.2.3 Cloning by Ring Technique

Cloning of cells also can be done by confining the cells by ring made up of ceramic or steel. This ring gives them a enclosed environment and thus separate clone of cells grown on the same plate. See Fig. 13.2

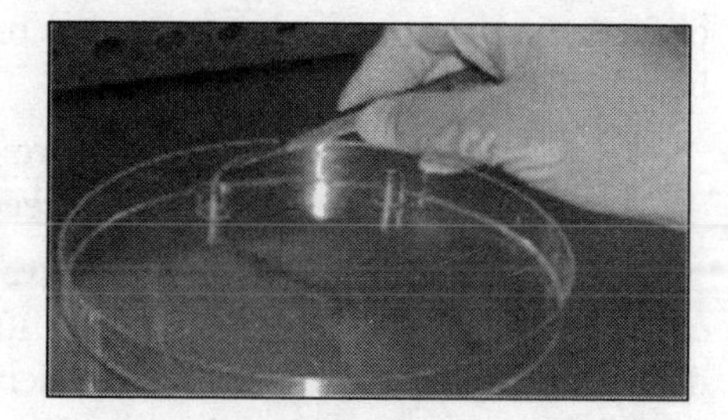

Fig. 13.2 Cloning by Ring method. Ring is usually made up of glass, stainless steel or cermic sealed with silicon grease

13.2.3 Cloning by Limiting Dilution

Dilution plating method was first of all devised by the Sanford and then by the Puck and marcus in 1955. In this method, cell are separated after subculturing, and then the cells are diluted several times before cloning.

In serial dilution cells are mixed with the distilled water for example for ten time dilution, 1 ml cell are mixed with 9 ml of distilled water and for 100 time dilution, 1 ml of the cell from ten time diluted cells are again mixed with 9 ml water etc. This method separate single cell easily. Single cell implanted in this way give rise to group of cells that has genetic similarity.

13.2.4 Roll-Tube Culture

It involves inoculating a tube of molten agar medium and rolling them while it is solidifying, and thus medium get dispersed in a thin layer on the inner surface of the tube. thus seperate clone of cells can be grown on single rolled bottle.

13.3 CLONING OF NON-ADHERENT CELLS

Cloning of hematopoietic stem cells or virally transformed (e.g. fibroblasts cells) can be done on growing them on semisolid agar culture but often these are harmful since they contains low concentration of agar or agarose which contains acidic and sulfated polysaccharides that are inhibitory to most of the growing cells. Alternative media is Methocel solution which is viscous in nature but do not contains any toxic agents like agar or agarose. Another advantage is that they are less dense than agar gel and thus cells would "fall" to bottom and lie at the methocoel /agar interface.

13.4 QUANTITATIVE MEASURES

Cloning can be quantified by either measuring plating efficiency or cloning efficiency.

13.4.1 Plating Efficiency

It is the percentage of cell seeded at subculture that give rise to colonies.

13.4.2 Cloning Efficiency

It is the percentage of Inoculums that attaches to substrate. It implies viability but not proliferative ability. Plating efficiency will determine how well single cells can survive and form colonies. Since the ratio of the volume of medium to the volume of the cells is very large, there is a minimal impact of the cells on their environment. Thus, there is little opportunity for the cells to metabolize and convert amino acids, to bind toxic components, or to secrete autocrine growth factors, they may need for growth. Plating efficiency is thus a more sensitive measure of the cell's response to its surroundings. This may be the method of choice for determining the nutritional requirements of cells or for testing and comparing serum lots or for toxicity testing of compounds.Plating efficiency can be calculated by following formulae.

$$\textit{plating efficiency} = \left(\frac{\text{No. of cell formed}}{\text{No. of colony seeded}}\right) \times 100 \text{ and } \textit{slope} = \mu \times \textit{specific growth rate}$$

Most of the primary cell cultures have low plating efficiency (around 0.1%), while established continuous cell lines shows better (over 10%) plating efficiency. Low Plating Efficiency may occur due to low cell density but at high cell density, cells secrete second messengers, which increase the survival and thus growth of the cells. Insulin and Dexamethosone addition often increase the plating Efficiency. Some media like serum, especially Fetal Bovine Serum is found to be better than Fetal Calf Serum or horse serum. Other factor that increases plating efficiency is proper concentration of CO_2, polylysine substrate and presence of Fibronectin.

13.5 ISOLATION OF CLONES

Cloning in multiwell plate is itself had no problem since the clones are separated from each other but if cloning has been done in agar or petri plate then it is necessary to separate the clone by special technique such as placing a stainless steel or ceramic ring around the colony to be isolated. (See Fig. 13.3.)

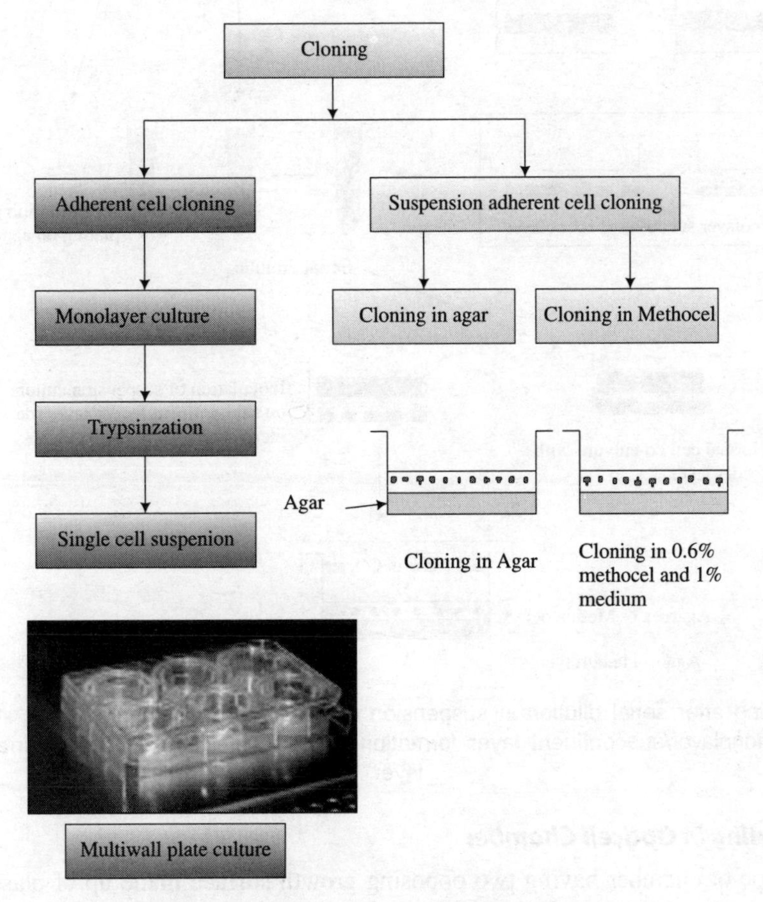

Fig. 13.3 Flow chart of Cloning Prcedure

13.5.1 Isolation Techniques for Monolayer Clones

13.5.1.1 Capillary Technique of Sanford et al. [1948]

Here colonies are developed inside the capillary tube by withdrawing some cells inoculated fluid into the capillary. The clone can be developed inside media by breaking the one end of small piece of capillaries. This technique has been well applied for clonogenic assay of hematopoietic cells and tumor cells, for which the colony-forming efficiency can be quantified by scanning the capillary in a densitometer. (See Fig. 13.6)

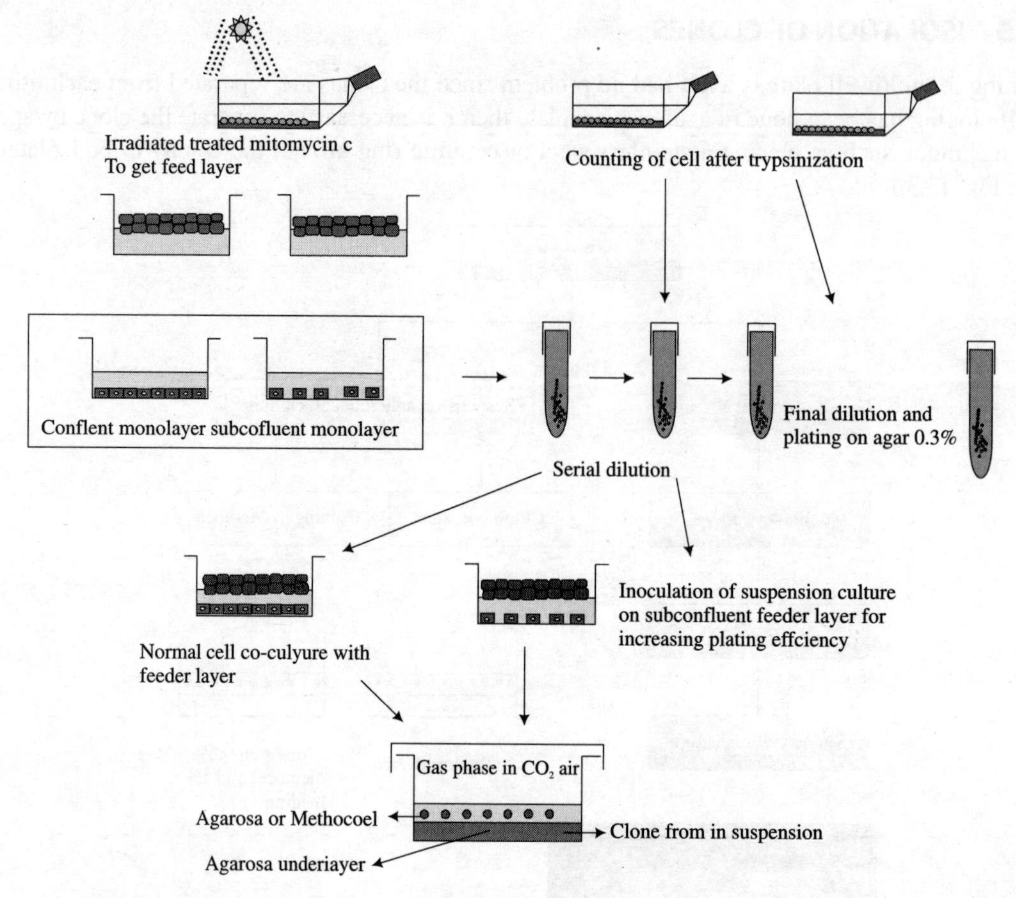

Fig. 13.4 Cloning after serial dilution in suspension or semibold agar culture. Basic step involve is trypsinization, monolayer/subconfluent layer formation and cell are concultured on irradiated feede layer

13.5.1.2 Cloning in Opticell Chamber

Opticell is a type of chamber having two opposing growth surface made up of plastics which is easy to cut and for further growth in multiwell plate or flask culture. (See Fig. 13.7).

13.5.2 Cloning By use of Selective Inhibitors

One can select the specific clone by using specific inhibitors that can selectively kills all the cell except the desired one e.g. Gilbert and Migeon [1975, 1977] replaced the L-valine in the culture medium with D-valine and demonstrated that cells possessing D-amino acid oxidase would grow preferentially. Kidney tubular epithelia [Gross et al. 1992], bovine mammary epithelia [Sordillo et al. 1988], endothelial cells from rat brain [Abbott et al. 1992], and Schwann cell cultures [Armati & Bonner, 1990] have been selected in this way. However, this technique appears not to be effective against human fibroblasts [Masson et al. 1993]. Fry and Bridges [1979] found that phenobarbitone inhibited fibroblastic overgrowth in cultures of hepatocytes, and Braaten et al. [1974] were able to reduce the fibroblastic contamination of neonatal pancreas by treating the culture with sodium ethylmercurithiosalicylate. Fibroblasts also tend to be more sensitive to geneticin (G418) at 100 μg/ml [Levin et al. 1995].

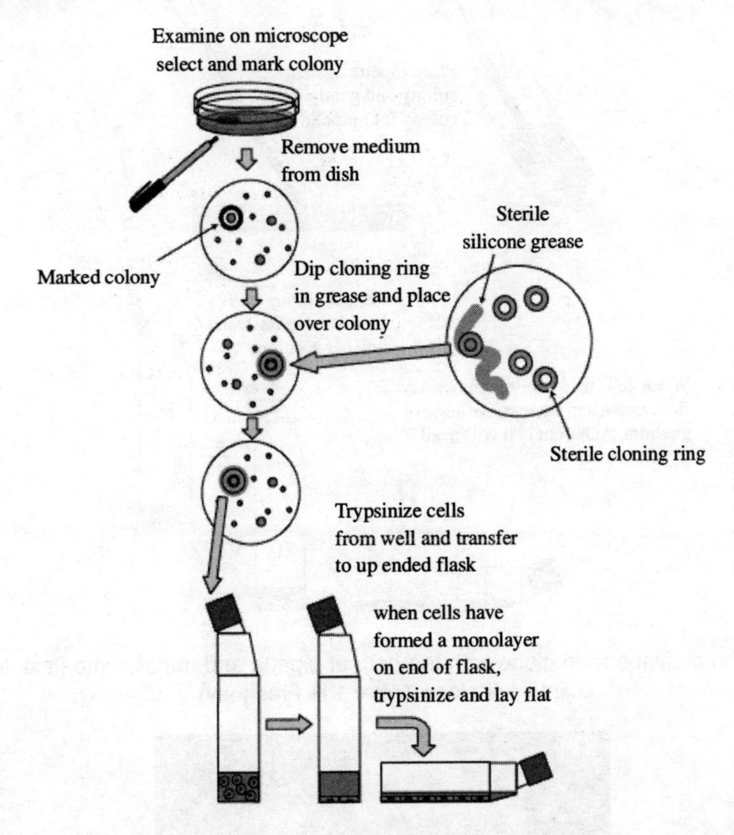

Fig. 13.5 Isolation of monolayer clone with the help of sterile silicone grease which is used to mark the colony in form of ring which is later in separated and cultured in flask. (Modified from freshney 2005)

One of the more successful approaches was the development of a monoclonal antibody to the stromal cells of a human breast carcinoma [Edwards et al. 1980] but when used with complement proteins, this antibody proved to be cytotoxic to fibroblasts from several tumors and helped to purify a number of malignant cell lines.

Cells may also be killed selectively with drug- or toxin-conjugated antibodies [e.g., Beattie et al. 1990]. However, selective antibodies are used more extensively in "panning" or magnetizable bead separation techniques. Selective media are also commonly used to isolate hybrid clones from somatic hybridization experiments. HAT medium a combination of hypoxanthine, aminopterin, and thymidine, selects hybrids with both hypoxanthine,glutanine phosphoribosyltransferase and thymidine kinase from parental cells deficient in one or the other enzyme [Littlefield, 1964a].

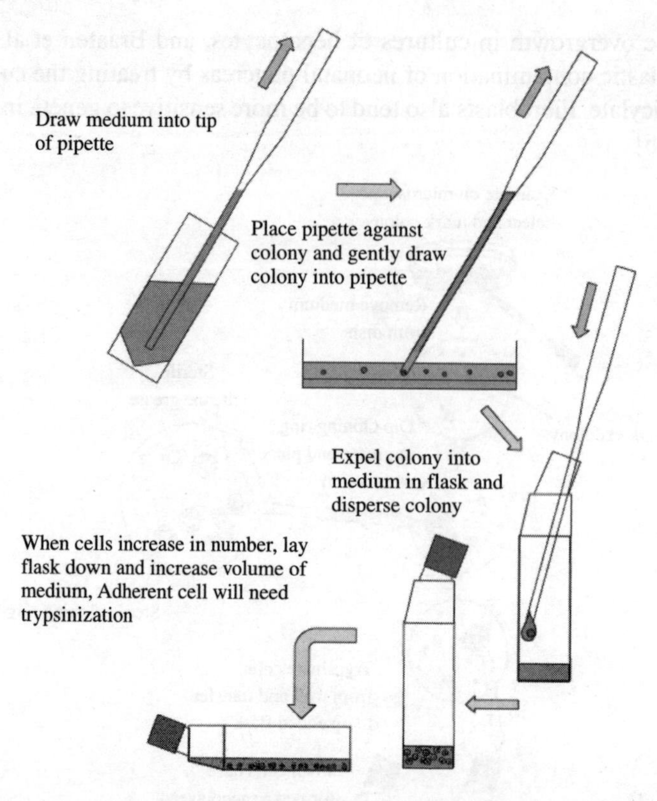
Draw medium into tip
of pipette
Place pipette against
colony and gently draw
colony into pipette
Expel colony into
medium in flask and
disperse colony
When cells increase in number, lay
flask down and increase volume of
medium, Adherent cell will need
trypsinization

Fig. 13.6 Isolation of suspension clone with the help of pipette and transfer into flask for development into full clone (After I R Freshney)

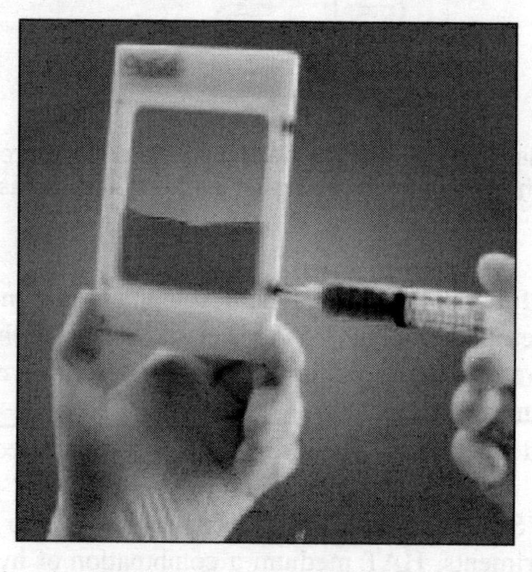

Fig. 13.7 An optical chamber in which medium has been filled with the help of syringe from one end

Transfected cells are also selected by resistance to a number of drugs, such as neomycin, its analog geneticin (*G*418), hygromycin, and methotrexate, by including a resistance conferring gene in the construct used for transfection [e.g., neo (aminoglycoside phosphotransferase).

BRAIN QUEST

1. How epithelial cell can be grown. Name the different precursor cells used for the epithelial cell culture.
2. Name the normal tissue that behave as cell line and grow in the suspension culture.
3. How Hematopoietic cell can be grown. Give the different type of cloning methods for the culture of animal cells.
4. Describe the method of cloning from single cells?
5. Describe the factor that increases cloning efficiency?
6. Describe cloning by ceramic ring technique. How cells can be cloned by selectively cloning methods.
7. How plating efficiency is measured?

MAINTENANCE OF THE CELL CULTURE

14

The maintenance of the cell culture increases self life of normal cells since normal cells have limited life span and died after few generations. First culture of the cell is called as primary culture and its further culture is called as Subculture. Thus, suclulturing increases the life of the cell and thus cell can be maintained for longer period of time.

14.1 MAINTENANCE OF THE CELL

Maintenance of the cell involves increasing self life by adapting various techniques, such as subculturing and preservation. This can be done by studying the growth pattern of the cells, their doubling time, and split ratio (dilutions in case of Non-adherent cells). Method of maintenance of adherent and nonadherent cells differ significantly. Adherent cells needs separate maintenance characteristics than the transformed cells. Adherent cells needs substrate to grow, and follow density dependent growth pattern, and trypsinization for harvesting the cells and thus their maintenance requires a lot of labor in maintaining pH, correct cell density, proper cell morphology, fix ratio of volume to surface area for proper diffusion of the gases. Sometime maintenance medium has to be supplied for adherent cells, due to lack of some inducer needed for the cell division after the medium get exausted, while Nonadherent cells need not to maintain all these characteristic because of loss in contact inhibition properties and thus they can be maintained simply by dilution.

14.2 SUBCULTURING INCREASES LIFE SPAN OF CELLS

Subculture involves the dissociation of the cells from each other and then culturing on suitable substrate. Reseeding this cell suspension at a reduced concentration into a flask or dish generates a secondary culture, which can be grown up and subculture again of such grown cells give rise to a tertiary culture, and so on. Thus, subculture gives the opportunity to expand the cell population, and allows the generation of replicate cultures for characterization, preservation by freezing, and experimentation. In most cases, cultures dedifferentiate during serial passaging but can be induced to redifferentiate by cultivation on a 3D

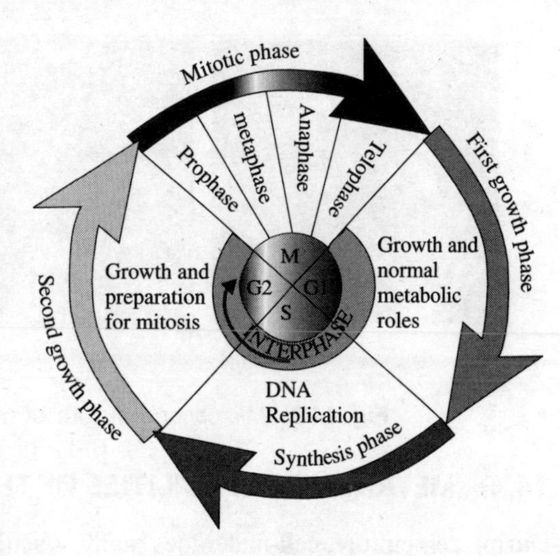

Fig. 14.1 Cell Cycle phases

scaffold in the presence of tissue-specific differentiation factors (e.g., growth factors, physical stimuli). However, the cell's ability to redifferentiate decreases with passaging. It is thus essential to determine (for each cell type) the source and a suitable number of passages before subculture. Protocols for subculture of specific cell types are given in later chapters, and a more general protocol is available in Chapter 13, Freshney [2005].

Monolayer culture is convenient for cytological and immunological observations, cloning, cell synchronization, chromosome preparation and in-situ-extraction without centrifugation. Also, the substrate might or might not provide nutrients to the cells. In vivo, most of the cells are bathed in "liquid matrix" that bathes cells attached to a surface. There are cells that require an "air-liquid interface" to grow properly; like cells grown on a "raft" of organic material that floats on the surface of a nutrient broth and acts like a wet sponge, feeding the cells from underneath while their tops are exposed to the air.

14.3 MAINTENANCE OF THE ADHERENT AND NON-ADHERENT CELLS

14.3.1 Adherent Culture

Adherent cultures can be maintained by simply by adding new medium, after removal of the old medium from culture plate. Exact time to replace medium for finite cell line can be calculated by their doubling time. For finite cell line, cell number determination helps in calculating doubling time. Cells are chosen at phase where there is short lag period and has short population doubling time and from where after seeding cells reaches to exponential phase. For maintenance of adherent cell line or finite cell line following steps should be followed. (a) Removal of medium. (b) Harvesting of cells (c) Counting of cells (d) sub culturing of cell.

14.3.2 Nonadherent Culture

Non adherent culture grows often in suspension medium. Suspension cultures are often maintained by feeding fresh medium at regular interval. If the dilution is below the required level the cells will not show the required performance. So the dilution level must be defined before the subculture.

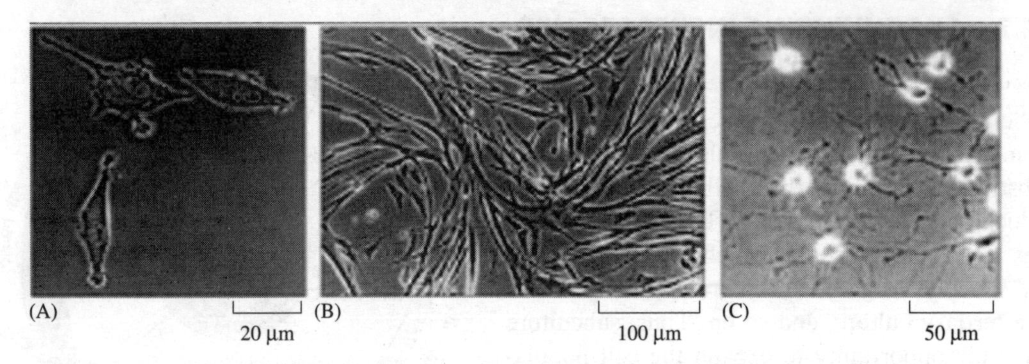

Fig. 14.2 Microscope picture of monolayer cells formed by neutrophils

14.4 METABOLIC CAPABILITIES OF THE FINITE CELLS

During cell culture, cell undergoes many subcultures. (All these steps are aimed towards increasing cell survival and their life-span.) As the cell grows, medium is continuously depleted due to utilization of the carbon source continuously. Normal cell utilizes glucose slowly, while cell line utilizes glucose rapidly. However, exact period of supplying glucose and other important factors can be determined by observing following factors. (a) cell density (b) cell morphology, (c) pH, (d) Ratio of medium volume to surface area.

14.5 GROWTH RESPONSE CURVE

Growth responce curve is similar to growth curve and follow the same pattern of growth characteristics. Growth responce curve gives exact idea about the lag, log and plateu phase (as we know that cells often undergo differentioation after the platue phase). (see Fig. 14.3.) Each time a cell undergoes subculture, it grow back to the original cell density. This process can be described by plotting a growth curve. The cells first enters a latent period of no growth, called the lag period, immediately after reseeding. This period lasts from a few hours to 48 h, but is usually around 12-24 h, and allows the cells to recover from trypsinization, reconstruct their cytoskeleton, secrete matrix to aid attachment, and spread out on the substrate, enabling them to reenter cell cycle. They then enter exponential growth in what is known as the log phase, during which the cell population doubles over a definable period, known as the doubling time and is fixed for each cells.

The doubling time of cell must be known for proper timing of changing the media, since after the confluency the cell growth and developments are checked by various factors like - depletion of the nutrients, limitation of the gases, and also due to lack of space to grow. Media changing may induce the cell growth because it may remove the several inhibitors secreted by the cell itself and also depletion of glucose. At large scale cell and tissue culture needs proper surface area to grow so several type of substrate have been utilized for increasing the growth of cell like microcarriers which increase the surface area thousand times.

As the cell population crowded, the cells become packed, spread less on the substrate, and eventually withdraw from the cell cycle. They then enter the plateau or stationary phase, where the growth drops to close to zero. Some cells may differentiate in this phase; others simply exit the cell cycle into G_0 but retain viability. Cells may be subcultured from plateau, but it is preferable to subculture before plateau is reached, as the growth fraction will be higher and the recovery time (lag period) will be shorter if the cells are harvested from the top end of the log phase.

Reduced proliferation in the stationary phase is due partly to reduced spreading at high cell density and partly due to exhaustion of growth factors in the medium at high cell concentration. In each case there are major changes in cell shape, cell surface, and extracellular matrix, all of which will have significant effects on cell proliferation and differentiation. A high density of cells will also limit nutrient perfusion and create local exhaustion of peptide growth factors. In normal cell population it leads to withdrawn of cell from the cell cycle whereas in transformed cells, cell cycle arrest is much less effective and the cells tend not to enter for apoptosis and thus they become immortal.

Cell Concentration per se, without cell interaction, will not influence proliferation, other than by the effect of nutrient and growth factor depletion. High cell concentrations can also lead to apoptosis in transformed cells in suspension, notably in myelomas and hybridoma, but in the absence of cell contact signaling is presumably a reflection of nutrient deprivation.

14.5.1 Cell Line Grow in Suspension Culture

In Suspension culture cell line grow by floating in culture medium i.e. Cells survive and proliferate without the attachment to culture container and they don't need solid substrate for their initiation of cell division. They already have lost the ability of cell division control. For example - Cells cultured from blood, spleen and bone marrow.

14.5.2 Normal Cell Requirements for Proper Growth

The key to success in culturing cells is to mimic the environment similar to their original one, before being transplanting to an artificial environment. Some cells like, Fibroblasts, Epithelial cells, and Neurons cell requires specialized surface for proper growth and development as cells get signal to divide and grow. Such cells are called as normal cells or adherent cells. Cells can be cultured for a longer time, if they had been split and cultured regularly. This environment is often bathed with an extracellular fluid derived from blood (Ham's tissue culture medium, a commonly used medium for mammalian cells.) Some normal cells naturally live without attaching to surface e.g. blood cells. Most cells derived from solid tissues require a surface, while others live either in solid or liquid condition and exhibit different phenotypes depending on whether or not they are attached to a surface, such as yeast and with the cells suspended in a nutrient broth.

14.6 FINITE AND CANCEROUS CELL LINE

Finite cell lines are obtained from normal cells after several subculture, while transformed cell (by viruses or any treatment) give rise to continuous cell line. Continuous cell line gives rise to cell strain. Cell line is of two type finite cell line and other is the infinite cell line. Finite cell line has definite life span and they can be grown up to 20 to 80 population doubling. The doubling time is the time required for the doubling of culture. The actual doubling time depends on various factors are written. If all line has the cell line name according to source origin, and some specific property out of total population formed as strain for example cell strain of normal human brain are named as NHB. The well characterized cell lines are: Human Diploid, fibroblastic cell line (WI-38), Human cervical adenocarcinoma epitheloid (HeLa), Mouse embryonic, fibroblastic (3T3) more name has been given in Table 14.1. Each cell line name is repesented by ATCC no.

Cell density log plot of cells at every two-hour interval gives idea about medium change life of the cell. Continuous cell line can be subculture at cell density of 10^4 cells /ml for fibroblast cell line and 10^5 cells/ml for the epithelial cell.

> ATCC is well known international patent culture repository for cell line. ATCC (American Type Culture Collection) was established in 1914 and have good repository of about 4000 human animal and plant cell lines.
>
> Terminology committee of the American Tissue Culture Association given a general recommendations that four letter acronyms to be used to designate specific lines or initial two letter to indicate the institution of origin followed by a number to specify identity, MCF-7 for the malignant cancer foundation., breast cancer line; and WI-38 denotes Wister Institute lung line number 38. Unfortunately, this system has not been adopted in uniform manner and many alternative are in common use today.

ATCC No	Designation	Utility
CCL-75	WI-38	Vaccine, Aging
CCL-171	MRC-5	Vaccine, diagnosis
CCL-10	BHK-21	Transformation studies and virology
CCL-81	VERO	Virology and vaccine
genetically modified cell line		
CRL-1650	COS-I	Host line for recombinant
CRL-8002	OKTG	SV40 propagation
CRL-1485	CCD-32 Lu	Normal human skin

14.6.1 Life Span of the Cell Lines

Finite cell have longer doubling time of about 24-96 hrs, while cancerous cell line have short doubling time of 12-24 hours. Most of the finite cell lines undergo a limited number of sub-cultures, or passages and the limit is determined by the number of doublings before it stops growing because of senescence. Therefore, cell density of continuous cell line become double in the time in which other cell has reached only half of density only. Increase in cell density means increase in cell population size in available surface area for finite cell line. Cell density effects cell morphology, pH, substrate in the medium and self-life of the cell itself. Addition of Phenol red in growing cells changes its colour from red to yellow if the cell increases which is also a good indication for changing of medium. A high cell density limits nutrient perfusion and create local exhaustion of peptide growth factors and thus leads to a withdrawal from the cycle, whereas in transformed cells, cell cycle arrest is much less effective. Finite cell line, have finite life spans for example MRC5, MRC9, W138. Normal chicken cells rarely are

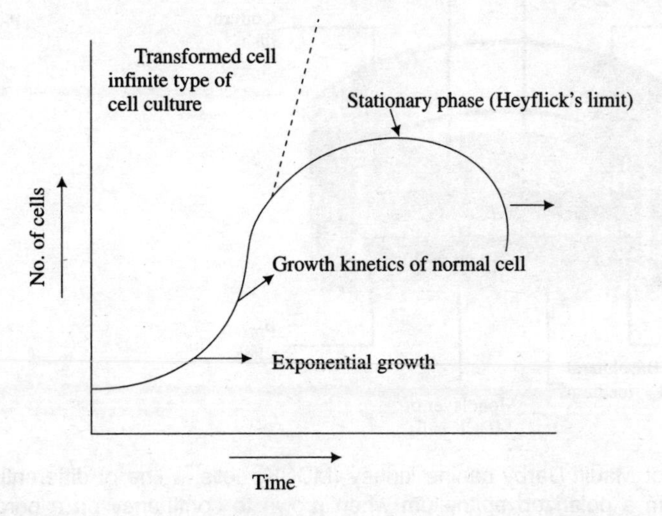

Fig. 14.3 Rate of growth of normal cell verses transformed cell. Note that normal cell growth decline after a certain period but transformed cell has limitation that growth is declined after a certain period.

transformed and die out after only a few doublings; even tumor cells from chickens almost never exhibit immortality while human cells, tumor cells grow indefinitely.

14.6.2 Continuous Cell Lines

They are derived from tumors or arise spontaneously in culture after transformation with viruses, radiation etc. The HeLa cell, the first human cell type, was originally obtained in 1952 from a malignant tumor (carcinoma) of the uterine cervix. This cell line had been proved invaluable for cancer research on human cells. In contrast to human and chicken cells, cultures of embryonic adherent cells from rodents routinely give rise to cell lines. During normal culture most of the cells die, but often a rapidly dividing variant grow forever if it is provided with the necessary nutrients.

Cells in spontaneously established rodent cell lines and in cell lines derived from tumors often have abnormal chromosomes. In addition, their chromosome number usually is greater than that of the normal cell, and it continually expands and contracts in culture. Such cells are said to be aneuploid (i.e., have an inappropriate number of chromosomes). One such line, Madin-Darby canine kidney (MDCK) cells, forms a continuous sheet of polarized epithelial cells one cell thick that exhibits many of the properties of the normal canine kidney epithelium from which it was derived (see Fig. 14.4). This type of preparation has proved valuable as a model for studying the functions of epithelial cells.

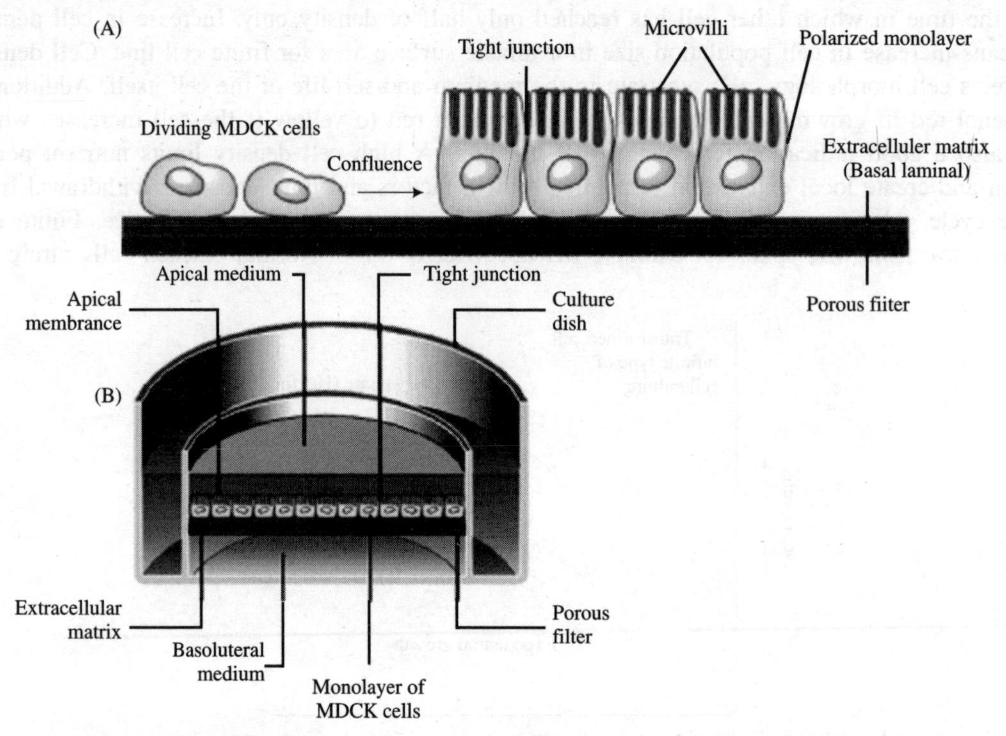

Fig. 14.4 Culture of Madin-Darby canine kidney (MDCK) cells, a line of differentiated epithelial cells. (a) MDCK cells form a polarized epithelium when grown to confluency on a porous membrane filter coated on one side with collagen and other proteins of the basal lamina, the extracellular matrix that supports an epithelial layer (b) Special culture dishes alow the cells to be bathed with an appropriate medium on each side of the filter; the apical surface faces the medium that bathes the upper side. [Modified from R. Van Buskirk, J. Cook, J. Gabriels, and H. Eichelberger.]

MDCK cells form a polarized epithelium when grown to confluency on a porous membrane filter coated on one side with collagen and other proteins of the basal lamina, the extracellular matrix that supports the epithelial layer. Special culture dishes allow the cells to be bathed with an appropriate medium on each side of the filter; the apical surface faces the medium that bathes the upper side. Note that the tight junctions connecting the epithelial cells form the physical barrier separating the basolateral from the apical extracellular space.

Properties of the Continous Cell Lines

They have following property.

1. Grow indefinitely in culture.
2. Usually have unstable complement of chromosomes number. (ancuploid)
3. Usually do not have fixed age in the culture means they have infinite capacity to grow therefore, they are 'immortal'. They often lose contact inhibition.
4. They often lose many normal characteristics and they are not dependent on growth factors. They may express 'large T-antigen' a p53 inhibitor.

Table 14.1 Finite cell line

All lines	*Source Morphology*	*Age*	*Tissue Characteristics*
IMR90 Human lung	Fibroblasts	Embryonic	Normal Infected by Human virus
MRC5 Human lung	Fibroblasts	Embryonic	Normal Infected by Human virus
MRC9 Human lung	Fibroblasts	Embryonic	Normal Infected by Human virus
WI38 Human lung	Fibroblasts	Embryonic	Normal Infected by Human virus

Table 14.2 Continuous cell line

Cell Line	*Source*	*Maturity*	*Characteristics*
EB, Human	Lymphocytic	Juvenile EB virus, + ve	
HeLa, Human	Epithelial	Adult	G6PD Type A
LS, Mouse	Fibroblastic	Adult	Grow in L929 suspension
P388D Mouse	Lymphocytic	Adult	Growinsuspension
S180, Mouse	Fibroblastic	Adult	Cancerchemotherapy
3T3A. Mouse	Fibroblastic	Embryonic	Feeder layer

Epithelial Cell Line:

1. NIH 3T3 Mouse embryo Fibroblast
2. L929 Mouse connective tissue Fibroblast
3. CHO Chinese Hamster Ovary Fibroblast
4. BHK-21 Syrian Hamster Kidney Fibroblast
5. HEK 293 Human Kidney Epithelial
6. HEPG2 Human Liver Epithelial
7. BAE-1 Bovine aorta Endothelial.

Table 14.3 Comparative difference of finite cell line and continuous cell line

FEATURE	FINITE CELL LINE	CONTINOUS CELL LINE
1. Ploidy	diploid	heteroploidy
2. Anchorage dependent	Yes	No
3. Transformation	Normal	Transformed
4. Tumoroginicity	Non	Yes
5. Contact Inhibition	Yes	No
6. Density limitation of growth	Yes	No
7. Mode of growth	Monolayer	suspension
8. Maintenance	Cyclic	steady state
9. Serum requirement	high	Low
10. Cloning efficiency	Low	High
11. Marker	Tissue specific	Chromosomal, enzymatic
12. Growth rate	slow (24-96 hrs)	Rapid (12-24 hrs)
13. Yield	Low < s10^6 cells /ml	High >10^6 cells /ml
14. Control feature	Generation number in vitro marker	Strain characteristics

Suspension Cell Lines

1. Mouse myeloma Lymphoblastoid- U937
2. Human Hystiocytic Lymphoma
3. Lymphoblastoid Namalwa Human Lymphoma
4. Lymphoblastoid HL60
5. Human Leukaemia Lymphoblastoid-like WEHI 231
6. Mouse B-cell Lymphoma
7. Lymphoblastoid YAC 1
8. Mouse Lymphoma
9. Lymphoblastoid U 266B1
10. Human Myeloma Lymphoblastoid SH-SY5Y
11. Human neuroblastoma Neuroblast

Table 14.5 Some commonly used cell lines

Properties	MONOLAYER	SUSPENSION
1. MAINTENANCE	Follow lag, log stationary and the death phase. Cyclic pattern of propagation. Growth dependent on availability of the substrate.	Cell may maintain in the steady state. Dependent on medium volume.
2. Differentiation	In differentiation of cell into monolayer cell there occur continuous cell to cell interaction and the cell develop communication among each other, as a result cell form a uniform layer on the available substrate, this is called as confluency.	Since they grow in homogenous composition, cell density is limited by nutrient but growth is not affected by confluency of the cell since they are independent of substrate surface. The cells are not rounded in shape.

Table contd...

Table contd...

	Thus confluency of the cell reaches after sometime of the growth. This confluency inhibits further growth of the cell. Cell is thus affected by the available substrate and the available food in the medium due to nutrient competition among each other. The cell growths are ceased after some time after the confluency reached. This may be due to end product inhibition or due to accumulation of some toxic substances.	
3. Nature of cell	Most of the cells are normal one.	Most of the cells are transformed one.

Many of these cell lines were derived from tumors. All of them are capable of indefinite replication in culture and express at least some of the special characteristics of their cell of origin. BHK21 cells, HeLa cells, and SP2 cells are capable of efficient growth in suspension; most of the other cell lines require a solid culture substratum in order to multiply.

14.7 CHARACTERIZATION OF CELLS

Most of the animal cell culture laboratories shall have the facility for the characterization of the culture since species identification is necessary before proceeding to their applications.This can be done by DNA profiling, chromosome analysis or isoenzyme electrophoresis to confirm cell line identity [Hay et al. 2000]

*CELL LINE**	*CELL TYPE AND ORIGIN*
3T3	fibroblast (mouse)
BHK21	fibroblast (Syrian hamster)
MDCK	epithelial cell (dog)
HeLa	epithelial cell (human)
PtK1	epithelial cell (rat kangaroo)
L6	myoblast (rat)
PC12	chromaffin cell (rat)
SP2	plasma cell (mouse)
COS	kidney (monkey)
CHO	ovary (chinese hamster)
DT40	lymphoma cell for efficient targeted recombination (chick)
R1	embryonic stem cells (mouse)
E14.1	embryonic stem cells (mouse)
H1, H9	embryonic stem cells (human)
S2	macrophage-like cells (Drosophila)
BY2	undifferentiated meristematic cells (tobacco)

14.7.1 Cell Lineage

Cell lineage can be identified by utilizing cell surface antigens, intermediate filament proteins, differentiated products/functions, enzymes, unique markers (MHC, HLA, DNA fingerprinting), and study of morphology (easiest and most direct way).

14.7.1.1 Chromosome Analysis

Chromosome analysis is used to identify cell lines species and even sex of original tissues. It also distinguishes normal vereses malignant cell. Chromosome analysis can be done by arresting cells in Metaphase with Colcemid and then staining which gives banded appearance (**Karyotyping).** normal cell can be identified since their number remains constant and aneuploidy occurs in malignant cells.

14.7.1.2 Chromosome Painting

Other technique is the chromosome painting which utilises specific fluorescent labeled DNA probe that hybridize specific DNA and thus can be identified visually or via southern blotting. Total DNA content of normal lines are fixed, while transformed lines are aneuploid or heteroploidy so always show increase number of total DNA content. Therefore, chromosome painting is a good technique for identification of the transformed lines.

14.7.1.3 Flow Cytometry

Use of fluorescent dye helps in detection of various molecules such as analysis of RNA and protein with the help of Antigenic Markers via flow Cytometry.

14.7.1.4 Use of Antibodies

The lineage or tissue origin can also be determined by using the antibodies against various intermediate filament proteins present in different tissue, for example, cytokeratins for identifications of epithelial cells, vimentin for mesodermal cells; fibroblasts, endothelium, myoblasts, and desmin for myocytes, neurofilament for neuronal and some neuroendocrine cells; and glial fibrillary acidic protein for astrocytes. Antibodies can be used against some cell surface markers that are lineage specific, for example, EMA in epithelial cells, A2B5 in glial cells, PECAM-1 in endothelial cells, and N-CAM in neural cells. Spontaneous transformation is unlikely in normal cells of human origin but sometime they show all the feature of transformed cell e.g. Nude or SCID mouse. Where transformation is detected, it is more likely to be due to cross-contamination.

14.7.2 Disadvantages of Cell Line

1. A cell line characteristic is hard to maintain.
2. Cell line grow in very small amount of tissue at high cost
3. Cell line can undergo dedifferentiation
4. Cell line is unstable, and is aneuploidy in nature

14.8 CROSS-CONTAMINATION

14.8.1 Hela Cell Line Cross-Contamination

It is estimated that about 20% of human cell lines are not the kind of cells they are generally assumed to be. The reason for this is that some cell lines exhibit vigorous growth and thus cross-contaminate cultures of other cell lines, in time overgrowing and displacing the original cells. The most common contaminant is the **Hela cell line**, while this may not be of significance when general properties such as cell metabolism are researched, it is highly relevant e.g. in medical research focusing on a specific type of cell.

This is less of a problem with short-term cultures, but the risk remains, if there are other cell lines present in the laboratory, since they can cross-contaminate even a primary culture, or misidentification can arise during subculture or recovery from the freezer. If a laboratory focuses on one particular human cell type, superficial observation of lineage characteristics will be inadequate to ensure the identity of each line cultured. Following precautions must be taken to avoid any cross-contamination:

(i) There should be different culture vessels and medium bottles for more than one cell line and the same pipette should not be used for different cell lines.

(ii) Media or other reagents should not be shared among different cell lines.

(iii) A tissue or blood sample from each donor should be retained and identity of each cell line should be confirmed by DNA profiling: (a) for seed stocks before frozen, (b) and the Cell line before experimental work.

(iv) Keep a panel of photographs of each cell line, at low and high densities, and consult this regularly when examining cells during maintenance. This is particularly important if cells are handled over an extended period, and by more than one operator.

(v) If continuous cell lines are in use in the laboratory, handle them after handing other, slower-growing, finite cell lines.

14.9 DEVELOPMENTS OF CONTINUOUS CELL LINE

14.9.1 Transformation

Transformed cell lines forms continuous cell lines. The transformation can be done by introduction of genes or Transfection by many viruses as described below.

14.9.1.1 Simian Virus 40 large T-Antigen

The cells that are immortalized by large T-antigen retained their original phenotype, however growth rates gets accelerated. Mechanism is not much clear but it thought to operate through inactivation of p53.

14.9.1.2 Papilloma Virus E6 and E7 Proteins

It is a DNA virus that can induce transformation of epithelial cells. Two proteins E6 & E7 are the transforming proteins, which binds to p53 and inactivate it, beside this E7 have additional mechanism of transformation for transformation by operating on Cyclin-A, p130, p33, cdk2, and histone h1 kinase, It also activates telomerase.

14.9.1.3 Adenovirus E1A Protein

Here transforming protein is E1A. E1A gene leads to the expression of 12S and 13S protein which are their immortalizing proteins, These proteins inhibits Rb genes. They also operates through HSP70, c-fos, proliferating cell nuclear antigen, p34, and cdc25.

14.9.1.4 Epstein Barr Virus-Nuclear Antigen 2

Often Lymphoma cells and B lymphocytes are immortalize by EBV nuclear antigen which acts through p53 inactivation.

14.9.1.5 HTLV-1, Tax and Rex genes

CD_4^+ cells and T lymphocytes are immortalized by using gene *tax* and *rax*, and some other cell like NFkB had been also immortalized by these genes.

14.9.1.6 Herpes Virus (Saimiri)

CD_4^+ ve and CD_8^+ ve T cells are transformed in the absence of exogenous cytokine or any antigenic stimulation, mechanism is not well understood

14.9.1.7 Oncogenes - myc, c-jun, c-ras, v-src, mdm-2, Hp53

Hp53, and p53 is tumor suppressive gene, but Hp53 can immortalize many rodents cells by reactivation of telomerase gene.

14.9.2 Telomerase Induced Immortalization

Telomerase / terminal transferase enzyme consist of two subunits: First is RNA subunit called as ***htr*** which is expressed in both normal and malignant cells and other is Protein subunit htrt which can be expressed in tumor cells, germ line cells and activated T-lymphocytes. If the cells are transfected with hTERT then they may or may not become immortalize but proliferation is certainly increased many folds.

14.9.3 Transient Transfection of Cells

Various methods of transient transfection are described in the literature (Calcium Phosphate, Electroporation, Ballistic Particles, DEAE Dextran, Cationic Matrix, and Lipofection) and they differ in their efficiency of transfection and cell toxicity. The primary decision for assay development involved the choice of a transfection method that should yield high efficiency, with minimal toxicity to the cells. There are following method of transformation.

14.9.3.1 Lipofection

It is a technique by which genetic material are injected into cell by means of lipid rich molecule called as liposomes. The Lipofection protocol is a relatively simple that has been used for high throughput screening. Various lipid and Plasmid is mixed which transfects membrane Lipofectamine mediated transfection and detection by enhanced green fluorescent protein (EGFP) and β galactosidase marker is another meths of lipofection. **Fugene-6** has been observed to work best in the presence of serum and resulting in little or no toxicity. (Fig. 14.5)

14.9.3.2 Transfection by Using Cationic Matrix Reagents

Superfect (Qiagen), Lipofection, Fugene 6 (Boehringer), Transfectam (Promega), TransFast (Promega), Tfx (Promega), CLONfectin (ClonTech), Lipofectamine Plus (GIBCO) are commercial name of some of cationic matrix reagent. Once the optimal transient transfection method and cationic matrix reagent have been selected then additional factors can be considered in developing a method for optimizing a transient transfection protocol.

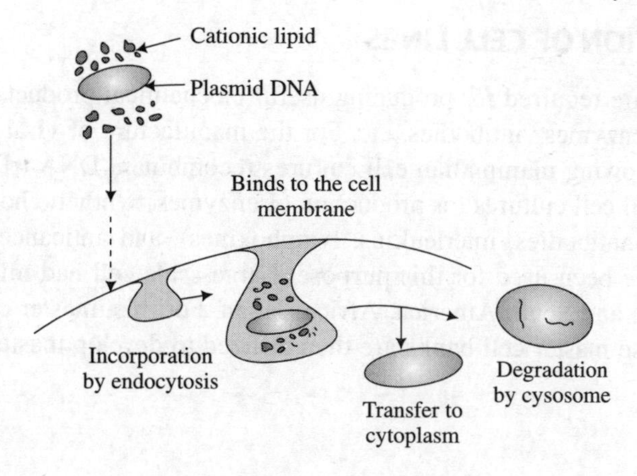

Fig. 14.5 Gene introduction by using cationic liposome

14.10 FACTOR AFFECTING GENE EXPRESSION

Gene expression is usually measured between 24 to 72 hours after transfection depending on the cell type. Induction or stimulation is usually done 48 hours after transfection, and then determine expression levels. Method for determining expression will also depend on type of expression- like GFP can be monitored by visual inspection; luciferase is measured by cell lysis, addition of substrate and measurement of light output. Note that following variables should be tested like DNA amount, Reagent amount, DNA to Reagent ratio, serum concentration, medium type, time of transfection. These variables can be optimized according to cell type.

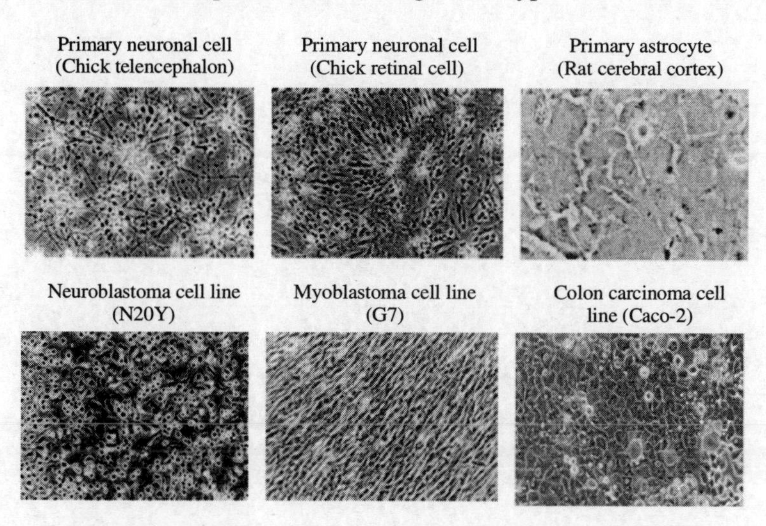

Fig. 14.6 Example of normal and continuous cell line

14.11 APPLICATION OF CELL LINES

Large-scale cultures are required for producing useful biochemical products like interferon, hormones, interleukins, enzymes, antibodies, etc. For the manufacture of viral vaccines now a days due to high cost of growing mammalian cell cultures recombinant DNA (rDNA) technology has been applied in animal cell cultures for production of enzymes, synthetic hormones, immunobiologicals (monoclonal antibodies, interleukins, lymphokines), and anticancer agents. Fermenters up to 10,000 liter have been used for this purpose. Large-scale cell had millions of dollars turn over in Europe, North and South America, Africa, Japan. For this, master cell banks (MCB) are first established. These master cell banks are then utilized to develop master working cell banks (MWCB).

BRAIN QUEST

1. How culture is maintained? Describe one method for its long term maintenance.
2. Define and compare primary, secondary and tertiary culture.
3. How subculturing is done and what is purpose of subculturing.
4. How cell line is obtained by transformation in genes? Also write different methods of transformation.
5. How maintenance of adherent and suspension cell is done?
6. How cell is harvested for the maintenance of the culture?
7. Write down the difference between the finite cell line and continuous cell line.
8. Write down the property of continuous cell line.

CRYOPRESERVATION OF ANIMAL CELLS

15

Here we will learn the method of preserving cells for longer durations without damaging. Cell storage in liquid nitrogen has many benefits but it is also harmful for the cell if not preserved methodologically.

15.1 CRYOPRESERVATION

The cryopreservation enables storage of cell at highly low temperature for the future stock. Under very low temperature (–80° to –196°C) cells remain dormant without death and can again be retrieved in functional state. After controlled freezing in the presence of cryoprotectants, cell lines can be cryopreserved in a suspended state for indefinite periods provided a temperature of less than –135°C is maintained. Such ultra-low temperatures can be attained in liquid or vapor phase. In liquid phase nitrogen allows the lowest possible storage temperature to be maintained with absolute consistency, but requires the use of large volumes (depth).

15.1.1 Cryopreservant

Cryopreservant are added to prevents the formation of ice crystals inside the cell. Ice crystal forms as a result exosmosis takes place in the cell. Therefore, generally DMSO (dimethylsulfoxide) or glycerol is added (cryopreservant) before cryopreservation. DMSO is used more frequently than glycerol because it penetrate slowly inside the cell and encourage greater dehydration of the cells prior to intracellular freezing and also glycerol is less toxic than DMSO. Ice crystals leads to cell death at lower temperature (–4°C to 80°C). At ultra low temperature no ice crystals forms, thus cells did not dies.

15.2 CRYOPRESERVATION IN LIQUID NITROGEN

Cryopreservation can be done by utilizing

1. Liquid phase nitrogen
2. Vapor phase nitrogen

15.2.1 Cryopreservation in Liquid Nitrogen

Liquid N_2 is used to preserve tissue or cells, (either in the liquid phase (–196°C) or in the vapor phase (–156°C)). Storage in liquid phase nitrogen allows the lowest possible storage temperature, but they requires the use of large volumes (depth) of liquid nitrogen in sealed glass ampules. Both of these requirements create potential hazards. There have also been documented cases of cross contamination by virus pathogens via the liquid nitrogen medium. For these reasons ultra-low temperature storage is done most commonly in vapor phase nitrogen.

15.2.2 Storage in Vapor Phase

In vapor phase nitrogen storage, the ampoules are positioned above a shallow reservoir of liquid nitrogen, the depth of which has to be carefully maintained. A vertical temperature gradient will exist through the vapor phase, the top of which contains temperature around –135°C and bottom of which maintains temperature around –170°C. Temperature variations in the upper regions of a vapor phase storage vessel can be extreme if regular maintenance is not carried out.

15.3 METHODS OF PRESERVATION

A cell contains approx. 70-80% of water and water is an important component of cellular metabolism which freezes due to ice formation. Therefore, glycerol or DMSO addition is essential which stops ice crystal formation during preservation. Ice crystal penetrate the cells membrane and thus cells dies.

15.3.1 Ampoules

Ampoules are available in Plastic as well as in glass forms. Plastic ampules have advantage over glass in respect they are unbreakable, are of a larger diameter and taller than equivalent glass ampoules. Therefore, plastic ampules are preferred for the average experimental and teaching laboratory, as they are safer and more convenient, but repositories and cell banks generally prefer glass ampules, since the long-term storage properties of glass are well characterized and, when correctly performed, gives successful results. The ampules should be labelled with the cell strain designation and, preferably, the date and user's initials, to keep the record.

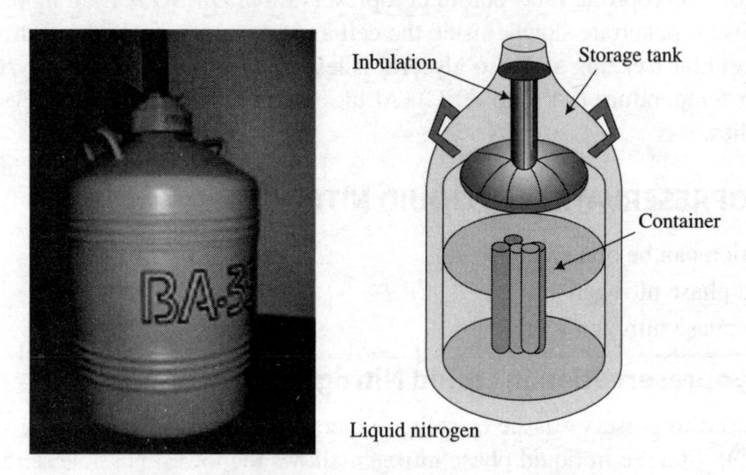

Fig. 15.1 Liquid Nitrogen can and interval design

15.4 REVIVAL OF CELLS

After cryopreservation the cells can be revived by putting the cryovials in the water bath at 37°C and as soon as the medium melts the cells are transfered in the medium. (The cell pellet are obtained after centrifugation at 4°C which is suspended in the fresh medium) and residual DMSO can be removed by washing.

15.5 FACTORS FAVORING GOOD SURVIVAL

1. A high cell density usually between $1 \times 10^6 - 1 \times 10^7$ cells/ml are used for preservatioin
2. Presence of a preservative at 5-10%.
3. Slow cooling at 10°C/min, down to –70°C and then rapid transfer to a liquid nitrogen.
4. Rapid thawing and slow dilution, 20 fold, in medium to dilute out the preservative.

15.6 BANKING OF CELL LINES

All established cell lines have to be cryopreserved for future use.

The cells in the bank should contain the same number and distribution from the total population immediately. Before using cells from bank or for checking feasibility of cells in banking the plating efficiency of the thawed cells should be determined immediately. The cells should contain at least 90% viability or greater than 75% of the cells can be used. If the banked cells do not meet these criteria, the conditions for freezing should be optimized and a new bank should be produced since frozen cells will slowly lose viability over time even in the optimal storage conditions. The purpose of a cell bank is to have cells available over years, it is important to start with a cell bank of high viability or the cells.

It is often wise to store a portion of the banked cells in a separate facility, or at least in two separate storage tanks, to prevent loss in case of accident. Alternatively, one can place new cell lines on deposit with the American Type Culture Collection (ATCC). They store the cells and provide them to other scientists for a minimal fee.

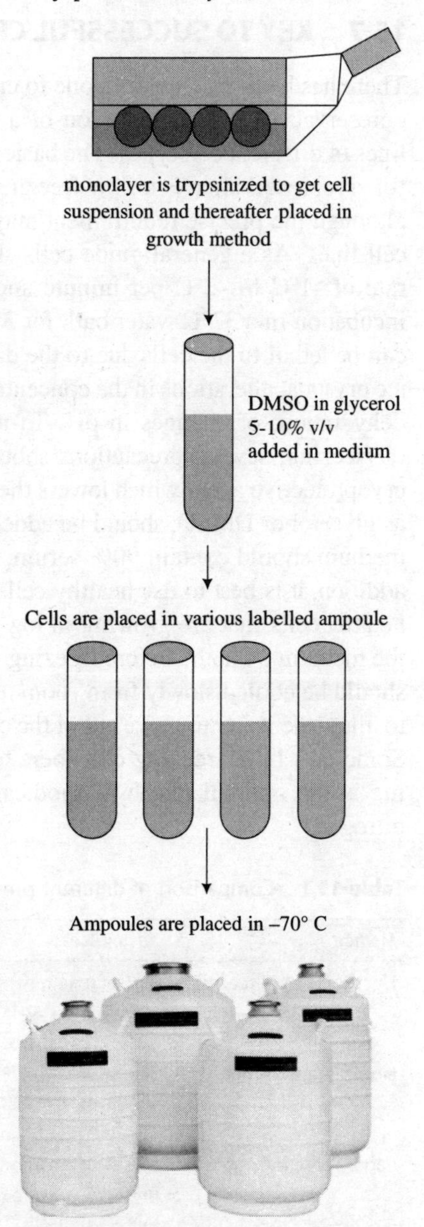
monolayer is trypsinized to get cell suspension and thereafter placed in growth method
DMSO in glycerol 5-10% v/v added in medium
Cells are placed in various labelled ampoule
Ampoules are placed in –70° C
Ampoules are placed in N_2 container

Fig. 15.2 Methodology of cryoprservation in liquid nitrogen

15.7 KEY TO SUCCESSFUL CRYOPRESERVATION

There has been much work done to ensure successful cryopreservation and resuscitation of a wide variety of cell lines of different cell types. The basic principle of successful cryopreservation is a slow freezing and quick thawing, although the precise requirement may vary with different cell lines. As a general guide cells should be cooled at a rate of –1°C to –3°C per minute and thawed quickly by incubation in a 37°C water bath for 3-5 minutes. Freezing can be lethal to the cells due to the damage caused by the ice crystals, alterations in the concentration of electrolytes, dehydration, or changes in pH. To minimize the effects of freezing, several precautions should be taken. First, a cryoprotective agent which lowers the freezing point, such as glycerol or DMSO, should be added. A typical freezing medium should contain 90% serum, and 10% DMSO. In addition, it is best to use healthy cells that it is best to use healthy cells that are growing in log phase and to replace the medium 24 hours before freezing. In addition, the cells should be cooled slowly from room temperature to –80°C to allow the water to move out of the cells before it freezes. Some labs have freezing chambers to regulate the freezing at the optimal rate by periodically pulsing in liquid nitrogen.

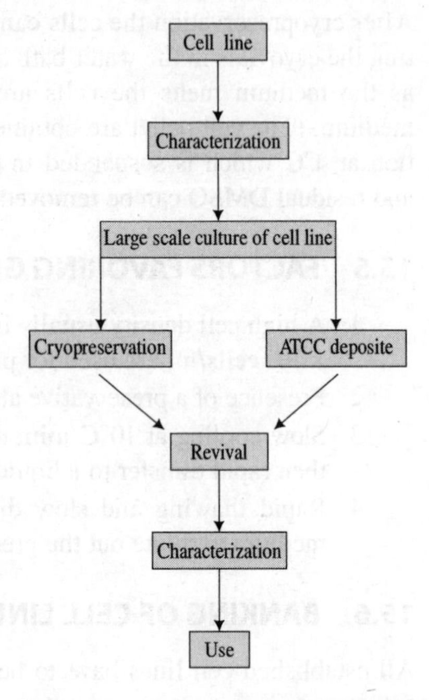

Fig. 15.3 Procedure of cell banking

Table 15.1 Comparison of different preservation methods (for detail see Sigma Aldrich protocol)

Method	*Advantages*	*Disadvantages*
Electric (135°C) Freezer	Ease of maintenance Steady temperature Low running costs	Requires liquid nitrogen back-up Mechanically complex Storage temperatures high relative to liquid nitrogen
Liquid Phase Nitrogen	Steady ultra-low (196°C) temperature Simplicity and mechanical reliability	Requires regular supply of liquid nitrogen High running costs Risk of crosscontamination via the liquid nitrogen
Vapor Phase Nitrogen	No risk of crosscontamination from liquid nitrogen Low temperatures achieved Simplicity and reliability	Requires regular supply of liquid nitrogen High running costs Temperature fluctuations between –135°C and 190°C

15.8 ADVANATAGES OF CRYOPRESERVATION

The main advantages of cryopreservation are:

1. It reduces risk of microbial contamination,
2. It reduces risk of cross contamination with other cell lines,

3. It reduces risk of genetic drift and morphological changes,
4. It work conducted using cells at a consistent passage number, reduced costs (consumable and staff time)

BRAIN QUEST

1. What is the different method for cryopreservation of cell?
2. Write short note on Cryopreservation.
3. Describe the four phases of growth during cell culture.
4. Why normal cells decline in their growth after certain period of time?
5. Define exponential period.
6. In which phase cells are added for the culture ________
7. Cell lines are characterize by ________

Describe the term

(a) Differentiation

(b) Dedifferentiation

(c) Explants

(d) DMSO

(e) Pluripotent

CULTURE OF SPECIFIC CELLS 16

Here we will learn about specific cell culture (from different organs). Different organs have different functions, therefore they need different type of stimulus and the medium compositions for example blood cells are bathed naturally in the plasma, therefore they never needs solid substrate to grow.

16.1 CULTURE OF SPECIFIC CELLS

As we have learn earlier primary cell cultures starts from the explant culture. Most of the animal cell types, (except fibroblasts and epithelial cells embryo fibroblasts) grow under appropriate conditions of temperature, pH growth factors cytokines and hormones. Several growth factors have been identified that stimulates cell division e.g. **skin fibroblasts** rapidly proliferate in order to repair damaged tissue like a cut or wound. Their division is triggered by released of several growth factor from platelets, (during blood clotting), which stimulates proliferation in the neighborhood of the damaged tissue. The identification of individual growth factors has made possible the culture of a variety of specific cells in serum-free media.

Many studies shows that, only a few differentiated cells grow well in culture for e.g. fibroblast cells are highly polarized. The epithelial cells that lines the intestine (a simple columnar epithelium) facing the lumen of the intestine and the apical surface is specialized for absorption; while the rest of the plasma membrane, specially the basolateral surface mediates transport of nutrients from the cell to the blood and forms junctions with adjacent cells and the underlying extracellular matrix called the basal lamina. Certain other cells of blood, spleen, or bone marrow adhere poorly to a culture dish, but they grow well in vivo (inside the body) since they are bathed in blood suspension and attached loosely to bone marrow or body part such as spleen.

Thus explants from adult grown in culture only for a limited period of time before they cease growing. Such a culture dies after few doublings, even if it is provided with fresh supplies of nutrients like serum.

16.2 EPITHELIAL CELL CULTURE

Skin culture is important in many application e.g. for cytotoxicity assay, inflammation study, and tumor development study and in organogenesis phenomenon. As we know every cell culture requires explant to start. Explant are named differently depending on the source e.g. epidermis explant is Keratinocyte, Liver cell explants is hepatocytes. Keratinocyte can be isolated from mature skin by Collagenase or Trypsin either alone or in combination with EDTA. Collagenase helps in dispersion of stromal cell but leaves epithelial cell attached in culture, while Trypsin do the opposite function, therefore, often combined treatment proves to be much helpful in complete separation of cells from the tissue.

Generally Epithelial cell culture requires many hormones either alone or in mixture such as epidermal growth factor EGF, hydrocortisone, and insulin. Specific hormone confers better growth of cells such as like Insulin and hydrocortisone for epithelial cells; estrogen, progesterone, and prolactin for breast epithelial growth. There are several commercial media available compiling all these requirements for specific cell culture e.g. for epidermis culture MCDB-153 media is available, for breast epithelial culture MCDB 170, and for liver epithelial culture, phenobarbitone is available. Buffer used during the culture is phosphate buffer saline (PBS), or HEPES (pH 7.65, (0.25%)).

Serum free media generally used for epithelial culture is KSM (Keratinocyte serum free media). It is reported that Ca^{+2} play very important role in differentiation of the epithelial cell culture. It had been observed that low calcium favors monolayer culture while high Ca^{+2} favor multilayer culture. Another important support (for epithelial culture) is addition of feeder layer that provides suppliments for the growth and development. Therefore, often irradiated 3T3 fibroblast, or collagen gels is preferred for Epidermal and Dermal cell culture. Combine culture of epidermis and stroma mounted on filter mesh are commercially available and can be used for several assay and toxicity.

16.2.1 Corneal Culture

For the growth of corneal epithelia ocular surface is used as explants, which is grown on fibronectin -collagen bed. Cell line explants of corneal epithelia can be obtained after transfection with SV40 early gene, in which large T-antigen integrates and confers them longer life. They are also able to maintain its phenotypic characteristic both in primary culture and in human corneal epithelial cell lines. They can be utilized in study of basic cell biology mechanism and in study of toxicological phenomenon.

16.3 STEM CELL CULTURE

A true stem cell is capable of self-renewal. Hematopoietic cells originate from totipotent stem cells, they undergoes lymphoid, myeloid, and erythroid cell developments while pluripotent stem cells may differentiate only into a subset of lineages (See Fig. 16.1). The intermediate-stage cells are capable of significant proliferation and so, known as **progenitor cells**. Lymphoid lineages give rise to T-cell and B-cells; while myeloid lineages give rise to erythrocytes or red blood cells, granulocytic (production of neutrophils, infection fighting cells), macrophages, dendritic (production of dendritic cells, which are antigen-presenting cells) and megakaryocytic (production of platelets, blood clotting component) lineages.

16.4 HEMATOPOIETIC CELLS

Hematopoietic cells are the blood forming cells and helps in maintenance of every cell type of the body relentlessly. The stem cells that form blood and immune cells are known as hematopoietic stem cells (HSCs). They are ultimately responsible for the constant renewal of blood- the production of billions of new blood cells each day. The first evidence and definition of blood-forming

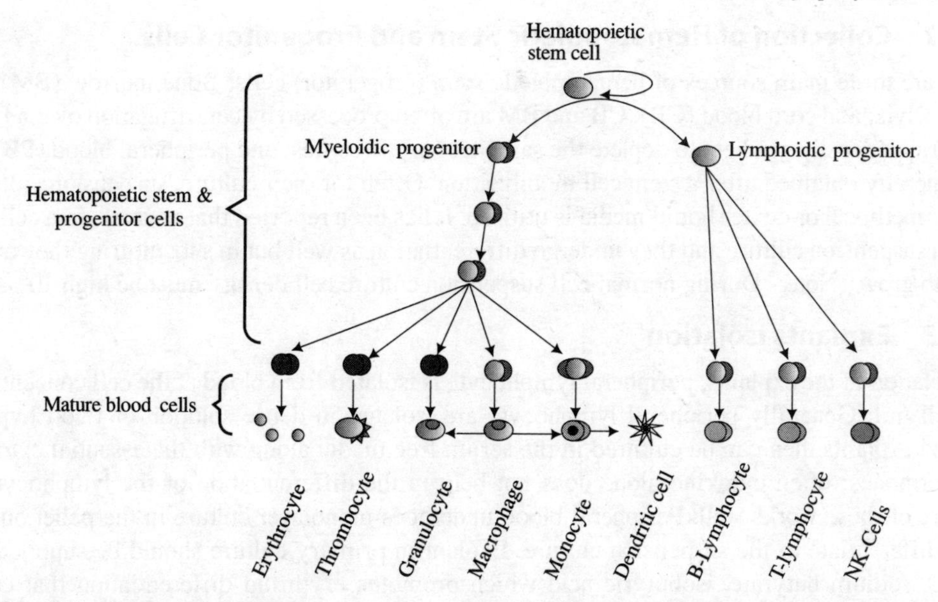

Fig. 16.1 Stem cell differentiation in two different directions (a) erythroid and (b) myeloid

stem cells came from studies of people exposed to lethal doses of radiation in 1945. They are considered as long-term stem cells and can immediately regenerate in all types of blood cells, while other cannot renew themselves, are referred to as short-term progenitor or precursor cells.

Progenitor (or precursor) cells are relatively immature cells that give rise to a fully differentiated cell of the same tissue type. They are capable of proliferating, but have a limited capacity to differentiate into other cell type as HSCs do. For example, a blood progenitor cell may only be able to make a red blood cell.

16.4.1 Origin of Hematopoietic Cells

A hematopoietic stem cell can differentiate to a variety of specialized cells, and can mobilize out of the bone marrow into circulating blood, or undergo programmed cell death, called **apoptosis**- a process by which detrimental cells undergo self-destruction. One initial primary culture can give rise to more than 10^{20} cells, which is equivalent to the weight of about 105 people. Thus a very high number of cells can be obtained from single lineage.

In vivo hematopoiesis occurs in the red bone marrow (BM), (in the sternum, pelvic and femur bones). The complex hematopoietic microenvironment (in the BM cavities between blood vessels and sinuses) is potentiated by a variety of nonhematopoietic supportive cells collectively known as **stroma**. Stromal cells primarily consist of macrophages, endothelial, adventitial, reticular and adipocyte cells and provide both soluble and membrane-bound growth factors to developing hematopoietic cells. Stromal cells produces extracellular matrix (ECM) - which consists of several adhesion molecules and proteoglycan such as fibronectin, thrombin and heparan sulfates, which also affects proliferation and differentiation. ECM binds and sequesters growth factors to provide adhesion sites for hematopoietic cells. Thus their culture requires same composition and environment.

16.4.2 Collection of Hematopoietic Stem and Progenitor Cells

There are three main sources of hematopoietic stem (progenitor) cells: Bone marrow (BM); sternum / pelvis; and cord blood (CB). CB and BM are often processed by centrifugation over a Ficoll-Histopaque density gradient to deplete the sample of erythrocytes; and peripheral blood (PB) cells are generally obtained after a stem cell mobilization. Often for their culture, suspension culture in agar or methocel or dexter liquid media is utilised. It has been reported that normal stem cell grow well in suspension culture and they undergo differentiation as well but in subculturing they are difficult to grow. (Note:- During normal cell suspension culture cell density must be high $10^6 - 10^9$).

16.4.3 Explants Isolation

For isolation of the explants, peripheral lymphocyte is isolated from blood at the cell concentration 10^6 cell /ml. Generally peripheral lymphocyte are isolated in dense solution of ficoll hypaque. Isolated explants then can be cultured in the serum free media along with the essential cytokines and hormones; often cytokine alone does not help in the differentiation of the lymphocyte but mixture of these works well. Peripheral blood undergoes monolayer culture in the pellet but they often differentiate in the suspension culture. Explant in primary culture should be supplied with DMSO, sodium butyrate, isobutyric acid which promotes erythroid differentiation that can be identified by the nuclear condensation, cell size reduction, hemoglobin accumulation that takes stain of benzidin.

16.4.4 Hematopoietic Cytokines

A cytokine may produce no effect when used alone, but may have dramatic effects when combined with certain other cytokines. Expansion of one lineage may be at the expense of another lineage. Cells of one lineage may secrete cytokines that promote or inhibit the differentiation into another lineage. Hematopoietic cytokines are members of a large and growing family of glycoprotein factors that are indispensable in hematopoietic cultures. The three cytokines that are essential for survival and proliferation are interleukin 3 (IL-3), stem cell factor [SCF; also known as c-kit ligand (KL) and mast-cell growth factor (MGF)] and fit3 ligand (FL). IL-3 prevents programmed cell death (apoptosis) of committed progenitor cells and SCF and FL prevent apoptosis of early progenitor cells. FL is structurally and functionally similar to SCF, except that it does not act on the erythroid lineage. IL-6 acts in concert with other cytokines to induce cycling of additional stem and progenitor cells. Thus, SCF IL-3 and IL-6 in combination for broad expansion across multiple hematopoietic lineages.

A number of cytokines also inhibit hematopoietic cell proliferation and differentiation: e.g. transforming growth factor [(TGF-] strongly suppress the proliferation of stem and progenitor cells; tumor necrosis factor (TNF) has both positive and negative effects depending on the target-cell population and the other cytokines employed; and macrophage inflammatory protein (MIP) inhibits differentiation of-the most primitive cells and stimulates proliferation of more mature cells. These cytokines may be useful in the ex vivo maintenance and expansion of stem cells by slowing down their differentiation in the presence of other cytokines.

16.4.5 Hematopoietic Cell Assays

Colony-forming cell (CFC) assays is done commonly *in vitro* for isolating hematopoietic progenitor cells. A small number of sample cells (1000-20, 000 cells ml^{-1}) are placed in a semisolid medium (usually 1.2% methylcellulose) containing a suitable mixture of growth factors, and are incubated for approximately two weeks. Cultures are then assess for the presence of colonies (>50 cells that originated from a single cell). Colony types are assessed by color and marked as CFU: for example colonies containing white cells are termed CFU-GM (colony forming-unit granulocyte-macrophage), colonies containing red cells are termed BFU-E (burst-forming-unit erythroid), and colonies containing both cell classes are termed CFU-Mix (colony-forming-unit mixed cells). CFU-DCs (colony-forming-unit dendritic cells), which are a subset of CFU-GM, are identified by their unique morphology when cultured in the presence of granulocyte-macrophage colony- stimulating growth factor (GM-CSF) and tumor necrosis factor (TNF-α). CFU-Mk (colony forming- unit megakaryocyte) are defined as a cell that produces three or more CD41 cells when cultured for 14 days.

These assays measure the presence of intermediate-stage hematopoietic progenitor cells. The long-term-culture initiating-cell (LTC-IC) assay is the most common assay for more primitive cells. For this assay, sample cells can be placed on a supportive stromal cell layer and maintained in culture for 5-8 weeks. At the end of this period, cells are placed into a colony assay to measure colony-forming cells. After 5-8 week culture period, the more primitive LTCICs (believed to be a subset of restricted stem cells) differentiate into CFCs, which can then be detected by the colony assay.

The stage of cell differentiation can be examined using flow cytometry, which has the advantage of providing immediate results. Its major disadvantages are that relatively large quantities of cells are required, and that the quantitative accuracy in measuring very small populations (such as stem cells) is limited.

16.4.6 Cloning of Hematopoietic Cells in Suspension

The choice of medium in a hematopoietic culture system depends on the aim of the expansion and on the culture system used. In general, serum-containing formulations (fetal bovine serum and horse serum) are superior for the expansion and maturation of the granulocyte and macrophage lineages. Serum is also important for the growth and maintenance of a stromal layer for use in a stroma-containing culture system. Serum-free media promote greater expansion of progenitor and total cells of the erythroid and megakaryocytic lineages, primarily because TGF-β (which is typically found in serum) is extremely inhibitory to the expansion of these lineages. Serum-free media have supported excellent expansion of highly purified primitive hematopoietic cells stimulated by multiple cytokines.

BRAIN QUEST

1. Describe the different method of
   (a) Hematopoietic cell culture
   (b) Epithelial cell culture
   (c) Stem cell culture

ORGAN CULTURE

17

In this Chapter we will learn about the technique of induction of embryonic cell and their control of morphogenesis and development into complete specific 3-D structure.

17.1 HISTORY AND DEVELOPMENT

A small tissue fragments or any embryonic portion grown in proper environment for development into specific portion of the organ on 3-D mesh constitutes **organ culture**. Organ culture is heterogeneous in nature and can be propagated into various tissue individually and then each tissue arranged in multilayered organization (not real tissue) constitutes a fully functional structure.

The beginning of the organ culture was credited to Harrison (1907) who cultured tadpole spinal chord in lymph drop and thus he was regarded as the **father of animal cell culture**. But the first organ culture was started by the **Loeb** in 1897 who cultured fragments of adult rabbit liver, kidney, thyroid, and ovary on small plasma clots inside a test tube and reported that the cells retained their normal histological structure for 3 days. **Loeb and Fleischer** showed that the tube filled with oxygen, prevents central necrosis of the explants. **Parker** also emphasized the necessity of oxygen and grew fragments of several organs in a shallow layer of medium in a flat bottomed flask filled with 80% oxygen. All these methods were unsatisfactory as most of the tissue sank to the bottom of the flask except skin which floated on the surface of the medium. Skin is not wettable, and this property was used to grow slices of rabbit ear skin on a serum-saline mixture in a flask filled with 70% oxygen. However, a disadvantage of growing skin in fluid medium was that the fragments tend to curl up and the epithelium migrates and covers the dermal surface. It had been recognized that, (with the exception of skin) most of the organ rudiments or organs could better be maintained and grow on a solid support rather than in the fluid medium.

17.2 HISTOTYPIC CULTURE

In Histotypic culture several different types of cell are grown in monolayer at high density and then they are recombined on a suitable three dimensional matrix to have the look and feel of tissue (See Fig. 17.1)

There are many ways in which cells have been recombined to simulate tissue, ranging from simply allowing the cells to multilayer by perfusing a monolayer [Kruse et al., 1970] to highly complex perfused membrane ([Klement et al., 1987]) or capillary beds [Knazek et al., 1972]. These are termed **as histotypic cultures**. Based on the type of matrix used following techniques are used for histotypic cultures. (See Fig. 17.1)

17.2.1 Gel and Sponge Technique

In this method, the gel (collagen) or sponges (gelatin) are used as matrix for inducing the morphogenesis, adhesion and cell growth. The cells penetrate these gels and sponges while growing and thus developed into complete histological similar to original one structure.

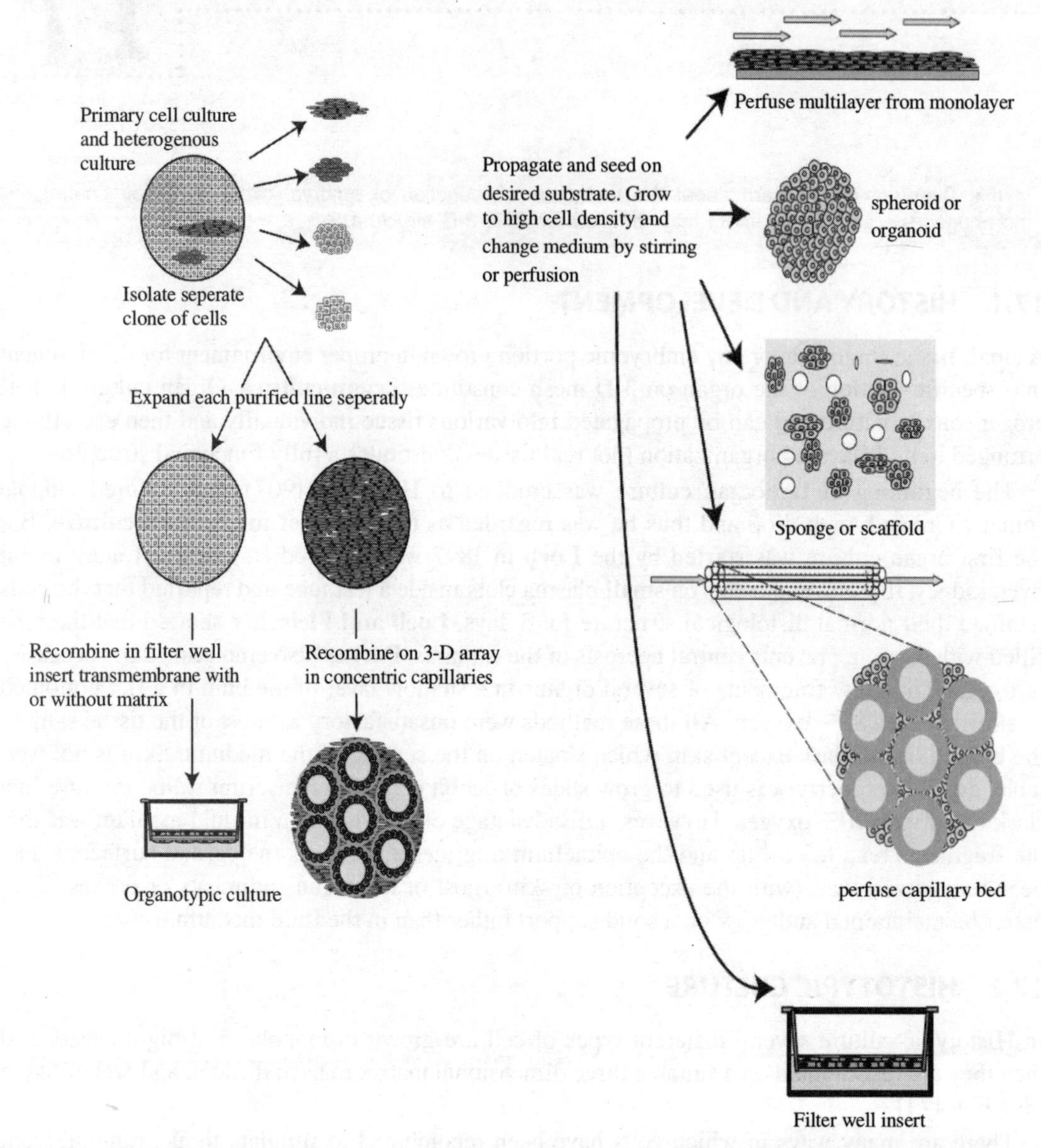

Fig. 17.1 Histotypic and organotypic culture. Expansion of purified populations and recombination can generate organotypic cultures, in filter well inserts or an concentric microcapillaries (modified from freshney)

17.2.2 Hollow Fibers Technique

In this method, hollow fibers are used as matrix which helps in more efficient nutrient and gas exchange. In recent years, perfusion chambers with a bed of plastic capillary fibers have been developed to be used for histotypic type of cultures that helps in better exchange of gases. The cells get attached to capillary fibers and increases cell density to form tissue like structures. See Fig. 17.2.

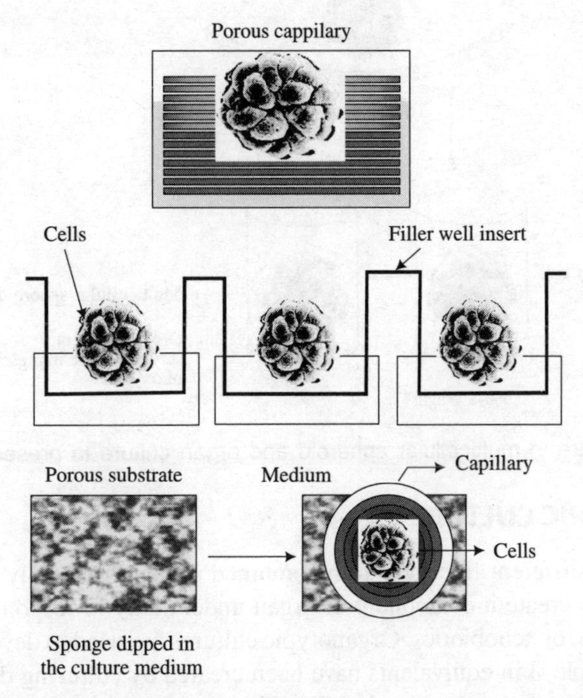

Fig. 17.2 Different type of substrate for organotypic culture 1. Perfusion culture, 2. On porous materials or 3. In sponge dipped in culture medium

17.2.3 Spheroids

The re-association of dissociated cultured cells leads to the formation of cluster of cells called **spheroids**. It is similar to the reassembling of embryonic cells on specialized structures. The principle followed in spheroid cultures is that the cells in heterotypic or homotypic aggregates have the ability to sort themselves and form groups which form tissue like architecture. However, there is a limitation of diffusion of nutrients and gases in these cultures.

17.2.3.1 Multicellular Tumour Spheroids

To form a multi cellular spheroid the monolayer of cells is treated with trypsin to obtain a single cell suspension. The cell suspension is inoculated in the medium in magnetic stirrer flasks or roller tubes. After 3-5 days, aggregates of cells representing spheroids are formed. Spheroid growth can be quantified by measuring their diameters regularly. The multicellular spheroids have a three dimensional structure which helps in performing experimental studies related to drug therapy, penetration of drugs regulation of cell proliferation, immune response, cell death, and invasion. They are used as models for a vascular tumour growth, to study gene expression in a three-dimensional configuration of cells, to study the effect of cytotoxic drugs, antibodies, radionucleotides, and rheumatoid arthritis. (See Fig. 17.3).

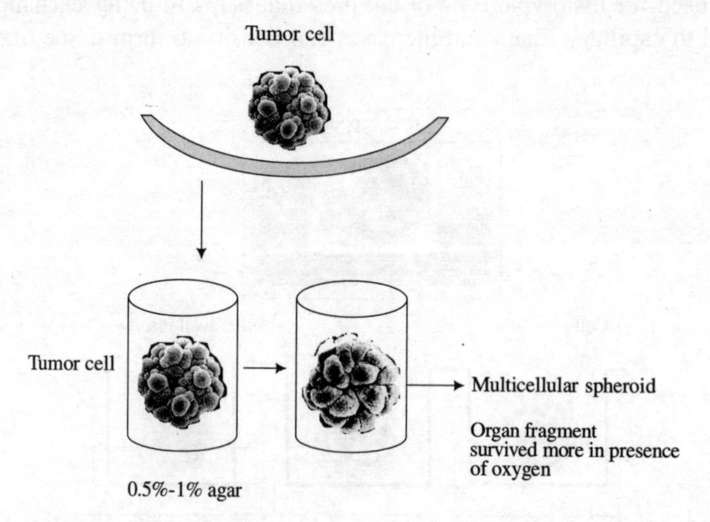

Fig. 17.3 Culture in multicellular spheroid and organ culture in presence of oxygen

17.3 ORGANOTYPIC CULTURE

In organotypic culture different lineages are recombined in experimentally determined ratio and spatial relationship to recreate a component of organ under study in response to external stimuli like hormones, nutrition, or xenobiotics. Organotypic cultures are used to develop certain tissues or tissue models for example skin equivalents have been created by culturing dermis, epidermis and intervening layer of collagen simultaneously. (Fig. 17.4)

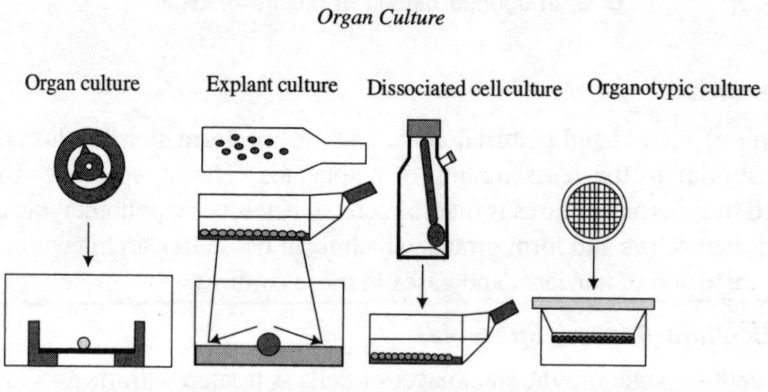

Organ Culture

Organ culture
Explant culture
Dissociated cellculture
Organotypic culture

Fig. 17.4 Illustrations of organ sulture, Explant culture dissociated cell culture and organotypic culture (redrawn from freshney 2005)

Similar model have been developed for prostrate, and breast cancer. The main objective is to maintain architecture of the tissue and direct it towards normal development. In this technique, it is essential that the tissue should never be disrupted or damaged. Media used for growing organ culture are generally the same as those used for tissue culture.

Steps in organotypic culture from suspension cell

1. Isolate individual cells.
2. Propagate them into various type of cells individually.
3. Recombined in experimentally determined ratio.
4. Finally grow them into medium to multilayered 3-D architecture.

17.4 ORGAN CULTURE

The pioneer work of Hardy (1949, 1951) for the growth of hair and hair follicles, followed up by Cleffman (1963) with studies of pigment formation in the hair-follicle melanocytes of agouti mice stimulated various workers for organ culture. *In vitro* culture and growth of organs or parts thereof in which their various tissue components (e.g. parenchyma and stroma) are preserved both in terms of their structure and functions so that the culture organs resemble closely the concerned organs in vivo is called **organ culture.** Therefore, in organ culture whole organ or its representative part are maintained as small fragments in the culture which retain their intrinsic, numerical, and spatial distribution of participating cells. It had been reported that retention of histologic structure, and its associated (differentiated) properties can be enhanced at the air/medium interface, where gas exchange is optimized and cell migration is minimized.

An alternative approach, is the amplification of the cell stock by generation of cell lines from specific cell types and their subsequent recombination in Organotypic culture. This allows the synthesis of a tissue equivalent or construct, for basic studies on cell-cell and cell-matrix interaction and for in vivo implantation.

17.5 TECHNIQUES OF ORGAN CULTURE

Organ culture may be done on variety of substrate viz (1) Semisolid agar, (2) Clotted plasma, (3) On raft of microporous filter, (4) Lens paper, (5) Rayon supported on stainless steel grid, or adherent to strip of perplex glass. The rolled culture bottles were successfully employed for the large scale culture.

17.5.1 Plasma Clot (Watch Glass Technique)

Fell and Robison introduced the 'watch glass technique' in which organ rudiments are grown on the surface of a clot (consisting of chick plasma) and chick embryo extract, contained in a watch glass. This became the classical standard technique for morphogenetic studies of embryonic organ rudiments. The method has later been modified to investigate the action of hormones, vitamins, and carcinogens in adult mammalian tissues.

Another type of culture vessel consist of an **embryological watch glass** containing a plasma clot and closed with a glass lid sealed on with paraffin wax (see Fig. 17.5). This was first introduced by **Rudnick** and later adopted by Gaillard.He used a clot consisting of two parts of human plasma, one part of human placental serum, and one part of human baby brain extract mixed with six parts of a saline solution. Since the clot liquefies in the vicinity of explants, they become partly or fully immersed in the medium. Biochemical analyses are not possible because of the complexity of the medium, and the duration of culture is short, usually less than 4 weeks.

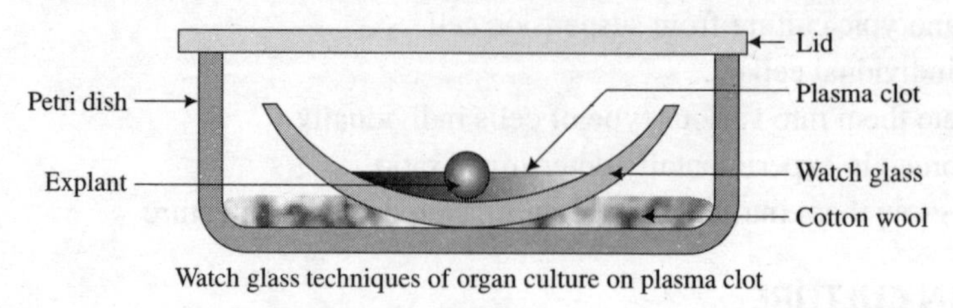

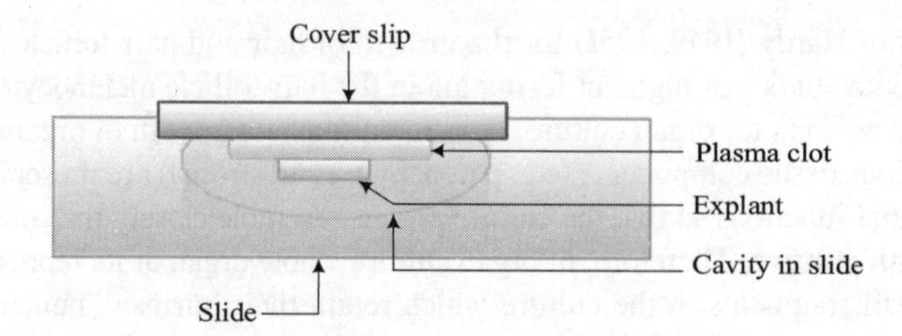

Fig. 17.5 Illustrations of plasma Clot on culture

17.5.2 Grid Method (Trowell, 1954)

The grid method utilises a metallic wire mesh or a perforated sheet with margins bent to form 4 mm high legs (see Fig. 17.6). Explants are placed on a raft of lens paper, rayon acetate or millipore filter membrane. Suitable wire mesh or perforated stainless steel sheet of size 25 × 25 mm is used as a grid, whose edges are bent to form four legs of about 4 mm height. The grids themselves are placed in a culture chamber filled with fluid medium up to the grid; the chamber is supplied with a mixture of O_2 and CO_2 to meet the high O_2 requirements of adult mammalian organs. A modification of the original grid method is widely used to study the growth and differentiation of adult and embryonic tissues.

The original Trowell technique was aimed at maintaining adult mammalian tissues which have a higher requirement for oxygen than fetal organs. To achieve this, the culture chambers are enclosed in containers, perfused with a mixture of carbon dioxide and oxygen. The method succeeded in preserving the viability and histological structure of the adult tissues, such as prostate glands, kidney, thyroid, and pituitary.

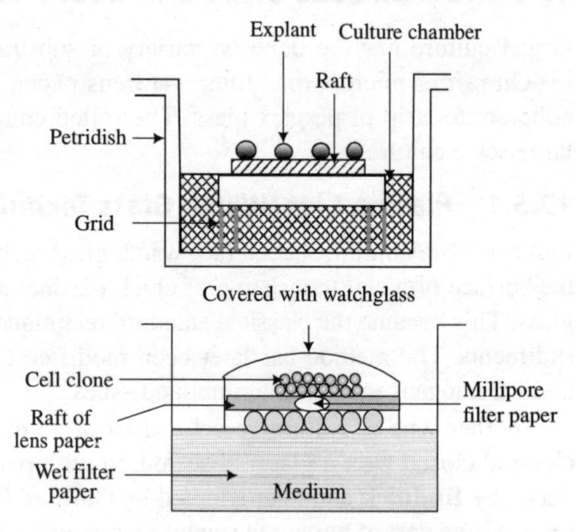

Fig. 17.6 Culture on (a) Grid and Raft (b) Wet filter

17.5.3 Raft Method

Chen (1979), reported that lens paper used in the cleaning of microscope lenses is non-wettable and can float on fluid medium. He explanted 4-5 cultures on a 25 × 25 mm raft of lens paper which floated on serum in a watch glass. Richter improved this technique by treating the lens paper with silicone which enhanced its flotation properties. Lash et al. (1968) combined the lens paper with Milli-pore filters. A small hole was created in the centre of the lens paper raft covered with a strip of Millipore filter which helps in exchange of gases.

Different types of tissues can be cultured on either side of the filter and their interaction with each other studied. Shaffer replaced the lens paper with rayon acetate strips. The rayon acetate strips were made to float on the fluid medium by treating the four corners with silicone. The rayon acetate has the advantage over lens paper that it is acetone soluble and can be dissolved during the histological procedures by immersing it in acetone. In raft and clot techniques, the explant is first placed on a suitable raft (floats), which is then kept on a plasma clot. This modification has made changing of media easy and also prevented the raft sinking into the liquefied plasma. (See Fig. 17.6).

17.5.4 Agar Gel Method

The problems encountered using plasma clots were eliminated by the use of agar gels. The agar gel technique was first introduced by **Spratt** (1956-57). Wolff and Haffen modified Gaillard's technique and used an agar gel contained in an embryological watch glass. The agar method has been successfully used for developmental and morphogenetic studies. Generally the explants need to be subcultured on fresh agar gels every 5-7 days. The agar gels are generally kept in embryological watch glasses sealed with paraffin wax.

This method have been used to study many developmental aspects of normal organs as well as tumors. Although the agar does not liquefy, but major difficulty is culture cant be analyzed without transplanting the cultures. This disadvantage were overcome by the use of fluid media combined with a support which prevented the cultures being immersed. The medium (consisting of suitable salt solution, serum, chick embryo extract or a mixture of certain amino acids and vitamins) gelled with 1% agar. This method avoids the immersion of explants into the medium and permits the use of defined media for e.g. media solidified with agar consist of 1% agar in BSS, or chick embryo extract and horse serum in with agar in the ratio 7:3:3 can be used. Defined media with or without serum may also be utilized with agar. The medium with agar provides the mechanical support for organ culture. It does not liquefy. Embryonic organs generally grow well on this medium; but cell from adult tissue is difficult to culture. The culture of adult organs or parts from adult animal is more difficult due to their greater requirement of O_2. A variety of adult organs (e.g., liver) have been cultured using special media with special apparatus.

17.5.5 Hydrogel Cell Culture

Hydrogel is the three-dimensional synthetic gels that can be used in cell culture to mimic the extra-cellular matrix for cell interactions studies. Bioengineers are moving forward in recapitulating endogenous factors in tissue to make hydrogels with characteristics similar to that of live tissues. Some challenges remain for utilizing hydrogels such as heterogeneities in the three-dimensional culture, oxygen availability in the culture environment, and technical difficulties in harvesting and analyze cells in the artificial 3-D matrix.

17.5.6 Intermittent Exposure to Medium and Gas Phase

More recently, a method which provides intermittent exposure to medium and gas phase has been successfully used for the long-term culture of human adult tissues, including bronchial and mammary epithelium, oesophagus, and uterine endocervix. In this technique, the explants are attached to the bottom of a plastic culture dish and covered with medium. The dishes are enclosed in an atmosphere-controlled chamber which is filled with an appropriate gas mixture. The chamber is placed on a rocker platform and rocked at several cycles per minute during cultivation.

17.6 ADVANTAGE OF ORGAN CULTURE

Organ cultures are suitable for physiological studies, since their results are comparable to the in vivo organs in both structure and function. Organ culture provides information on the pattern of growth, differentiation and development and the influence of various factors on. Sometimes, organ cultures may replace whole animals in experimenting, as the results from them are easier to interpret, however, result from organ cultures are often comparable to those from whole animal studies. A typical example could be the study of drug action, since drugs are metabolized in vivo, and not in vitro.

17.7 LIMITATIONS OF THE ORGAN CULTURE

1. Difficult to monitor the differentiation due to the sampling variation in preparing an organ culture than in propagated cell line.
2. Difficult to prepare than replicate the cultures from passage cell line, because it can be maintained only for few months.
3. Preparation is labor intensive as a result the yield is often too low to be used for monitoring assay.
4. Difficult to propagate and each time they require original donor tissue.
5. Essentially a technique for studying the behavior of integrated tissues rather than isolated cells.
6. Organ cultures can be maintained only for few months, while effects of certain factors require study over several months.

17.8 LIMITING FACTORS FOR 3D ORGANIZATIONS

Gaseous and nutrition exchange get hampered if size is greater than 250 mm (5000 cells) due to blood vessels blockage or O_2 and CO_2 exchange and thus their nutrition blocks occur which result in necrosis of internal area. In the absence of O_2 and nutrition supply from blood vessels then the polarity of cells is changed or get disturbed.

17.9 APPLICATIONS

Organ culture is used principally for (1) the maintenance of structural organization in tissues which are to be subjected to experimentally varied environments (e.g., to hormones, drugs, or radiation); (2) the study of morphogenesis, differentiation, and function in excised organs or presumptive organs; and (3) for comparison of the growth and behavior of explanted organs with the growth and behavior of similar organs in situ. (4) The most significant application of organ culture is the production of tissues for implantation in patients. This is called **tissue engineering**.

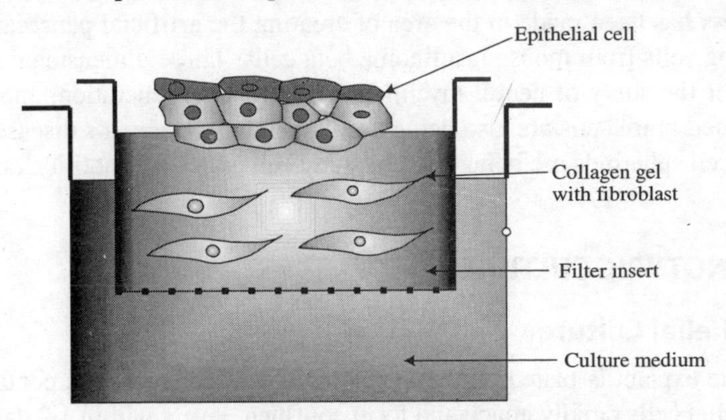

(A) Explat culture of epithelial cells on feeder layer

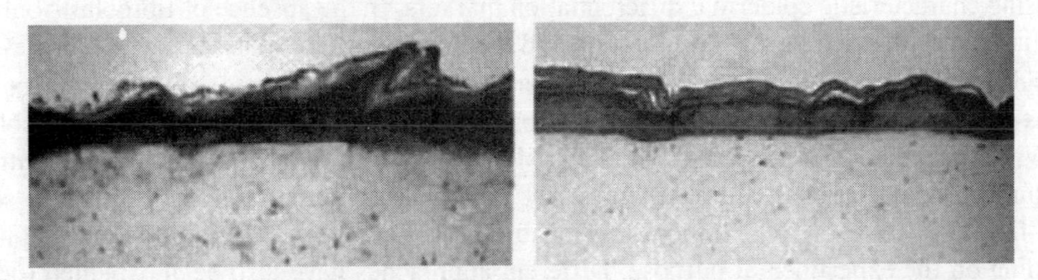

(B) Keratinocyte and fibroblast coculture 2 weeks after implantation in vivo 3 weeks later

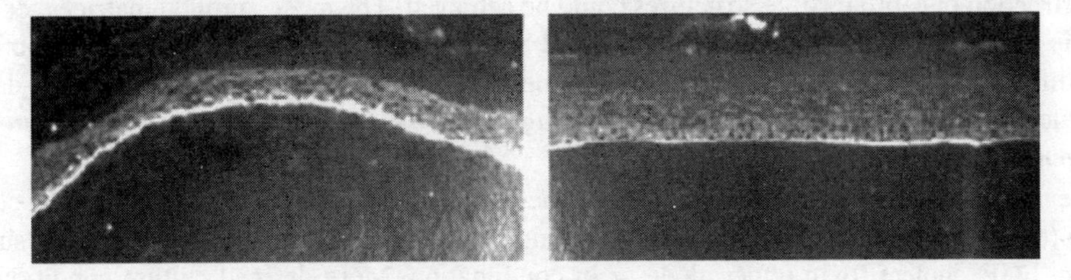

(C) Integrin and cytokeratine coculture in vitro and in vivo

Fig. 17.7 Organotypic culture of skin cells

Human skin has been successfully produced in vitro and used for transplantation in more than 500 cases of serious burns, ulcers etc. The ultimate objective of tissue engineering is to reconstitute body parts *in vitro* for use as grafts or transplants, and as models for studies on drug delivery and action. It is expected that cartilage tissue developed *in vitro* **(artificial cartilage)** will be available for human implantation in case of injuries, arthritis, etc. Experiment with rabbits have produced promising results. It is hoped that studies will permit the culturing and constitution of bones, liver, pancreas, etc. In the case of serious injuries and the arthritis artificial cartilage can be prepared.

The attempts has been done to create tissue model in the form of artificial organs using tissue engineering for example the artificial liver is being created using hepatocytes cultured as spheroids and held suspended in artificial support system such as porous gelatin sponges, agarose or collagen. Some progress has been made in the area of creating the artificial pancreas using spheroids of insulin secreting cells from mouse insulinoma beta cells. Three dimensional brain cell culture have been used for the study of neural myelination, neuronal regeneration, and neurotoxicity of lead. The aggregated brain cells are also being used to study Alzheimer's disease and Parkinson's disease. Thyroid cell spheroids are being used to study cell adhesion, motility, and thyroid follicle biogenesis.

17.10 ORGANOTYPIC CULTURE

17.10.1 Epithelial Culture

When keratinocyte explant is plated on the upper surface of collagen gels containing embedded fibroblasts, epithelial cells rapidly attach and form confluent layers within 1-2 days. Subsequently, keratinocyte reconstitutes into an epithelial tissue architecture resembling the epidermis and expressing characteristic epidermal differentiation markers. In the absence of fibroblasts only thin epithelia arise with rapid loss of proliferation and incomplete differentiation.

Besides collagen type I, other extracellular matrices have been utilized in organotypic cultures such as mixtures of collagen type I and III or basal lamina constituents (e.g., laminin, collagen type IV, fibronectin). A widely used material is Matrigel, a mixture of ECM components extracted from tumor stroma Dermal cell type mouse 3T3 fibroblasts, human dermal fibroblasts scleroderma fibroblasts or even SCC tumor-derived fibroblasts have been incorporated into the matrix depending on the experimental purpose. Different approaches have also been exploited (devitalized dermis, **skin equivalents**), to establish recombinant cultures with endothelial cells an *in vitro* reconstruction of capillary-like structures could be achieved. The most common matrices for organotypic cocultures are collagen type I gels used at different concentrations (14 mg/ml) obtained from different species and organs (e.g., rat, bovine, tendons, skin, placenta). Alternatively, dermal equivalents can also be reconstituted from *collagen type IV, Matrigel, soft agar, or mixtures of collagen and glycosaminoglycans.*

The majority of epithelial cell culture experiments have been performed in conventional culture media (e.g., DMEM or other formulations) containing serum and additional supplements (such as insulin, EGF, cholera toxin either alone or in combination. More defined culture conditions are now available by the development of serum-free culture media. Under those defined conditions it is possible to examine the intrinsic mesenchymal-epithelial interaction free of ill-defined serum constituents and to evaluate more precisely the responses of cells to addition or depletion of growth factors, hormones, vitamins, and other compounds in the culture medium. Cell-cell interactions via diffusible factors and cell-matrix interactions via integrins have been shown to modulate epithelial morphogenesis. Several growth factors and interleukins have been detected in skin as well as in keratinocyte and fibroblast cultures, such as IL-1, IL-6, IL-8, GM-CSF, TGF and NGF, and PDGF as well as several members of the FGF family is essential in proliferation and developemnt of complete tissue.

17.10.1.1 Artificial Skin (LSE)

It has become possible to produce the skin by utilizing the epidermis portion of the skin-in vitro. When this epidermis is reconstituted the complete skin-called living skin equivalent (LSE) is formed. This requires the addition of a collagen matrix as a support for tissue growth. The source of explant is either the patient itself or the prepuce of new born babies. The use of a synthetic polymer (PGA), allows the new born skin to grow without scars. This artificial skin is used to cover the wound until the patient's skin is cultured and artificial skin is obtained for grafting.

17.10.1.2 Procedure

The keratinocytes making up the bulk of the epidermis is trypsinized. These cells are cultured with underlying layer of irradiated 3T3 fibroblast cell line. Proliferation of keratinocytes is stimulated by certain products from fibroblasts. Keratinocyte forms colonies which is dissociated and cultured again. The process is repeated till a continuous sheet of pure epithelium is formed. This sheet is detached, cleaned and used for grafting. Rejection of implant can be avoided by taking the explant from the patient itself. A 1.7 m^2 artificial skin can be obtained from a 3 cm^2 skin in 34 weeks a two-thousand fold increase. Artificial skin grafts have been used to repair successfully several skin defects, skin ulcers, etc. after the grafting of the artificial skin.

17.10.1.3 Assay for Epithelial Cell Culture

Cell proliferation can be analyzed in tissue sections by established methods such as *mitotic index or immunolabeling.* Some of the conventional methods for studying cell proliferation in normal cell culture systems, such as cell counting and cloning efficiency, cannot be applied to these complex model systems, in which keratinocytes form keratinizing epithelia (since a complete dissociation into single cells is no longer feasible). Biochemically, proliferation can be assessed by measuring changes in total DNA content, which usually parallel to the number of nucleated cells. However, this technique is not reliable for determining the total cell number of the epidermis, because the DNA is degraded during the differentiation process in the stratum corneum. Measurement of protein content as a criterion of cell proliferation is also not reliable because of changing cellular protein levels during keratinocyte maturation.

Determination of proliferating cells has been done conventionally by incorporation of radioactive thymidine and analysis by autoradiography. The most frequently used method is labeling with the thymidine analogue BrdU (bromodeoxyuridine) and subsequent detection of the proliferating cells on paraffin or frozen sections by specific anti- BrdU antibodies. Positive nuclear reaction in cells that are in S phase during the BrdU pulse can be recorded relative to the total number of nuclei counterstained by a DNA dye.

The viability of the explants and macroscopic changes can be monitored during cultivation by viewing them in daylight or with the aid of a light source from the dissecting binocular. Healthy and growing tissues usually appear translucent with a shiny surface; opacity suggests loss of viability or the beginning of necrosis of the explanted tissue. Morphology, (the most important qualitative parameter of epidermal tissue reconstitution) is evaluated by light and electron microscopy.

17.11 TISSUE ENGINEERING

Tissue Engineering is the study of the growth of new connective tissues, or organs, from cells and a collagenous scaffold to produce a fully functional organ for implantation back into the donor host. Tissue engineering frequently involves the stem cells, (a kind of premature cell first isolated from the body in 1992). The explant can be as small as a 2 mm punch biopsy for some applications and can be grown on artificial mesh like structure called as **scaffold**. The graft produced can be used for patient to replace the old tissue (see Fig. 17.8 a). This technique allow organs to be grown from implantation (rather than transplantation) and hence free from immunological rejection. Implanting stem cells in the appropriate location can generate everything from bone to tendon to cartilage. Physical structure of scaffold may control cell function by regulating the diffusion of nutrients, waste products and cell-cell interactions by providing spatial and temporal control of biochemical cues, whereas scaffold surface chemistry indirectly affects cell adhesion, morphology and subsequent cellular activity by controlling adsorption of ions, proteins and other molecules from the culture medium. The term regenerative medicine (more emphasis on the use of stem cells to produce tissues) is often used synonymously with tissue engineering,

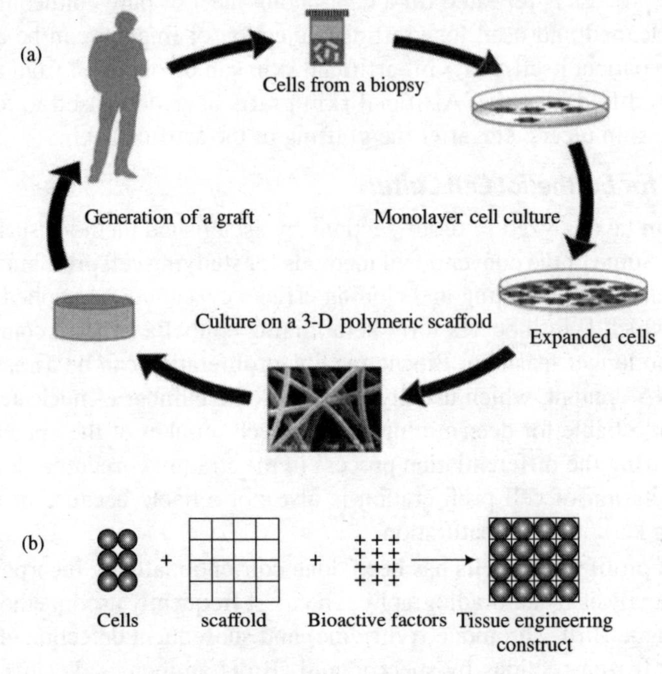

Fig. 17.8 Basic principle of tissue engineering

17.11.1 Design and Engineering of Tissues

The design and tissue engineering should essentially cause minimal discomfort to the patient. The damaged tissues should be easily fixed with the desired functions quickly restored. Another important factor controlling the designing of tissue culture is the source of donor cells. The cells from the patient himself, is always preferred as it considerably reduces the immunological complications. However under certain situations allogeneic cells (cells taken from a person other than the patient) are also used. The other important factors are the support material, it's degradation products, cell adhesion characteristics etc. It was demonstrated in 1975 that human keratinocytes could be grown in the laboratory in a form suitable for grafting. A continuous sheet of epithelial cells can be grown now however there is still difficult to grow the skin with the dermal layer with all the blood capillaries, nerves, sweat glands, and other accessory organs. Some of the implantable skin substitutes which are tissue engineering skin constructs with a limited shelf life of about 5 days are

(a) **Integra ™:** A bioartificial material composed of collagen-glycosaminoglycan is mainly used to carry the seeded cells.

(b) **Dermagraft™:** This is composed of poly glycolic acid polymer mesh seeded with human dermal fibroblasts from neonatal foreskins.

(c) **Apligraf™:** It is constructed by seeding human dermal fibroblasts over collagen gel with the placement of a layer of human keratinocytes on the upper surface. These tissue constructs integrate into the surrounding normal tissue and form a good skin cover with minimum immunological complications. The urothelial cells and smooth muscle cells from bladder are now being cultured and attempts are on to construct the urothelium. Some progress has also been made to repair injured peripheral nerves using tissue engineered peripheral nerve implants. The regeneration of the injured nerve occurs from the proximal stump to rejoin at distal stump.

The regeneration process requires substances like

(a) Conducting Material

The conducting material is composed of collagen- glycosaminoglycans, PLGA (poly lactic- co-glycolic acid), hyaluronan and fibronectin and forms the outer layer.

(b) Filling Material

The filling material contains collagen, fibrin, fibronectin and agarose. This supports the neural cells for regeneration. and

(c) Additives

A large number of additive is required for appropriate proliferation e.g. growth factors, neurotrophic factors such as fibroblast growth factor (FGF), nerve growth factor (NGF).

17.11.2 Scaffold

Scaffold is an artificial structure capable of supporting three-dimensional tissue formation. Scaffolds make balance between temporary mechanical function with mass transport to aid biological delivery and tissue regeneration. To achieve the goal of tissue reconstruction, scaffolds must meet some specific requirements. A high porosity and an adequate pore size are necessary to facilitate cell seeding and diffusion throughout the whole structure of both cells and nutrients. Biodegradability is often an essential factor since scaffolds should preferably be absorbed by the surrounding tissues without the necessity of a surgical removal. The rate at which degradation occurs has to coincide as much as possible with the rate of tissue formation: this means that while cells are fabricating their own natural matrix structure around themselves, the scaffold is able to provide structural integrity within the body and eventually it will break down leaving the neotissue, newly formed tissue which will take over the mechanical load. Injectability is also important for clinical uses.

Many different materials (natural and synthetic, biodegradable and permanent) have been investigated. Most of these materials have been known in the medical field before the advent of tissue engineering as a research topic, being already employed as bioresorbable sutures. Examples of these materials are collagen and some polyesters.

New biomaterials have been engineered that have ideal properties and functional customization: injectability, synthetic manufacture, biocompatibility, non-immunogenicity, transparency, nano-scale fibers, low concentration and resorption rates, etc. PuraMatrix, originating from the MIT labs of Zhang, Rich, Grodzinsky and Langer is one of these new biomimetic scaffold families which has now been commercialized and is impacting clinical tissue engineering.

A commonly used synthetic material is PLA - polylactic acid. This is a polyester which degrades within the human body to form lactic acid, a naturally occurring chemical which is easily removed from the body. Similar materials are polyglycolic acid (PGA) and polycaprolactone (PCL): their degradation mechanism is similar to that of PLA, but they exhibit respectively a faster and a slower rate of degradation compared to PLA. (See Fig. 17.9). Scaffolds may also be constructed from natural materials: in particular different derivatives of the extracellular matrix have been studied to evaluate their ability to support cell growth. Proteic materials, such as collagen or fibrin, and polysaccharidic materials, like chitosan or glycosaminoglycans (GAGs), have all proved suitable in terms of cell compatibility, but some issues with potential immunogenicity still remains. Among GAGs hyaluronic acid, possibly in combination with cross linking agents (e.g. glutaraldehyde, water soluble carbodiimide, etc.), is one of the possible choices as scaffold material. Functionalized groups of scaffolds may be useful in the delivery of small molecules (drugs) to specific tissues. Thus, Scaffolds have been investigated in different directions, from matrix components (e.g., collagen, synthetic polymers), to geometry (e.g., gels, fibrous meshes, porous sponges, tubes), structure (e.g., porosity, distribution, orientation, and connectivity of the pores), physical properties (e.g., compressive stiffness, elasticity, conductivity, hydraulic permeability), and degradation (rate, pattern, products).

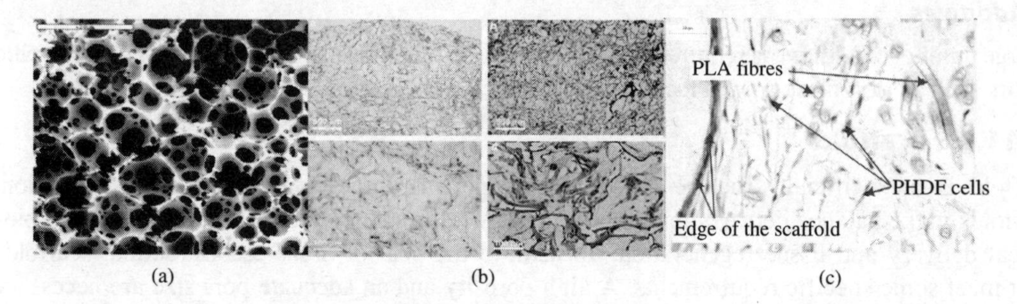

Fig. 17.9 Various type of scaffold (a) polystyrene, (b) collagen and (c) PLA scaffold

Scaffolds usually serve at least one of the following purposes:

(i) Allow cell attachment and migration
(ii) Deliver and retain cells and biochemical factors
(iii) Enable diffusion of vital cell nutrients and expressed products
(iv) Exert certain mechanical and biological influences to modify the behaviour of the cell

17.13 APPLICATIONS OF TISSUE ENGINEERING

1. In growth of cells in three dimensional systems.

2. Delivery systems for protein therapeutics.
3. Cell cultivation methods for culturing 'recalcitrant cells'.
4. Expression of transgenic proteins in transplantable cells.
5. To develop vehicles for delivering transplantable cells.
6. Development of markers for tracking transplanted cells.
7. Avoiding immunogenicity in transplantable cells.
8. Development of in vivo and ex vivo biosensors for monitoring cell behavior during tissue production.

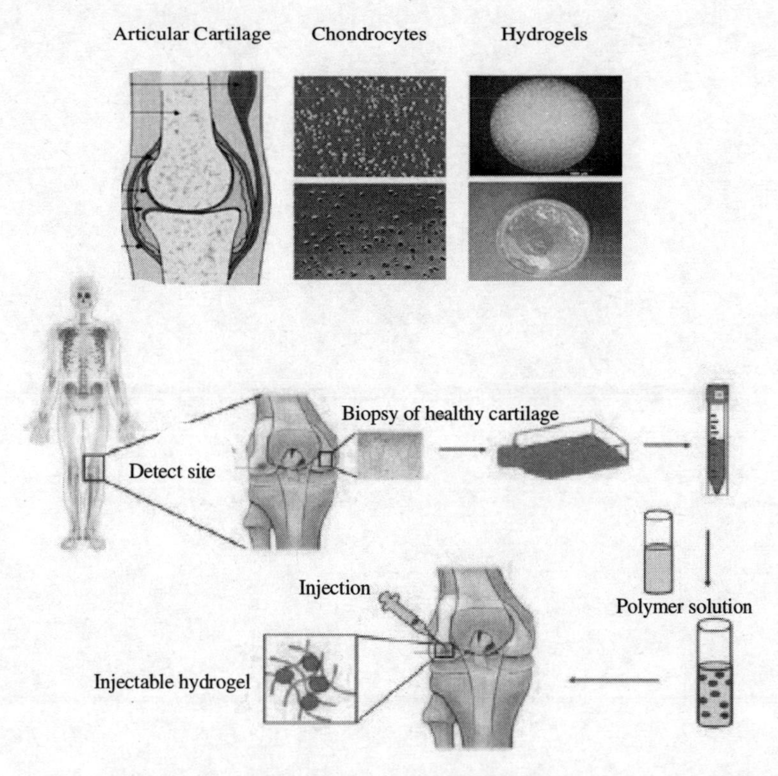

Fig. 17.10 Creation of new cartilage (controlled delivery of biological signals in addition to physical signals provided by a scaffold to design anisotropic, organized cartilage tissue)

BRAIN QUEST

1. Write short note on organ culture.
2. What is various type of organ culture.
3. What are the different type of 3-D organization for organ culture.
4. What are the differences between Histotypic culture and Organotypic culture?
5. What are the different substrates used for organ culture.
6. Write the limitation of the organ culture.
7. Write the application of organ culture.
8. Write the importance of scaffold in tissue engineering.

CELL TO CELL INTERACTIONS

18

In this Chapter, we will focus on role of different type of extracellular component and cell-adhesion molecule in cell to cell interactions, organizing the tissues and in understanding their role in proper growth and development of animal cell.

18.1 ROLE OF EXTRACELLULAR MATRIX

All multicellular organisms, establishes an elaborate communication network that coordinates the growth, differentiation, and metabolism of cells with the help of a complex network of proteins and carbohydrates, called as the extracellular matrix (ECM) which is present in the spaces between the cells and has specific role. The matrix helps in binding the cells and tissues together and is a reservoir for many signals molecule such as hormones which controls cell growth and differentiation. The matrix is also equipped with a lattice through which cells can move, particularly during the early stages of differentiation and with various integral membrane proteins, collectively termed as **cell-adhesion molecules** (CAMs), which enables the cells to adhere tightly and specifically with cells of the same type and helps in aggregation of the cells to elaborate specialized cell junctions that stabilize these interactions and promote local communication between adjacent cells. These interactions allow populations of the cells to segregate into distinct tissues. Defects in these connections lead to cancer and other developmental malformations in the cells. The extracellular matrix has three major protein components: (See Fig. 18.1)

1. Highly viscous proteoglycans, which cushion the cells; (protein and carbohydrates)
2. Insoluble collagen fibers, which provide strength and resilience; and
3. Soluble multiadhesive matrix proteins, which bind these components to receptors on the cell surface. e.g. integrins, cedherins and selectins

For proper cell differentiation certain adhesion molecules, external growth factor is required (PDGF, EGF FGF).

Adhesion is modified by surface receptors and the matrix (ECM) components

- ECM is secreted by normal cells and helps in adhesing cells to charged plastics, while receptor binds to matrix.
- ECM secretes various cell surface adhesion molecules either between the cells (Cadherins) or around the cells (CAMs)
- There are various other cell substrate molecule eg Integrin which helps in cell attachment via binding to fibronectin, entactin, laminin and collagen. Integrin is comprised of α and β subunit and they bind via specific motic (RGD) arginin, glycine, aspartic) — ECM is

dependent on cell type eg fibrocyte secrete collagen I and fibronectin while epithelial cell secrete laminin

- Other cell adhesion component is proteoglycan they bind not utilising RGD motif but by low affinity receptors
- Cell line make their own ECM thats why they grow indefinitly.

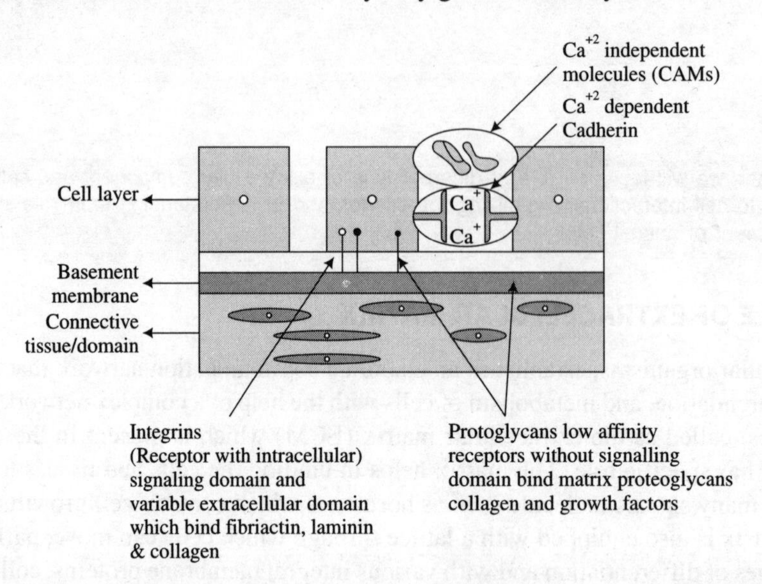

Fig. 18.1 Role of various adhesion molecule in attachment to cells or cells to basement membrane

Note that without CAMs or cadherins cell to cell attachment not possible while to basement membrane various ECM component like integrins and proteoglycans helps since differentiation requires, cell to all interaction cell to metrix interaction, since without this wrong lineage of cell may be expressed

Different combination of these components tailor the strength of the extracellular matrix for different purposes for example, strength in a tendon, cushioning in cartilage, or adhesion. In the case of smooth muscle (surrounding artery), the extracellular matrix must provide strong but flexible connections. Therefore, the extracellular matrix is not just an inert framework or cage that supports or surrounds cells but also communicates directly and indirectly with the intracellular signaling pathway that directs a cell to carry out specific functions for example, hepatocyte the principal cells in the liver, express liver-specific proteins depends on their association with a matrix of appropriate composition. Specific ECM components directly activate cytosolic signal-transduction pathways by binding to cell-adhesion protein receptors in the plasma membrane. This signal-transduction pathways either sequester these signals from cells or, conversely, present them to cells, thereby indirectly inducing or inhibiting signaling pathways due to binding of growth factors and other hormones. Morphogenes is the later stage of embryonic development (during which cell movements and rearrangements occurs) also is critically dependent on ECM components. Even in adults resynthesis of ECM components occurs in areas of wounding and degradation.

18.2 CELL JUNCTION

Animal cells have several specialized cell junctions which are responsible for cell to cell attachment. Cell junctions are of following three types: 1. Adhering cell junction 2. Tight junction and 3. Gap junction.

18.2.1 Adhering Junction

It includes 1. Adherence junction, 2. Desmosomes and 3. Hemidesmosome.

18.2.1.1 Adherence junction

Adherence junctions are connection sites for actin filaments; while other two junctions, Desmosomes and Hemidesmosome are the junction site of intermediate filaments. Adherence junctions are commonly found in the epithelial cell in the form of zonula adherence around each of interacting cells located near the apex of each cell just below the tight junction.

The adhesion belts are members of large family of Ca^{++} dependent cell-cell adhesion molecules and are called as **cadherins** which are present in adjacent epithelial cells. It is proteinaceous in nature and holds the plasma membrane together to provides strong anchorage to the adjoining cell. In basal surface of some epithelial cell, **hemidesmosome** are found which prevent underlying cell from the peeling off.

18.2.1.2 Desmosomes

Desmosomes are mostly found in the various types of epithelial cells. They are insoluble in the non ionic detergent against most the membranes that are soluble. Most of the epithelial sheet of the cells are held together by dozen of Desmosomes on their surfaces.

18.2.2 Tight Junction

Tight junction is present at the fusion point of the cell membrane, mainly at the point of contact tight junction as can be seen in Fig. 18.2 and leaves no space in between. They may appear as belt or band like, completely sealing off the area of contact. They offer a strong physiological barrier and prevent flow of material across the epithelium for example: Singulin ZO-1 is found as major protein encircling the cell such as spot Desmosomes and seal enough not to pass any molecule from them. Tight junction, are also known as **zonula occludens**, which is common in the epithelial cells, play a central role in the regulation of permeability in epithelia.

18.3 GAP JUNCTION

They are the simplest nexus for cellular interaction and have two main functions, one cellular adhesion and other is the intercellular communication. They allow up to 1,300 Daltons and other low molecular weight compound such as sugar, amino acids, and vitamins and its permeability is regulated by calcium -ion concentration. Normally the Ca^{++} is low in the cell hence the permeability of the channels is high. If the concentration of the Ca^{++} is high then permeability is low and vice versa. Gap junction present in the neurons has low resistance so it allows impulse conduction instantaneously. (Fig. 18.3)

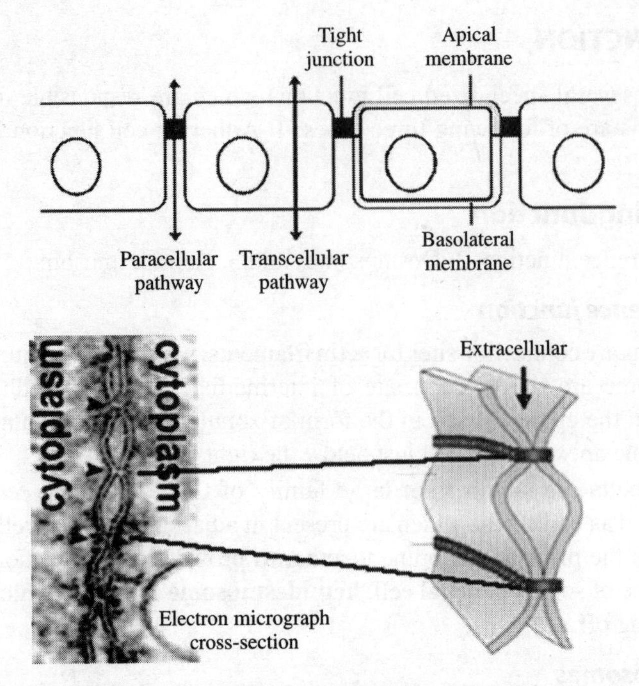

Fig. 18.2 Plasma membrane joined tightly at several places by tight junction that tightly seals the junction and does not allow anything to pass. (a) Outline of the tight junction (b) Joints of tight junction

18.3.1 Features of the Gap Junction

Gap junction occurs between the plasma membrane of two adjacent cells and is formed of a protein called as **connexin** of molecular weight- 32,000. The Gap junction channel is generally used for intracellular transfer of the molecules and ions and its function can be observed by using inhibitor chemical such as Oubain which abolishes the intracellular concentration of K^+ where as resistant cells maintain high K^+ levels. If cells having no gap junction are co-cultured, with cells having gap junction, the cells develops gap junction between them. (Fig. 18.3)

Gap junction usually regulates the signaling molecule such as c-AMP which at high concentration, arrest the development of the oocyte. In cardiac cells, gap junctions allow synchronous heart beat, during labor, uterine muscle contract simultaneously, this ability is accompanied by the presence of the gap junctions between the muscle cells. Many cancerous cells which do not have gap junctions, do not allow the metabolic regulation or ionic coupling.

18.4 CELL ADHESION MOLECULES (CAM)

Various glycoprotein's between cells are responsible for species-specific aggregation they were later recognized as CAM. Some of the best characterized CAM molecules in the vertebrate's cells and tissue shown in Table 14.1 and the structure is shown in Fig. 18.4. Isolation of CAM can be done by the treatment of Ca^{+2} and Mg^{++} free sea water; since cell are held together by the ionic forces.

18.4.1 Cadherins

The cadherins are general classes of adhesion molecules which requires Ca^{++} for its activity. They are of three types, uvomorulin (LCAM), P-cadherin and N-cadherins. Molecular cloning of the DNA for uvomorulin (LCAM), P-cadherin and N-cadherins shows that these proteins are quite similar in structure. Cadherins are composed of 723-748 amino acids, with a single peptide membrane spanning hydrophobic region and extracellular hydrophobic region, with cytoplasmic domains. The average amino acid is conserved in the cytoplasmic domain. The extracellular part has 3 cytoplasmic domains that bind to the Ca^{+2}.

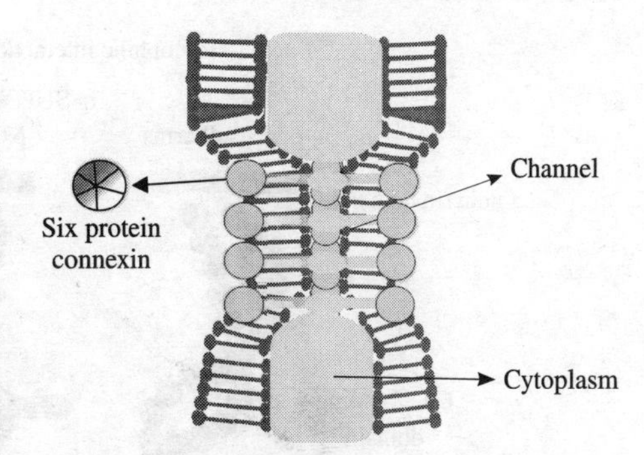

Fig. 18.3 Gap junction are having six subunit proteins Connexin to check and pass the molecule across the membrane.

Their expression in the cell increases the adhesion with other cells. It also found to promote the alteration in the cell shape. A cytoplasmic domain called catenins binds with the cadherins. **Plakoglobin** a component of Desmosomes and is a type of catenins. In addition, Desmocolin and Desmoglein two other desmosomal proteins are also cadherins.

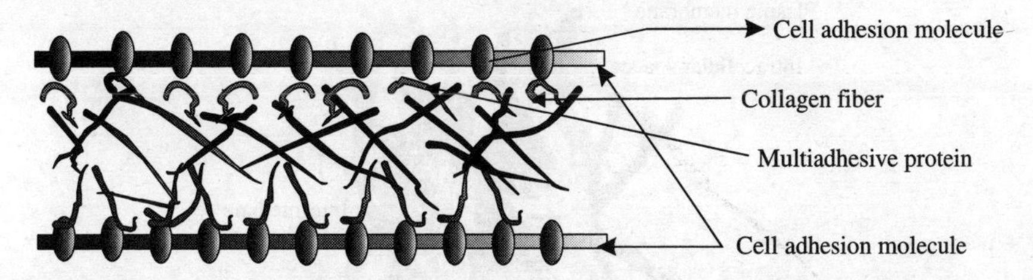

Fig. 18.4 Electron micrograph of smooth muscle in the wall of a small artery. The muscle cells are separated by relatively wide intercellular spaces that contain a group of glycoproteins, called proteoglycans, and collagen, which is the most abundant fibrous component of the extracellular matrix. Interactions between cell-adhesive integral membrane proteins and components of the extracellular matrix allow the cells to adhere to one another and give this tissue its strength and resistance to shear forces, while cell adhesion molecules resulates the passage of molecules.(Redrawn from Lodish, 2005).

Cadherins expression changes during tissue development and this apparently leads to separation of different tissue that has the same origin. Neuronal retina expresses N-cadherins equally on their surface and on differentiation retina looses expression of N-cadherins. The neuronal ectoderm expresses E- cadherins and some time they express N-cadherins. This difference allows these cells to detach from the ectoderm and bind homotypically, and then eventually form the neural tube.

The best studied of the Ca^{+2} independent CAM is NCAM. This molecule exist in the three major forms of molecular weights 180 K,140 K, 120 K in skin, lung, and the intestine which undergo continuous shape change throughout life. Elasticity is due to presence of the Elastin in collagen. Low sialic acid results in the low affinity. Although the NCAM and IgG has similarity in recognition, pattern one difference is that one recognizes self and other recognizes the nonself. See Figs. 18.5 and 18.6. They acts like the marker and thus helps in differentiation of neighboring cell.

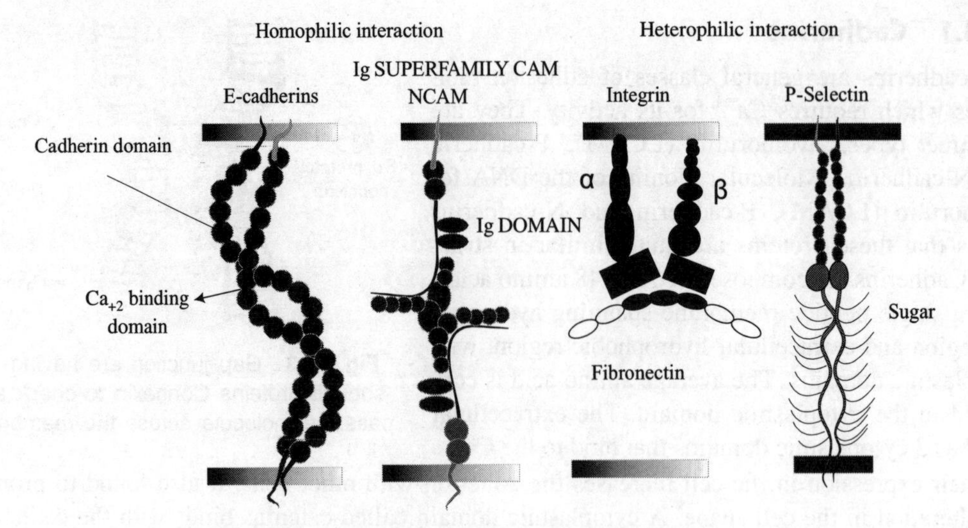

Fig. 18.5 Homotropic and Heterotropic interactions

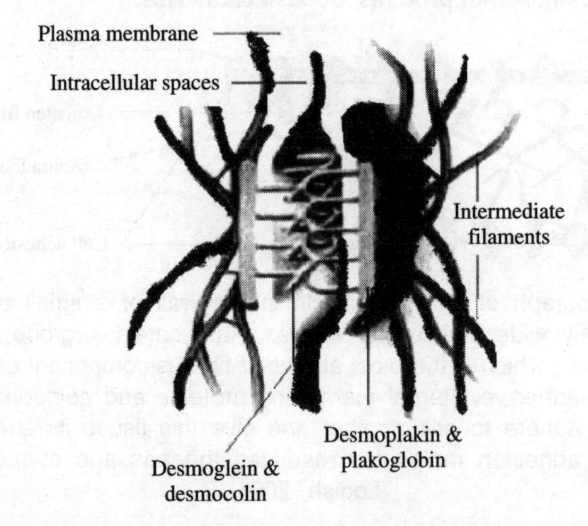

Fig. 18.6 Schematic structure of the desmosome. On the cytoplasm side of each of the two plasma membrane is a desmosomal plaque Attached to the plaque are intermediate filament that radiate into the cytoplasm, some time attached to cell organelle, or periphery. The core is glycoprotein in nature; also they have several proteins like Desmocollins (stick out from the plasma membrane), Desmoglein (member of the cadherins family) and plakoglobin (that interact with the intermediate filaments). The hemidesmosome in the epithelia indicate that two epithelia are attached to form complete Desmosomes where cell interacts with basal lamina. Trypsin treatment only, can separate the two membranes. The modified Desmosomes are the adhering band found in the intestine, cardiac muscles. These bands hold them together to form a spot Desmosomes.

Low sialic acid results in the low affinity. Although the NCAM and IgG has similarity in recognition, pattern one difference is that one recognizes self and other recognizes the nonself. See Figs. 18.5, 18.7 and 18.8. They acts like the marker and thus helps in differentiation of neighboring cell.

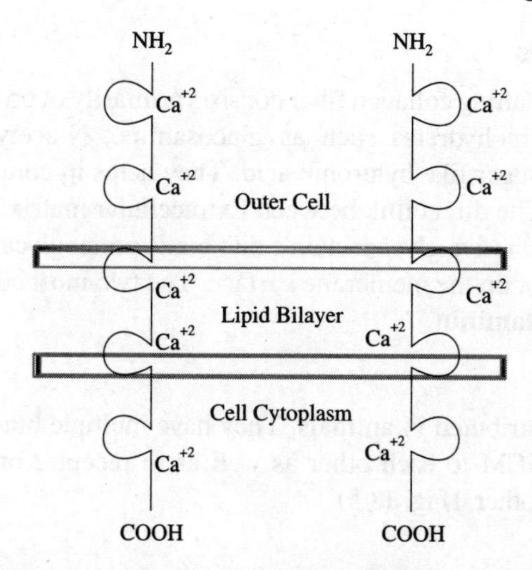

Fig. 18.7 Cadherins. There are three Ca^{+2} binding domains, a membrane spanning region, and a cytoplasmic domain

Table 18.1 Cell adhesion molecule and their specific tissue distribution

Molecule	*Ca^{++} dependence*	*Major tissue distribution*
Liver CAM	Yes	Epithelia
P-cadherin	Yes	Epithelia
N-cadherin	Yes	Neurons
E-cadherin	Yes	Epithelia
Uvomorulin	Yes	Blastomeres
Neural CAM	No	Neurons
Axonal	No	Axons
MAG	No	Neuron, Glia
ICAM	No	WBC, Endothelium
LEC-CAM	No	Leukocyte, endothelium

18.5 COMPONENTS OF EXTRACELLULAR MATRIX

18.5.1 Collagens

Collagen is the most abundant protein present in the animal cells and is secreted from fibroblasts cells (present throughout tissue). It is a prominent component of the bone cartilage, tendons and ligaments and they have high contents of glycine, hydroxylysine, and hydroxyproline amino acid. Smallest unit of collagen is Tropocollagen. Collagen provides support to the cells and tissue due to its fibrous rod like structure which provides elasticity to skin, lung, and the intestine (which undergo continuous shape change throughout life). Elasticity is due to presence of the Elastin in collagen. Collegen controls phenotypic expression by interacting integrin receptors on cell surface sites in the matrix. It promotes cell differentiation and improve survival of epithelial cells.

18.5.2 Proteoglycans

Hydrated gel like network bathing collagen fiber consists primarily of proteoglycans. Proteoglycans are formed of various carbohydrates such as glucosamine, N-acetyl Glucosamine, N-acetyl galactosamine and amine sugar like hyluronic acid. They helps in communicating to the cells by direct or indirect binding. The direct link between Extracellular matrix and plasma membrane are reinforced by a family of adhesive glycoprotein's that binds proteoglycans and collagen molecules to each other and to receptor on the membrane surface. The two most common adhesive glycoproteins are **Fibronectin and laminin**.

18.5.3 Fibronectins

Fibronectins are widely distributed in animals. They have multiple binding sites so they can link to various component of ECM to each other as well as to receptor on cell surface. Fibronectin molecule can bind to each other. (Fig. 18.5)

18.5.4 Integrins

There are various receptors presents on the surface of plasma membrane called as Integrin. They are under family of receptor protein called as Integrin. Integrin are made up of two subunit alpha and ßeta. Both subunits differ in specificity. Surrounding it other extracellular structure is secreted by the cell itself. The chemical nature of substrate differs among organism but all have common theme and common structure. A long rigid fiber embedded is amorphous hydrated matrixes of branched molecule that are usually are glycoprotein or polysaccharides. Therefore its presence in cell wall save the cells from infection and damage by the virus's bacteria and other infectious organisms. (Fig. 18.5)

Types of collagen

1. Type I collagen (skin, bone and cornea)
2. Type II collagen (blood vessel, uterine wall)
3. Type IV collagen (basal lamina and all connective tissue)

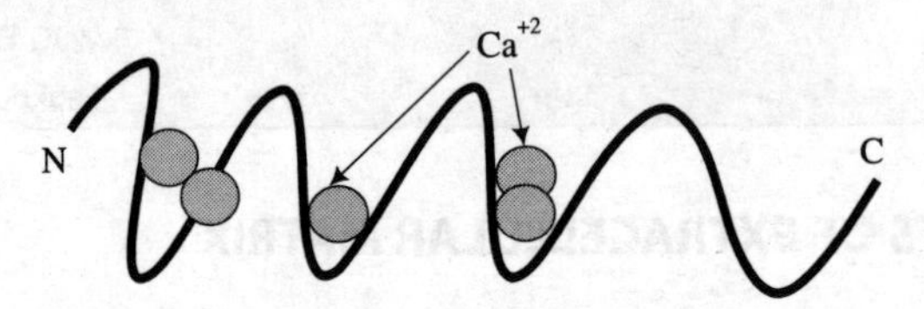

FIG 18.8 Calmoduline and Ca^{+2} binding domain

BRAIN QUEST

1. Give a detailed account of Extracellular matrix.
2. Write short notes on
   (a) Cadherins (b) Desmosome (c) Gap-junction
   (d) Ion-coupling (e) Adherence junction

3. Discuss the importance of adherence junction in cell to cell communication.
4. Discuss the importance of extracellular matrix in cell attachment to the matrix
5. Discuss the role of divalent ions in attachment of normal cells.
6. Discuss why transformed cell cannot attachment to the cells
7. Discuss the role of collagen in development of different types of cells and tissue.
8. Name some junction common found in epithelial cell.
9. Write short notes on
   (a) Desmosome
   (b) Extracellular matrix
   (c) Gap junction
   (d) Cadherins
   (e) Integrin

Fill in the blanks

(a) The most abundant protein present in the animal cell is _______.
(b) Proteoglycans are part of the _______.
(c) Desmosome structure is like _______.
(d) N-cadherins are found in the _______.
(e) Cadherins need the spread of viruses from one cell to another occur via _______.
(f) Communication among two cells occur by _______.
(g) Desmosome are found in various type of _______.
(h) Septets junction are widely found in _______.

Answers

(a) Collagen (b) ECM (c) Buttons (d) Neurons
(e) Plasmodesmata (f) Gap junction (g) Epithelial cell (h) Invertebrate tissue.

HYBRIDOMA TECHNOLOGY

19

In this Chapter we will learn about monoclonal antibody, their types such as Bispecific, chimeric, and humanized MAB, and the technique involved in formation, history, and their applications with recent MAB in use today.

19.1 INTRODUCTION

Hybridoma refers to the hybrid cells which specifically secretes monoclonal antibodies continuously. Kohler and Milestein in 1975 was first to combined the two desirable properties together for large scale production of antibody, i.e., property of specific antibody secretion plus continuous cell division into single fused cell called as hybridoma (derived basically from the union of myeloma cell and a normal spleen cell of mouse). Thus, Monoclonal antibodies (MABs), now is an extraordinarily important resource for medical research, diagnosis, therapy, and basic science. Furthure, specificity of antibody can be increased by change in CDR region of antibody by rearranging genes. Many MABs are in trial (see http://www.clinicaltrials.gov/) and till date at least more than 15 MABs has been approved by FDA and many more are near the approval and will be successfully applied for treatment of various type of cancers, transplantation of organs, asthma, psoriosis etc. Chimeric antibody contains variable chains of heavy and light chains from different animals and have been successfully applied in treatment of many diseases but the major disadvantages is, they are highly immunogenic thus humanized antibody was the another breakthrough with decrease immunogenicity, since they have less rodent consensus sequences of CDR region (6%) than chimeric antibody (more than 25%).

19.2 HISTORY

Harris and Watkins (1965), succeeded in somatic cell fusion between human HeLa cell and tumor mouse cells using inactivated Sendai virus as fusing agent. Weiss and Green (1967) noticed a peculiar feature in some somatic hybrid in which chromosomes of one of the parent cells gradually lost. In somatic hybrid of human and mouse cells, human chromosomes generally lost after fusion while in somatic hybrid of monkey and mouse cells the chromosomes of monkey were lost. Davis (1981) used polyethylene glycol as fusing agent and successfully obtained somatic hybrids of animal and plant cells separately belonging to two different species and even hybrid of animal and plant cells. **Kohler (1974)** successfully produced a hybridoma by fusing a P3 myeloma cell (resistant to azaguanine) and a lymphocyte (from the spleen of a mouse) immunized against sheep-red blood cells. The experiment consisted of immunizing mice against red blood cells (the antibodies produced against this antigen are easily detected in the serum by means of an assay developed in 1963 by Jerne and Nordin), then mixing mouse myeloma P3 cells with spleen cells from immunized mice in the presence of polyethylene glycol. These chimeric cells, were called as hybridoma which retained the property of immortality of the myeloma cells as well as property of secreting an antibody against specific antigen.

1847: Bence jones discovered urinary protein in myeloma.
1890: Von Behring and Kitaazato discovered antibodies.
1958: Okada discovered Cell fusion by Sandai virus.
1960: Barski discovered spontaneous fusion of cells.
1964: Littlefield discovered selective media for selection of hybrid.
1973 : Cotton and Milstein fusion of mice and rat myloma cells for MAB production.
1975: Kohler Milstein construction of hybridoma secreting antibodies.
1976: Pentecarvo PEG polyethylene glycol for cell fusion.
1980: Human MAB.
1981: Bispecific antibodies.
1987: Catalytic antibodies Abzyme.
1989: Transgenic production of MAB in plants by R-DNA technology.

19.3 CELL FUSION TECHNOLOGY

A cultured animal cell infrequently undergoes cell fusion spontaneously. Fusion of different cultured animal cells can yield interspecific hybrids. The fusion rate, however, increases greatly in the presence of certain viruses that have a lipoprotein envelope similar to the plasma membrane of animal cells. Cell fusion is also promoted by polyethylene glycol, which causes the plasma membranes of adjacent cells to adhere to each other and to fuse. Köhler and Milstein first used Sendai virus to create their revolutionary cells. The activities of the receptor binding hemagglutinin-neuraminidase protein are solely responsible for inducing close interaction between the virus envelope and the cellular membrane. However, one of many membrane fusion proteins (F protein) is triggered by local dehydration and there occurs a conformation change in the bound HN protein which actively inserts into the cellular membrane, which causes the envelope and the membrane to merge, followed shortly by virion entry. When the HN and F protein are manufactured by the cell and expressed on the surface, the same process may occur between adjacent cells, causing extensive membrane fusion and resulting in the formation of a syncytium. As most fused animal cells undergo cell division, the nuclei eventually fuse, producing viable cells with a single nucleus that contains chromosomes

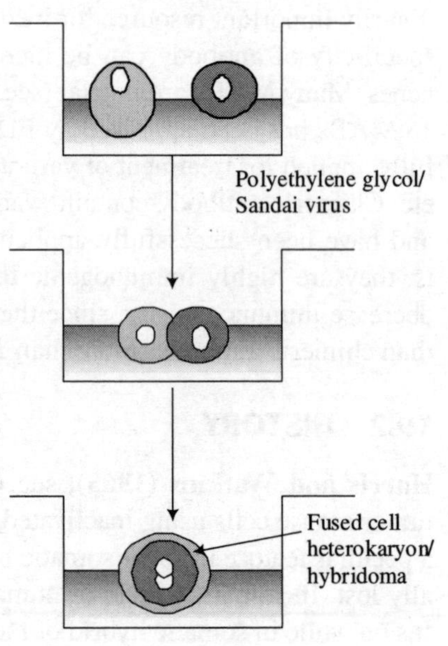

Fig. 19.1 Formation of hybridoma after fusion

from both "parents." The fusion of two cells that are genetically different yields a hybrid cell called a heterokaryon (See Fig. 19.1).

19.4 DISADVANTAGES OF CONVENTIONAL ANTIBODIES

1. The titers are low, and are seldom higher than 25.
2. The antibodies are heterogeneous, and not specific to one determinant.
3. The supply of antibodies is limited and also different from batch to batch.
4. It is impossible to reproduce a combination of specific antibodies in new animals.

19.5 ADVANTAGES OF MONOCLONAL ANTIBODIES

1. The titers are high, compared to conventional antibodies, can be up to 212.
2. Because the antibodies are secreted by clones of hybrids, they are homogeneous.
3. *In-vitro* or *In-vivo*, the cell line can keep on growing continuously in theory and the cells will be the same.

19.6 HYBRID CELL SELECTION ON HAT MEDIUM

Cultured cells from different mammals can be fused to produce interspecific hybrids e.g., hybrids from human cells and mutant mouse cells. As the human-mouse hybrid cells grow and divide, gradually they loose one of chromosomes in random order in a medium lacking the gene to utilise essential metabolite (lack in mouse but present in human) will not be retained.

19.6.1 Utilization of Metabolic Pathway For Hybrid (Cells) Selection

There are two pathways for synthesis of nucleotide in the cells; one is de-novo pathways and other is alternative pathway (salvage pathway) which operates only when De-novo pathway has been blocked by some inhibitor. Most animal cells can synthesize (the purine and pyrimidine) nucleotides de-novo, from simpler carbon and nitrogen compounds, rather than from already formed purines and pyrimidines). But the folic acid antagonist's (aminopterin) interfere with the donation of methyl and formyl groups (by tetrahydrofolic acid) in the early stages of de novo synthesis. These drugs are called as antifolates, since they block reactions involving tetrahydrofolate, an active form of folic acid. Many cells, contain enzymes that synthesize the necessary nucleotides from purine bases and thymidine (if they are provided in the medium;) which is part of salvage pathways to bypass the metabolic blocks imposed by antifolates.

Thus Cell lacking thymidine kinase (TK) are resistant to the toxic thymidine analog 5-bromodeoxyuridine. Since TK converts 5-bromodeoxyuridine into toxic 5-bromodeoxyuridine monophosphate and then into a nucleoside triphosphate which is incorporated by DNA polymerase into DNA, where it exerts its toxic effects and kills the cells. TK mutation prevents production of functional TK enzyme. Hence, TK mutants are resistant to the toxic effects of 5-bromodeoxyuridine. Similarly, cells lacking the HGPRT enzyme are resistant to the otherwise toxic guanine analog 6-thioguanine. As we will see next, HGPRT cells and TK cells are useful partners in cell selection. (See Fig. 19.2).

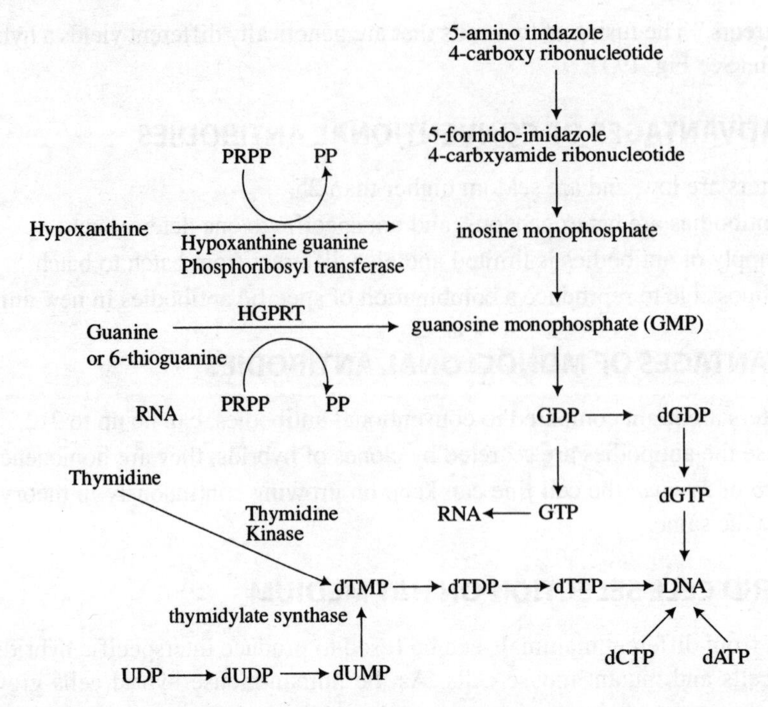

Fig. 19.2 Utilization of precursor and formation of nucleotides by different enzymes

19.6.2 Basics of Selection in HAT Medium

Normal cells are killed in HAT medium because of lacks of an enzyme for salvage pathway. However, hybrids (formed by fusion of these two cells) will carry a normal TK gene from the HGPRT parent and a normal HGPRT gene from the TK parent. The hybrids thus will produce both functional (salvage-pathway) enzymes and will survive on HAT medium. Likewise, hybrids formed by fusion of mutant cells and normal cells can grow in HAT medium.

19.7 TYPES OF ANTIBODY

Normal B lymphocyte is capable of producing antibody directed against a specific determinant, or epitopes, on an antigen molecule. Antigen-activated B lymphocyte form a clone of cells in the spleen or lymph nodes, producing antibody of seperate specificity against epitopes is said to be polyclonal.

Monoclonal antibodies are structurally similar to natural antibodies which consist of a variable region (Fab) that is responsible for antigen recognition, and a constant region (Fc) (see Fig. 19.3) that binds to receptors on immune effector cells, such as natural killer cells, monocytes, and macrophages. These effector cells function to elicit phagocytosis and cytotoxicity. The Fc portion of the immunoglobulin also functions to fix and activate complement, another important component of the host immune defense. The constant region of an antibody molecule is responsible for recruitment of a variety of effector functions. Methodologies have been developed to isolate switch variants in culture or genetically engineer an antibody molecule to express a desired isotype. There are different screening assays including fluorescence activated cell sorting (FACS) and ELISA.

19.7.1 Monoclonal Antibody

For many studies involving antibodies, monoclonal antibody is preferable to polyclonal antibody. However, biochemical purification of monoclonal antibody from serum is not feasible, in part because the concentration of any given antibody is quite low. For this reason, researchers looked to culture techniques in order to obtain usable quantities of monoclonal antibody. Because primary cultures of normal B lymphocytes do not grow indefinitely, such cultures have limited usefulness

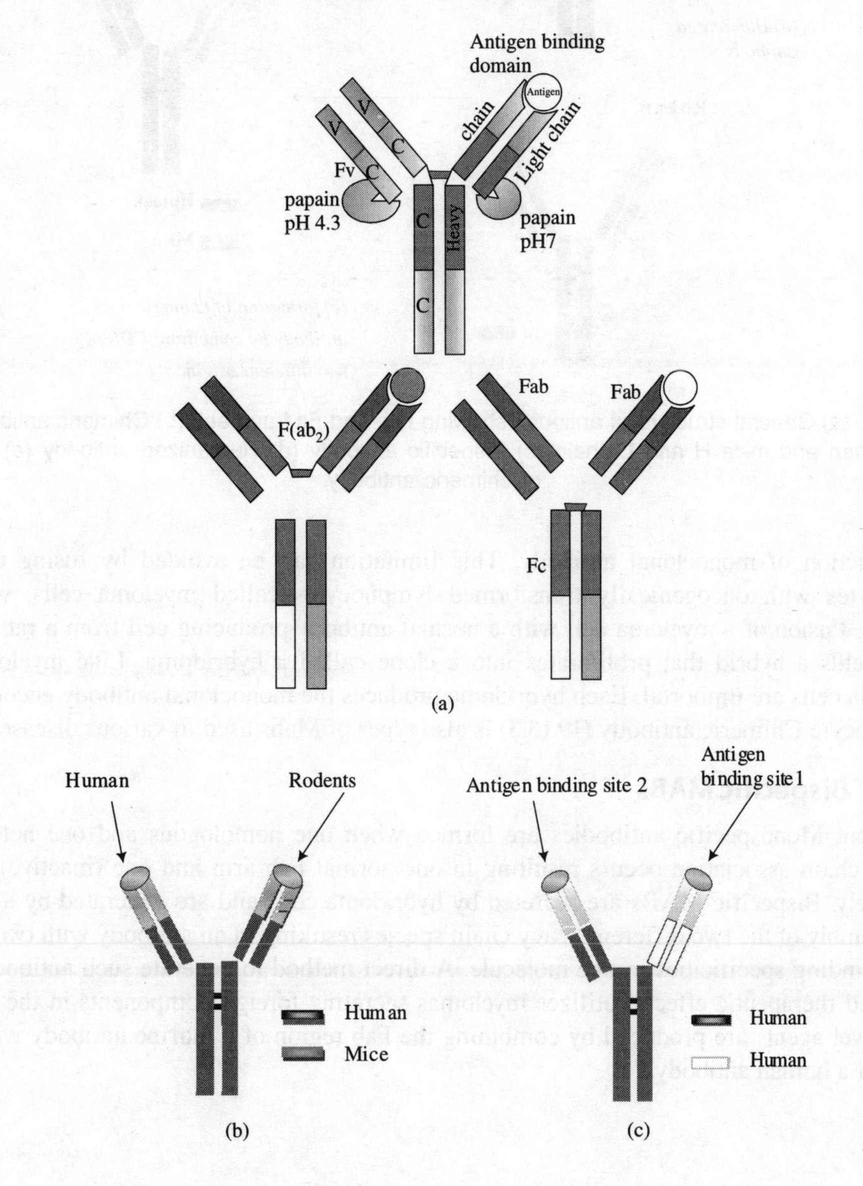

Fig. Contd..

Fig. Contd..

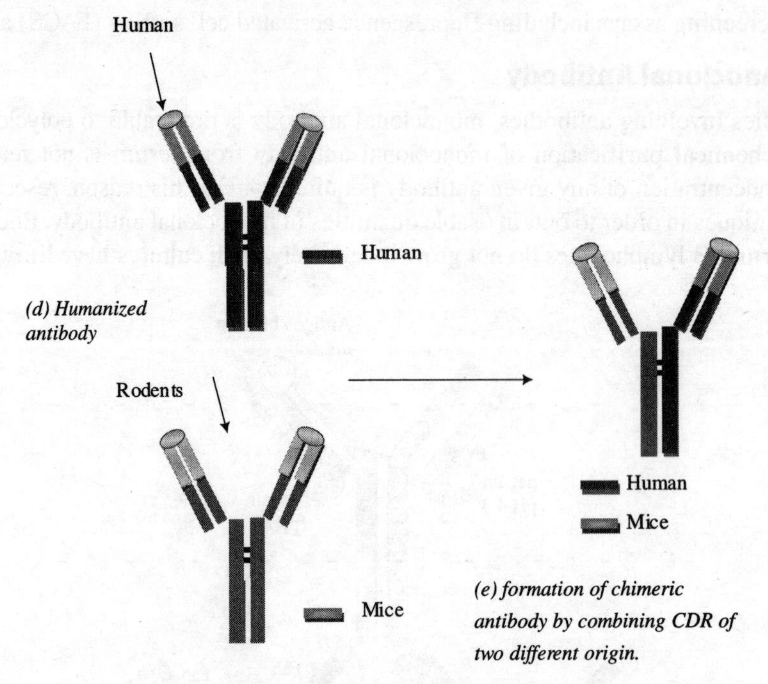

Fig. 19.3 (a) General structure of antibody showing Fab and Fc fragment. (b) Chimeric antibody made up of human and mice H and L chain (c) Bispecific antibody (d) Humanized antibody (e) formation of chimeric antibody.

for production of monoclonal antibody. This limitation can be avoided by fusing normal B lymphocytes with oncogenically transformed lymphocytes called myeloma cells, which are immortal. Fusion of a myeloma cell with a normal antibody-producing cell from a rat or mouse spleen yields a hybrid that proliferates into a clone called a hybridoma. Like myeloma cells, hybridoma cells are immortal. Each hybridoma produces the monoclonal antibody encoded by its B-lymphocyte Chimeric antibody (19.13.3) is also types of Mabs used in various disease therapy.

19.7.2 Bispecific MABs

Monovalent Monospecific antibodies are formed when one homologous and one heterologous H and L chain association occurs resulting in one normal Fab arm and one (inactive) Fab arm respectively. Bispecific MABs are secreted by hybridoma cells and are generated by a heterologous assembly of the two different heavy chain species resulting in an antibody with two separate antigen binding specificities in one molecule. A direct method to generate such antibodies (with augmented therapeutic effects) utilizes myelomas secreting foreign components in the antibody. These novel agents are produced by combining the Fab region of a murine antibody with the Fc portion of a human antibody.

The advantage of human monoclonal antibodies over murine monoclonal antibodies is that human recipients are less likely to develop antibodies against them (although antiidiotypic and possibly anti-allotypic antibodies may still be produced) and also human antibodies have the full range of biological functions, because of Fc region. There may be other advantages such as selection of a subclass of antibody with particular properties. (See Fig. 19.3d).

Murine monoclonal antibodies have been prepared using cell lines (hybridoma) made by fusion of lymphocytes from an immunized donor with myeloma cells. A number of alternative strategies have been devised for production of human monoclonal antibodies. These are:

(a) **Fusion of human lymphocytes** (usually peripheral blood or lymph-node derived) with a murine myeloma or hybrid human-murine myeloma line. This procedure is essentially similar to the hybridoma technique used to produce murine monoclonal antibodies, but presents some technical problems in that a lower fusion efficiency results and human chromosomes are lost preferentially. This procedure may be regarded as a compromise due to the absence of a suitable human myeloma fusion partner.

(b) ***Transformation of human lymphocytes with Epstein-Barr virus (EBV).*** This procedure has been used for many years to produce continuous, rapidly growing human B cells.

(c) Fusion of human B-lymphocytes with a human lympho-blastoid B-cell line.

(d) Fusion of an EBV-transformed human B-lymphocyte line with a mouse myeloma cell line.

A chimeric monoclonal antibody has about a 30% to 35% murine component, a humanized monoclonal antibody possesses less than 10% of the murine constituent. A chimeric molecule can be identified in human circulation for two weeks, while a murine molecule is identified for only two days. The faster clearance of the more foreign murine molecule by the immune system causes this. Murine, chimeric, and humanized monoclonal antibodies are all available anticancer agents.

19.8 NOMENCLATURE

The United States Adopted Name (USAN) Council has outlined specific guidelines for the naming of monoclonal antibodies. These guidelines provide knowledge about a specific monoclonal antibody just by looking at the generic name. Specifically, all monoclonal antibodies end in the suffix -MAB. Important safety information can be determined by the syllable preceding -MAB because it identifies the animal source of the product, which can elicit host antibodies against that source (Table 19.1). The general subclass is identified immediately before the product source by use of a code syllable. The last consonant of this target syllable is often dropped to make the name more pronounceable (e.g., -tum- changed to -tu-). All of the approved monoclonal antibodies in the treatment of cancer use the -tum- identifier for "miscellaneous tumors." The starting prefix is a distinct syllable that creates the unique monoclonal antibody name. If the product is radiolabeled or conjugated to another chemical such as a toxin, a separate word is used to identify the conjugate. For toxin conjugates, the -tox suffix is included within the toxin name.

Table 19.1 Nomenclature of MABs

Source	*Identifier*	*Human*	*-u*
Mouse	-o	Rat	-a
Humanized	-zu	Hamster	-e
Primate	-i	Chimera	-xi
Tumors	Identifier	Colon	-col
Melanoma	-mel	Mammary	-mar
Testis	-got	Ovary	-gov
Prostate	-pr (o)	Miscellaneous	-tum
Disease	Identifier	Viral	-vir
Bacterial	-bac	Immune	-lim
Infectious lesions	les	Cardiovascular	-cir

19.9 LARGE SCALE PRODUCTION OF ANTIBODIES

Most of the MAB are produced in very small quantities (less than 0.1 g) at bench scale for research purposes. Commercial interests consider production scales of 0.1-10 g as small, 10-100 g as medium, and over 100 g as large. The complete flow chart has been given below for large scale production of antibody. Following steps involved in production of MAB. (See Fig. 19.4.)

1. Immunization (in vivo or in vitro)
2. Fusion of spleen cell and myloma cell
3. Screening of fused cell in HAT medium
4. Isolation and cloning of selected cells
5. Characterization of fused cells
6. Propagation in ascites
7. Purification
8. Storage

19.9.1 Immunization

The adequate amount of antigen for mouse is about 1 to 50 µg in 50 to 200 µl volume for proper immunization. For rat, higher amount of antigen is needed but the emulsion should be thick and creamy. After the first injection, other booster immunization should be in incomplete Fruend's to prevent possible hypersensitivity reaction to the bacteria. Typically, 1 to 3 boosters may be given at intervals of 2 to 8 weeks and the final booster is best given 2 to 4 days prior to fusion in IP injection. For intact living cells, no adjuvant is needed and 10^7 of cells are suitable for each injection. The first injection can be S.C. (subcutaneous) and the others may be I.P. or I.V. Preparation of the antigen in Freund's adjuvant, equal volumes of antigen and adjuvant can be made together.

19.9.2 Growth of Myeloma Cells

Myeloma cells can be induced by mineral oil or pristine (2, 6, 10, 14-tetramethy-pentadecane) by intraperitoneal injection (IP). Normally, the myeloma cells are maintained in the laboratory with 10% fetal calf serum. Myeloma cells cannot growth in high density, and cells will die very soon,

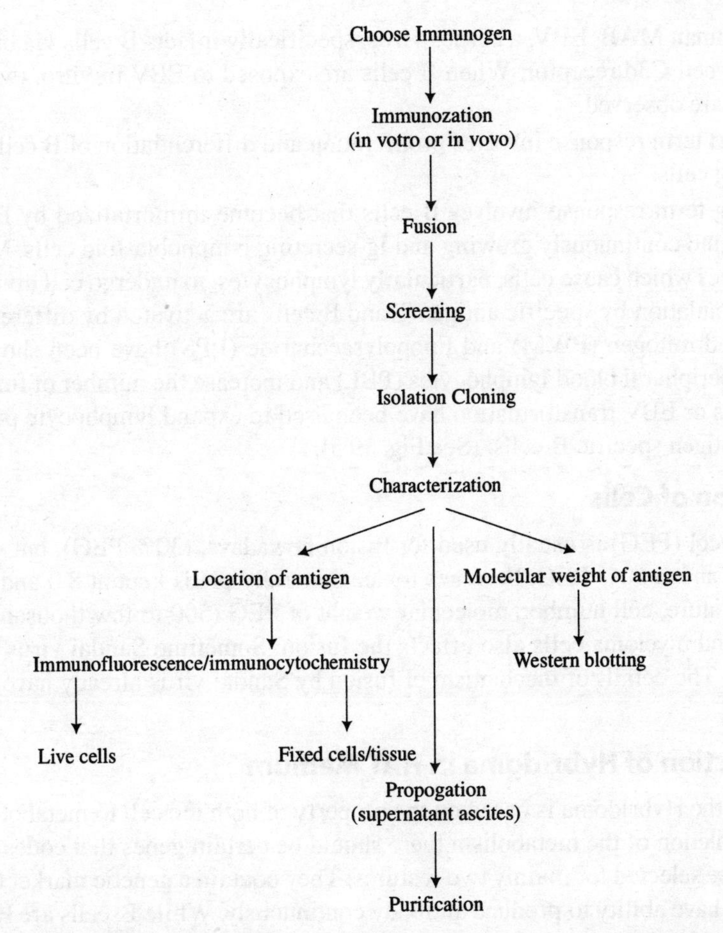

Fig. 19.4 Flow chart of complete step involved in production of monoclonal antibody

10^5 to 10^6 /ml of cell number can be grown in stationary phase. Some of the culture medium used are as follows:

1. Dulbecco's modified Eagle,s medium (DMEM), (no asparagines), and RPMI-1640, (no pyruvate).
2. CO_2 concentration is kept maintaining in between 4 to 10%, buffering with $NaHCO_3$.
3. Glutamine is usually added (2 mM).
4. Antibiotics, streptomycin and penicillin (100 μg/ml) are added.
   (Culture medium must be protected from direct sunlight which can cause toxic photoproducts.)

19.9.3 Transformation of Antibody Secreting B Cells

Sometime transformation of antibody secreting B cells is done by using Epstein bar virus. Human B lymphocyte transformation with Epstein-Barr virus (EBV) was one of the first trials for the production of human MAB. EBV, a herpes virus, specifically infects B cells via binding of a viral protein to the B cell C3d receptor. When B cells are exposed to EBV in vitro, two types of polyclonal response are observed:

(1) The short term response involves proliferation and differentiation of B cells into antibody-secreting cells.

(2) The long term response involves B cells that become immortalized by EBV, i.e.., transformed into continuously growing and Ig-secreting lymphoblastoid cells. Mitogens are the substances which cause cells, particularly lymphocytes, to undergo cell division. Following their stimulation by specific antigen T and B cells are activated by different mitogens e.g. Pokeweed mitogen (PWM) and lipopolysaccharide (LPS) have been shown to stimulate human peripheral blood lymphocytes (PBL) and increase the number of functional B cells. Mitogens or EBV transformation have been used to expand lymphocyte population, especially antigen specific B cells. (See Fig. 19.5).

19.9.4 Fusion of Cells

Polyethylene glycol (PEG) is usually used for fusion nowadays, (30% PEG), but very few hybrid are formed, and above 50% PEG, becomes toxic. Generally pH is kept at 8.0 and 8.2. Other factors, like temperature, cell number, molecular weight of PEG (500 to few thousand) and the ratio of spleen cells and myeloma cells also affects the fusion. Sometime Sandai virus is also used for fusing two cells. The details of mechanism of fusion by Sandai virus already have been described (See Fig. 19.3).

19.9.5 Selection of Hybridoma in HAT Medium

The selection of the Hybridoma is based on the property of both the cell to metabolize certain substance. For completion of the metabolism there should be certain genes that codes for an enzyme. Myeloma cells are selected for mainly two features. They contain a genetic marker e.g. ($HGPRT^-$), TK+ gene. They have ability to produce antibody continuously. While B-cells are $HGPRT^+$ (TK^-). On mixing both cells the two genes get mixed producing hybridome. HAT medium are used for the selection in which only those cells remain live having both gene HGPRT+ and TK+. HAT contains three antimetabolite hypoxanthine, aminopterin and thymidine. Antimetabolite aminopterin blocks the cellular biosynthesis of purine, Pyrimidine, from simple sugars and amino acid. So, alternative pathways of the metabolism become active called as salvage pathway. In salvage pathway two metabolites are used for conversion of the purine and pyrimidine. These two metabolites are hypoxanthine and thymidine. HGPRT contains hypoxanthine guanine phosphoribosyl transferase which converts hypoxanthine into guanine .While thymidine is phosphorylated by thymidine kinase (TK). Thus in this way only those cells are capable of surviving that contains both TK or thymidine kinase and HGPRT or hypoxanthine guanine phosphoribosyl transferase. (See Fig. 19.6).

19.10 CLONING

Cloning of hybrid cells has a advantage that it reduces the risk of overgrowth by non-producer cells and it ensure that the antibodies produced are truly monoclonal. For cloning, cells should be plated at 10, 3 and 0.5 cells per 200 μl in 96-well plates. Efficiency of cloning can be improved

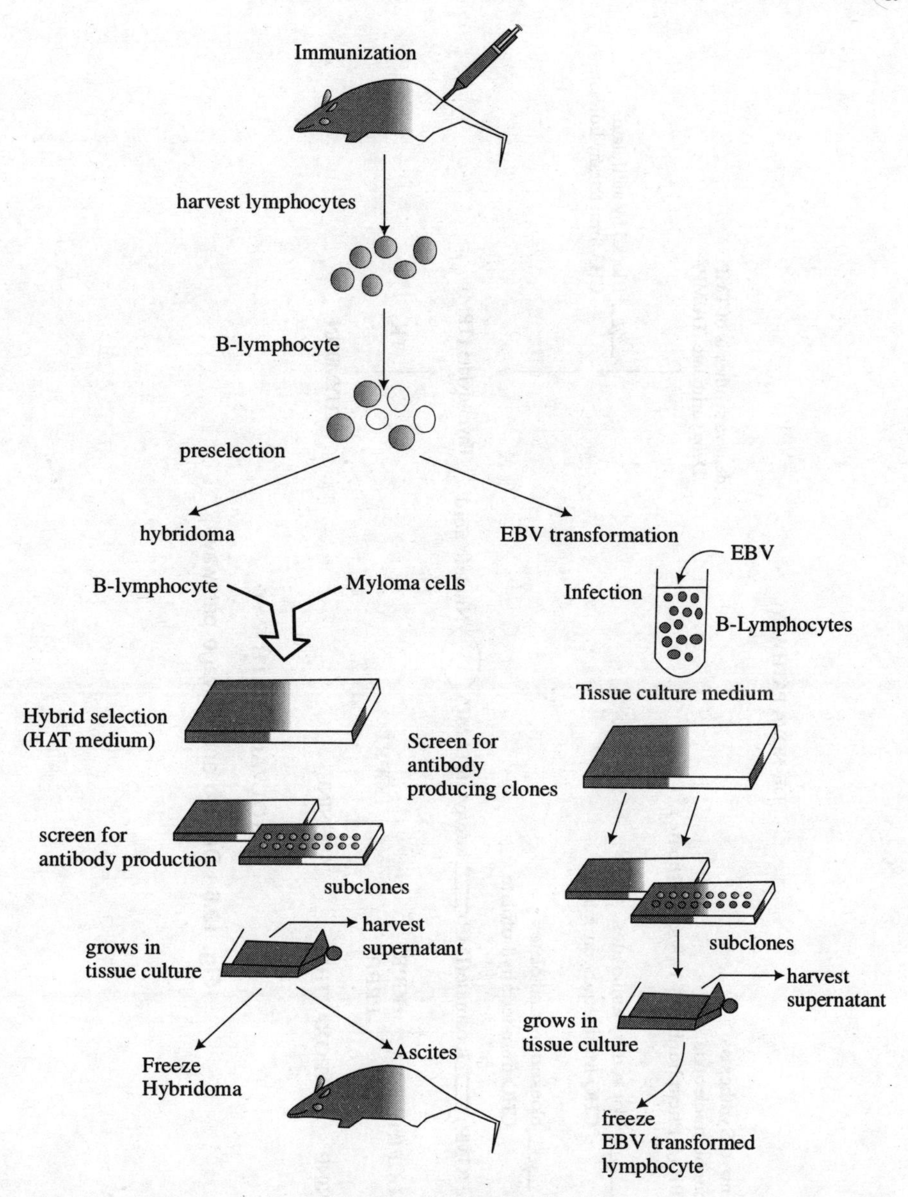

Fig. 19.5 Production of MAB by fusion of myloma and B-cells or by transformation by EBV

in presence of feeder cells. The colonies can be visible after 1 to 2 weeks time. Clones should be assayed as soon as possible after they become visible. At least two cloning method are used for making sure of the production of monoclonal antibodies. Two methods are used for the cloning:

1. Soft agar (less used) consist of two layers of agar, a firm underlayer containing of 0.5% (w/v) agar in culture medium and a second layer on top called "soft" of 0.3% (w/v) with the fused cells.
2. **Limiting dilution:** Most commonly used method.

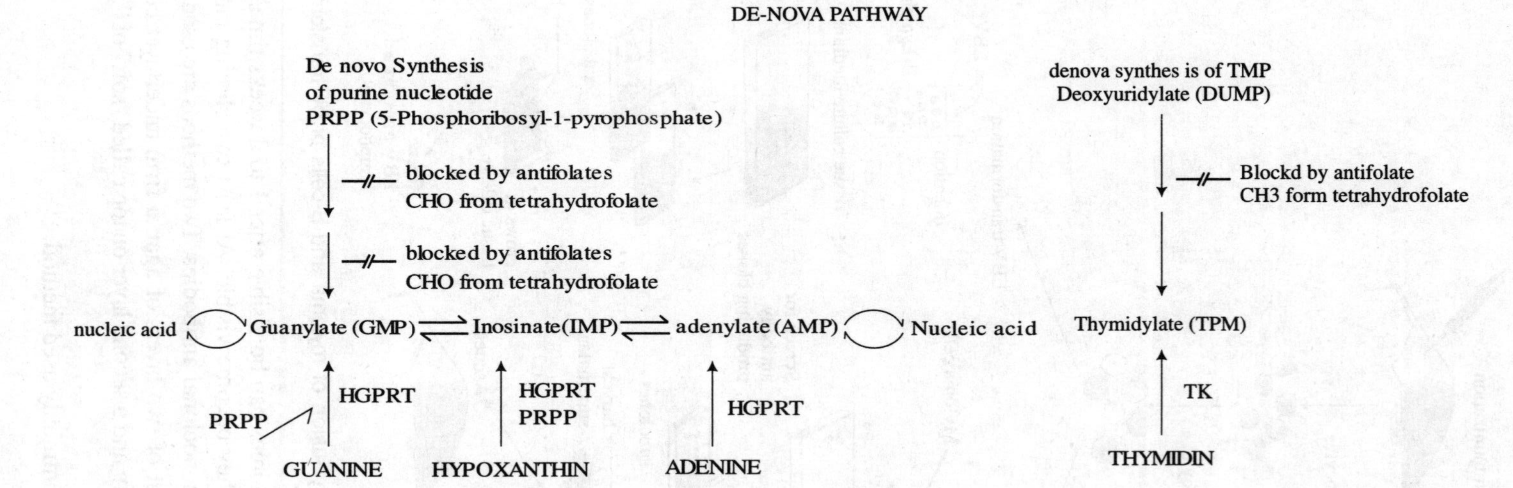

Fig. 19.6 Denovo and Salvage pathway

19.10.1 Ascite Formation

Monoclonal antibodies can be produced by growing hybridoma cells within the peritoneal cavity of a mouse (or rat). When injected into a mouse, the hybridoma cells multiply and produce fluid (ascites) in its abdomen. This fluid contains a high concentration of antibody which can be harvested. This usually provides higher antibody yields than hybridoma cell culture. Injection of 10^6 to 10^7 histocompatible hybrid cells into mice or rats l give results in tumor formation after 2 to 4 weeks (normally IP injection). The mice or rats can form ascites after injection of hybridoma cells. In ascitic fluid, amounts of monoclonal antibodies can be up to 5 to 15 mg/ml of serum, about 1,000 times of normal medium culture. A single mouse can produce up to 2 to 5 ml each time and about 20 ml for the whole life. To obtain better reaction, pristane can be used to inject a few days before injection of hybridoma for tumor production.

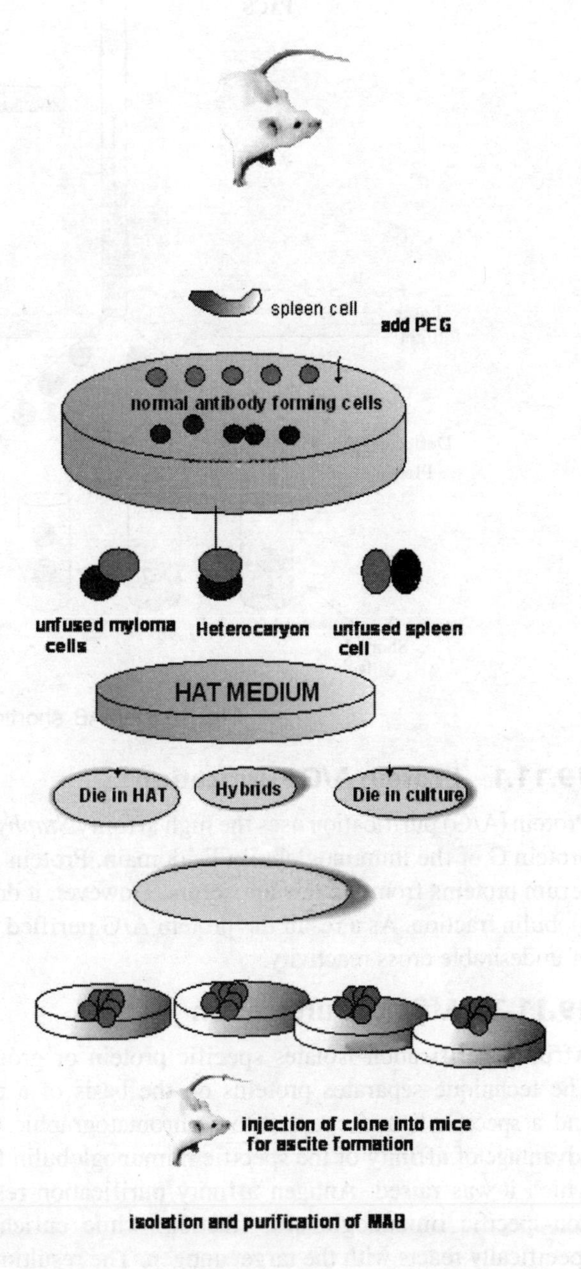

Fig. 19.7 Hybridoma selection in HAT medium

19.11 PURIFICATION

Purification of the monoclonal antibodies is similar to the purification methods for conventional antibodies. The method commonly used is

1. Precipitation with ammonium sulphate
2. Separation by ion exchange chromatography
3. By electrophoresis
4. Gel filtration
5. By FACS

Precipitation with ammonium sulfate (Ammonium sulphate precipitation is a further preparation step often used with ascetic fluid to concentrate the immunoglobulins.)

Monoclonal ascites fluid/tissue culture supernatant is commonly purified by one of two methods: (See Fig. 19.8).

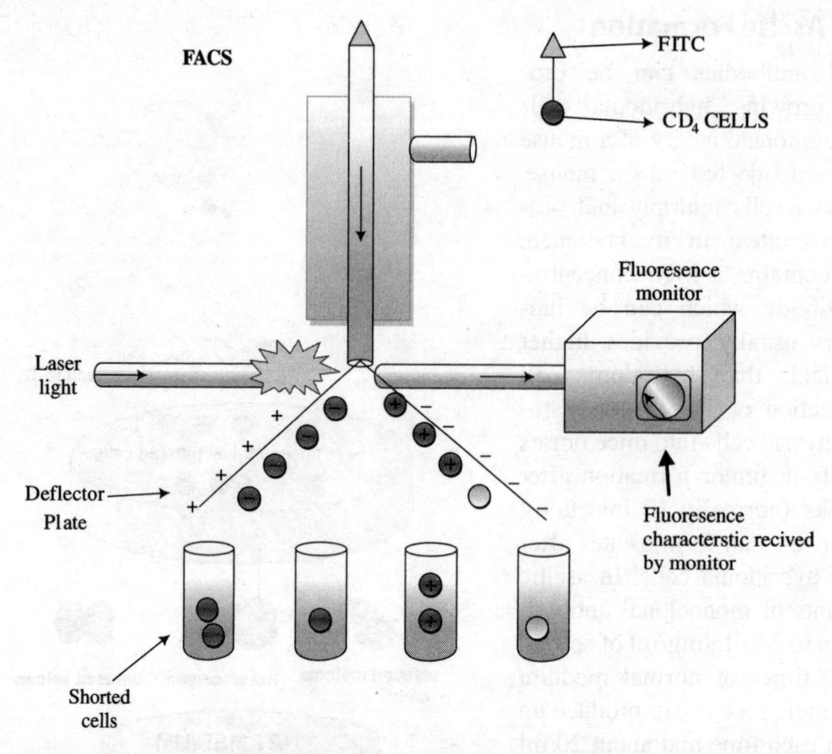

Fig. 19.8 MAB shorting by FACS

19.11.1 Protein A/G Purification

Protein (A/G) purification uses the high affinity *Staphylococcus aureus* protein A or Streptococcus protein G of the immunoglobulin Fc domain. Protein A/G purification eliminates the bulk of the serum proteins from the raw antiserum. However, it does not eliminate the non-specific immunoglobulin fraction. As a result the protein A/G purified antiserum may still possess a small amount of undesirable cross reactivity.

19.11.2 Affinity Purification

Affinity purification isolates specific protein or group of proteins with similar characteristics. The technique separates proteins on the basis of a reversible interaction between the proteins and a specific ligand coupled to a chromatographic matrix. Antigen affinity purification takes advantage of affinity of the specific immunoglobulin fraction for the immunizing antigen against which it was raised. Antigen affinity purification results in the elimination of the bulk of the non-specific immunoglobulin fraction, while enriching the fraction of immunoglobulin that specifically reacts with the target antigen. The resulting affinity will contain primarily the immunoglobulin of desired specificity.

19.11.3 Solid Phase Radioimmunoassay

$Iodine^{125}$ labeled anti-immunoglobulin or $Iodine^{125}$ labeled protein (A) allows separation of antigen which can bind to IgG antigen tightly. The process detail is given in Fig. 19.9. The primary

and secondary antibody is used to bind and locate the antigen. Secondary antibody is linked with Iodine^{125} isotope.

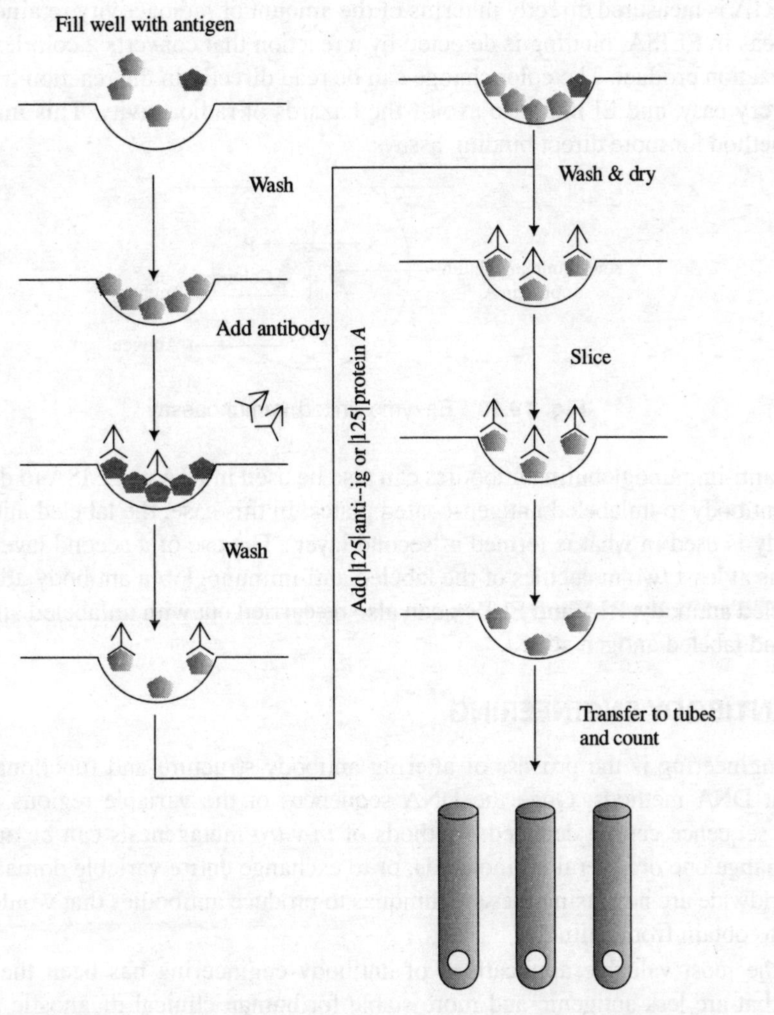

Fig. 19.9 Solid phase radioimmunoassay for Hybridoma screening

Radioimmunoassay (RIA) and **enzyme-linked immunosorbent assay (ELISA)** are direct binding assays for antibody or antigen and both work on the same principle, but the means of detecting specific binding is different. Radioimmunoassay is commonly used to measure the levels of hormones in blood and tissue fluids, while ELISA assays are frequently used in viral diagnostics. For both these methods one needs a pure preparation of a known antigen or antibody, or both, in order to standardize the assay (See Fig. 19.10). The principle is similar if pure antigen is used instead. In RIA for an antigen, pure antibody is radioactively labeled, usually with I^{125}; while for ELISA, an enzyme is linked chemically to the antibody. The unlabeled component, antigen, is attached to a solid support, (such as the wells of a multiwell plate), embaded with any protein. The labeled antibody is allowed to bind to the unlabeled antigen, under conditions where nonspecific absorption is blocked, and any unbound antibody and other proteins are washed away. Antibody binding to RIA is measured directly in terms of the amount of radioactivity retained by the coated wells, whereas in ELISA, binding is detected by a reaction that converts a colorless substrate into a colored reaction product. The color change can be read directly in the reaction tray, making data collection very easy, and ELISA also avoids the hazards of radioactivity. This makes ELISA the preferred method for more direct binding assays.

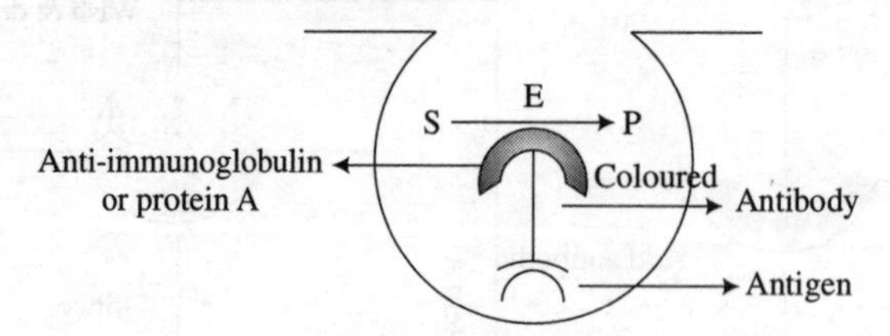

Fig. 19.10 Enzyme linked immunoassay

Labeled anti-immunoglobulin antibodies can also be used in RIA or ELISA to detect binding of unlabeled antibody to unlabeled antigen-coated plates. In this case, the labeled anti-immunoglobulin antibody is used in what is termed a 'second layer'. The use of a second layer also amplifies the signal, as at least two molecules of the labeled anti-immunoglobin antibody are able to bind to each unlabeled antibody. RIA and ELISA can also be carried out with unlabeled antibody stuck on the plates and labeled antigen added.

19.12 ANTIBODY ENGINEERING

Antibody engineering is the process of altering antibody structure and functional properties by recombinant DNA methods. Once the DNA sequences of the variable regions are known, the amino acid sequence can be deduced. Methods of *in-vitro* mutagenesis can be applied to insert, delete, or change one or several amino acids, or to exchange entire variable domains. Many laboratories worldwide are now using these techniques to produce antibodies that would be difficult or impossible to obtain from animals.

One of the most valuable applications of antibody engineering has been the preparation of antibodies that are less antigenic and more stable for human clinical diagnostic and therapeutic applications. Many MABs that are useful for medical imaging or therapy were derived in rodents and evoke an undesired immune response in patients. Human MABs have proven to be very difficult to produce by conventional hybridoma methods. The alternative has been to "humanize" mouse MABs using recombinant antibody techniques.

In the production of human monoclonal antibodies (in the production of so called chimeric molecule) recombinant DNA (rDNA) technology is utilised. In chimeric antibodies the variable heavy and light chain domains of a human antibody are replaced by those of a rodent (usually murine) antibody, which possesses the desired antigen specificity. In humanized antibodies only three short hypervariable sequences (complementarity determining regions or CDR's) of the rodent variable domains (for each chain) are engineered into the variable domain frame work of a human antibody producing mosaic variable regions. Humanized antibodies contain a minimum of rodent sequence. Suitable cells for expression of the rDNA monoclonal antibody genes are mammalian cell lines such as immunoglobulin non-producing myeloma cell lines that are capable of high level expression of exogenous heavy and light chain genes and the glycosylation, assemblage and secretion of functional antibodies. Engineered monoclonal antibodies may have the advantages of decreased immunogenicity, enhanced *in-vivo* circulating half life in combination with optimized specificity and effecter functions.

19.12.1 Procedure involve in Genetic Manipulation

Recombinant antibody technology involves recovering the (antibody) genes from the source cells, amplifying and cloning the genes into an appropriate vector, introducing the vector into a host, and achieving expression of adequate amounts of functional antibody. Recombinant antibodies can be cloned from any species of antibody-producing animal, if the appropriate oligonucleotide primers or hybridization probes are available.

Table 19.2 MAB and possible contaminations with their quantity

MAB	*Contaminations*	*Quantity*
Hybridoma cell culture supernatant with 10% fetal calf serum	Phenol red, water albumin, Albumin, transferrin, bovine IgG, alpha-2 microglobulin, other serum protein and viruses	1 mg/ml
Hybridoma cell culture supernatant serum free	Albumin transferrin	Up to 0.05 mg/ml
Ascite fluids	Lipid, Albumin transferrin, lipoprotein, endogenous IgG and other host proteins	1-15 mg/ml

The host in which most recombinant antibody methods developed was the bacterium *Escherichia coli*. Since growth of bacteria is rapid and inexpensive, a number of vectors are other available for expression and manipulation of cloned genes. DNA can be introduced directly into *E. coli* (the process known as transformation) or by infectious bacteriophage (transfection). Genetic constructions of antibody fragments (Fab and ScFv) can be quickly assessed and various selection methods can be applied. In addition it is easier to scale up the production of antibody fragments in *E. coli* than in mass culture of mammalian cells such as hybridomas. See Fig. 19.11. The mRNA population contains messages for a vast number of genes in addition to antibody genes. The Fab or Fv sequences of antibodies can be selectively amplified from the cDNA by PCR using specific complementary oligonucleotide primers. The products of PCR amplification are populations of different H and L sequences. The versatile phagemid vectors normally replicate in *E. coli* as an extrachromosomal element (plasmid). However, when the host is infected with an M13 helper phage, the phagemid DNA is packaged into complete, infective bacteriophage.

Bacterial systems proved to be very efficient for expression of some proteins but sometimes give disappointing results with recombinant antibodies. Efficiencies of recombinant antibody expression in *E. coli* vary with the antibody. Two known problems are misfolding of the antibody chains and incompatibility with bacterial secretory pathways. Despite these problems, some laboratories have successfully produced usable amounts of antibodies. One strategy is to optimize the export of Fabs to the periplasmic space where they are more likely to fold correctly. Functional antibody fragments have also been expressed in other prokaryotic and eukaryotic hosts, including *Bacillus subtilis* WB600 (Wu et al. 1993), the yeast strains *Saccharomyces pombe* (Davis et al. 1991) and *Pichia pastoris*, Xenopus (toad) oocytes (Biocca et al. 1993), insect cells (Page and Murphy, 1990), and plants (Hiatt et al. 1989).

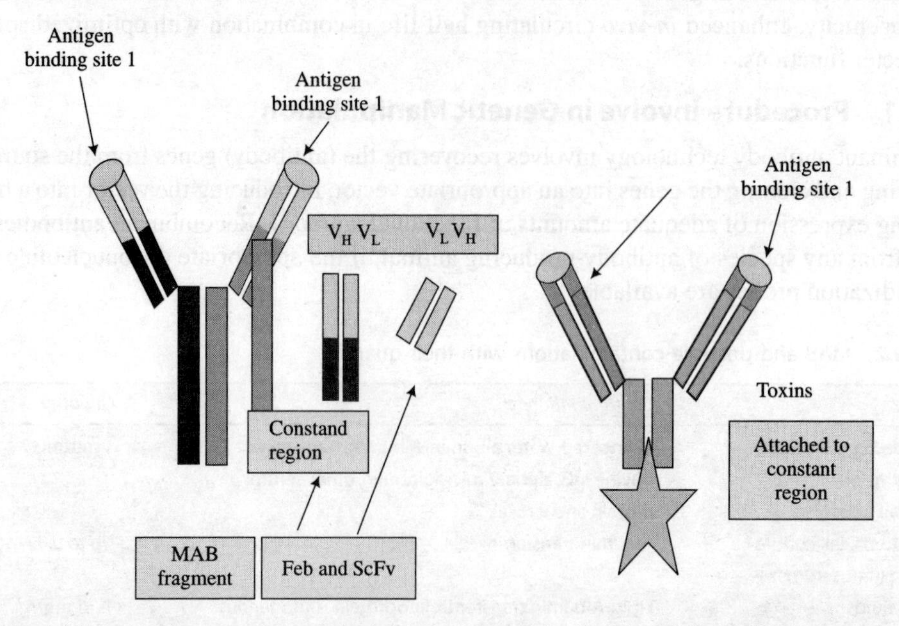

Fig. 19.11 (a) Fab (involve complete region of heavy and light chain) and Sc Fv (inovolve only variable region of both heavy and light chain) fragment used for preparation of Engineered MAB production (b) Toxin attached to Fc region are used to targeted killing of specific cells.

19.13 APPLICATION OF MAB

Generally large scale production of MAB is done by culturing the cell into mouse ascites or in peritoneum. Later on these cells can be used in the Air Lift fermenter for the large scale production of the antibody. Monoclonal antibody are much useful than the polyclonal antibody, because they are more specific for single antigen. In this way they can be employed against certain specific disease that is difficult to cure. MAB can be attached to the toxin for killing certain tumor cells. Now a day ricin toxin isolated from the *Ricinus Communis* are attached with the MAB for successful treatment of the cancerous cells.

19.13.1 Disease Diagnosis

Other application of the MAB is in the disease diagnosis. Monoclonal antibodies have become important diagnostic and therapeutic tools in medicine. Monoclonal antibodies that bind to and inactivate toxic proteins (toxins) secreted by bacterial pathogens are used to treat diseases caused by these pathogens. Other monoclonal antibodies are specific for cell-surface proteins expressed by certain types of tumor cells; chemical complexes of such monoclonal antibodies with toxic drugs are being developed for cancer chemotherapy. Generally our bodies don't produces monoclonal antibody, but it produces the polyclonal antibody. For the disease identification MAB are attached in the wall of the microtiter plate and then microorganism specific antigen are detected with the help of secondary antibody. This method is called as ELISA or enzyme linked Immunosorbant assay. MABS are also used for identification of certain type of tumors like benign tumor and early cases of the metastasis.

19.13.2 Therapeutic Monoclonal Antibodies

Therapeutic monoclonal antibodies may be conjugated to radioisotopes, toxins, or chemotherapy agents, using these antibodies to deliver cytotoxic treatments directly to targeted tumor cells. Immunotoxins consist of monoclonal antibodies attached with a lethal cellular toxin, such as **calicheamicin**, the plant toxin ricin, and the Pseudomonas exotoxin that is delivered to the cell and internalized by the receptor causing cytotoxicity. Radioimmunoconjugates are monoclonal antibodies labeled with a beta-emitting, radioactive isotope such as **Yttrium-90 or Iodine-131**. These agents bind receptors of cell surface antigens on tumor cells and deliver localized radiation to both those cells and surrounding tumor cells. Because of the effect on neighboring cells, radioimmunoconjugates have the advantage of not requiring either receptor internalization or a functional immune system to be effective.

The first radioimmunoconjugate approved as an anticancer agent was **Ibritumomab Tiuxetan** (Zevalin). Ibritumomab is similar to Rituximab in that it binds specifically to the CD20 surface antigen on B-cell precursors and mature B-cells; however, Ibritumomab is a murine monoclonal antibody and is attached to the linker Tiuxetan to provide a chelator site for the beta-emitter yttrium-90. The beta emission from the yttrium-90 forms free radicals and induces cellular damage in both the target cells and surrounding cells. In this case, the murine component of ibritumomab has an advantage because it allows for more efficient clearing of the radioactivity from the body. **Ibritumomab tiuxetan** has the same indication as Rituximab for the treatment of patients with relapsed or refractory, low-grade or follicular, B-cell NHL, including those patients with Rituximab-refractory follicular NHL

19.13.3 Chimeric Antibody in Therapy of Different Cancer

Rituximab (Rituxan) was the first monoclonal antibody approved in the United States for the treatment of a malignancy. Rituximab is a chimeric human/mouse antibody that targets the CD20 surface antigen

(i) Breast Cancer

The HER2/neu protein is a proto-oncogene that is overexpressed in approximately 25% to 30% of patients with breast cancer and is associated with a poor prognosis. Trastuzumab (Herceptin) is a humanized monoclonal antibody that selectively binds to the extracellular domain of the HER2/neu protein, causing down-regulation of HER2/neu and inhibiting the growth of HER2/neu-overexpressing cells. Trastuzumab is indicated as monotherapy for patients with refractory metastatic breast cancer and in combination with Paclitaxel for patients with metastatic breast cancer who have not received chemotherapy for their metastatic disease. (See Fig. 19.12).

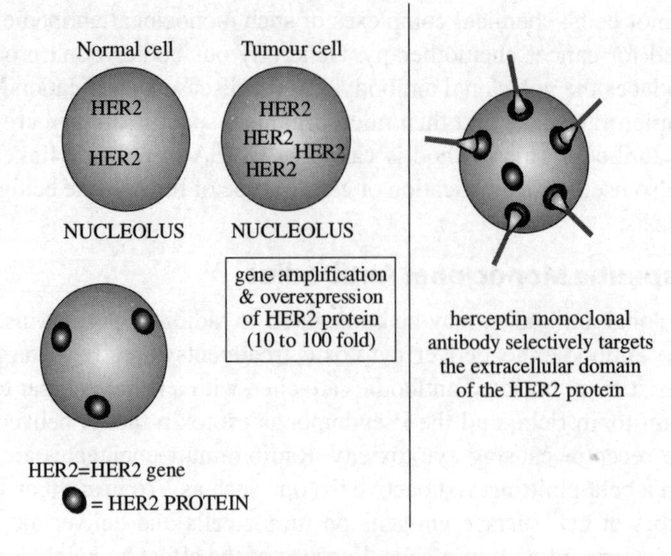

Fig. 19.12 Application of herceptin MAB in Breast cancer

(ii) B-Cell Chronic Lymphocytic Leukemia

Alemtuzumab (Campath) is a humanized monoclonal antibody targeted against the CD52 surface antigen that is expressed on virtually all normal and malignant lymphocytes and most natural killer cells, monocytes, macrophages, and tissues of the male reproductive system. **Alemtuzumab** is indicated for patients with B-cell chronic lymphocytic leukemia (B-CLL) who have been treated with alkylating agents and have failed treatment with fludarabine.

(iii) Anticancer MAB

Gemtuzumab Ozogamicin (Mylotarg) is a humanized monoclonal antibody that is targeted to the CD33 surface antigen and conjugated with Calicheamicin, an antitumor antibiotic. Upon internalization of the CD33-antibody complex, the Calicheamicin portion is released due to changes in pH and then exerts its anticancer effect by cleaving double-stranded DNA. Gemtuzumab Ozogamicin is approved for the treatment of patients older than 60 with CD33-positive AML who are in first relapse and are not candidates for cytotoxic chemotherapy.

(iv) MAB in Treatment of Multiple Solid Tumors

Cetuximab (Erbitux), a chimeric monoclonal antibody that targets the epidermal growth factor receptor (EGFR), has shown activity in multiple solid tumors. Bevacizumab (Avastin) is a novel humanized monoclonal antibody directed against the vascular endothelial growth factor, inhibiting angiogenesis. Angiogenesis is the formation of new blood vessels that are necessary for tumor growth and metastases. **Rituximab** has chimeric structure. Binding regions from original murine anti-human CD20, consisting of variable regions of immunoglobulin heavy and light chains, are fused to human IgG1 heavy-chain and human kappa light-chain constant regions. Fc portion from human IgG1 was selected for its ability to fix complement and activate antibody-dependent cellular cytotoxicity. (See Fig. 19.13).

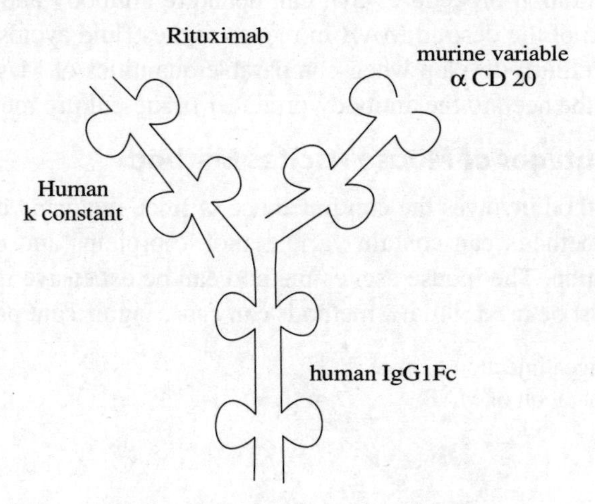

Fig. 19.13 Structure of Rituximab

19.14 ADVANTAGES AND LIMITATIONS

19.14.1 Advantages of *In-Vitro* Methods

In-vitro methods reduces the use of mice at the antibody-production stage, therefore it can be used for large-scale production by the pharmaceutical industry because of ease in culture for production, and because of economic considerations. It also avoids or decreases the need for laboratory personnel experienced in animal handling. Another advantage is by using semipermeable-membrane-based systems produce MAB in concentrations often as high as those found in ascitic fluid and are free of mouse ascitic fluid contaminants.

19.14.2 Disadvantages of *In-Vitro* Method

Some hybridoma do not grow well in culture or are lost in culture and generally require the use of FCS, which limits some antibody uses which is also a concern from the animal-welfare perspective. Antibody product is unsuitable for *in vivo* experiments because of increased immunogenicity, reduced binding affinity, changes in biologic functions, or accelerated clearance *in vivo.* Another disadvantage, batch-culture supernatants contain less MAb (typically 0.002-0.01) per milliliter of medium than the mouse ascites method. In batch tissue-culture methods, MAb concentration tends to be low in the supernatant; this necessitates concentrating steps that can change antibody affinity, denature the antibody and add time and expense. adequate concentration of MAb might be obtained in semipermeable membrane based systems. In this method most of the antibody are conatminated with dead hybridoma products that might require costly purification steps. MAb produced *in-vitro* may yield product with poorer binding affinity than those obtained with ascites method and also is more expansive comparatively.

19.14.3 Advantages of Mouse Ascites Method

The mouse ascites method usually produces very high MAB concentrations that often do not require further concentration procedures that can denature antibody and decrease effectiveness. The high concentration of the desired MAB in mouse ascites fluid avoids the effects of contaminants in-vitro (batch-culture) develop when comparable quantities of MAB are used. The mouse ascites method avoids the need to the antibody producer tissue-culture methods.

19.14.4 Disadvantages of Mouse Ascites Method

The mouse ascites method involves the continued use of mice requiring daily observation. MAB produced by in-vivo methods can contain various mouse proteins and other contaminants that might require purification. The mouse ascites method can be expensive if immunodeficient mice in a barrier facility must be used. *In-vivo* methods can cause significant pain or distress in mice.

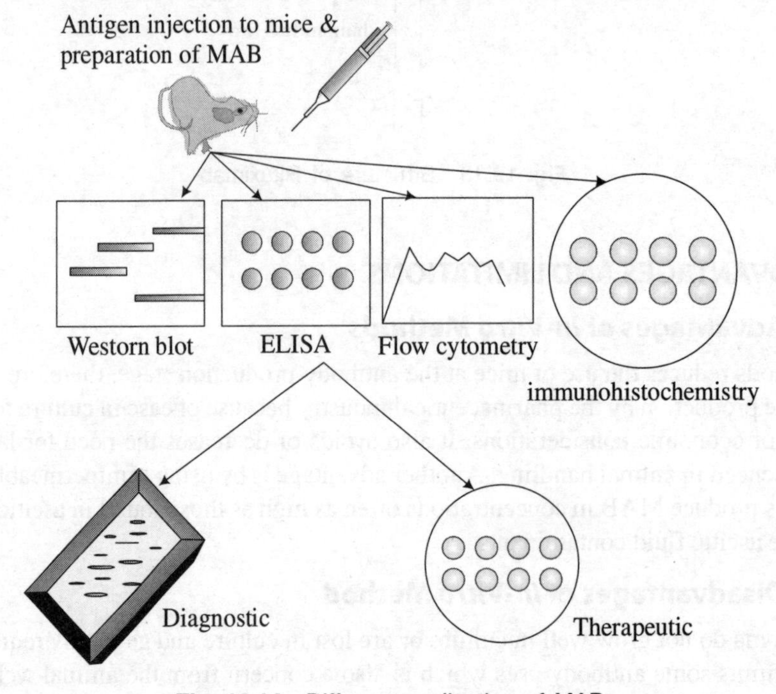

Fig. 19.14 Different application of MAB

BRAIN QUEST

1. Define application of MAB in the medical field.
2. Discuss in detail the selection principle of MAB in HAT medium.
3. What is the basic difference between Salvage pathway and De-novo pathway?
4. Why monoclonal antibody can divide for longer period of time?
5. How monoclonal antibody can be produced at large scale.
6. How MAB can be utilized for cancer treatment?

GENE THERAPY

20

In this Chapter we will discuss the complete process and mechanism of gene therapy. The dream of treating disease at genetic level is in almost completion stage. The gene therapy is going to treat almost many genetic diseases and the cancerous cells successfully in next five years.

20.1 INTRODUCTION

Gene therapy is the insertion of genes into an individual's cells and tissues to treat a disease, (such as a hereditary disease) in which a deleterious mutant allele is replaced with a functional one. Gene therapy involves the modification of genetic material with the aim of curing a disease. These modifications includes correct insertion of gene, its entry into the nucleus, and its expression with the help of right promoter. Many times antibodies form against the new DNA, and thus results in severe killing of the cells. Many time FDA trials results in death of patient, which many time, discourage the application of gene therapy.

20.2 HISTORY OF GENE THERAPY

The First Gene Therapy Trial was done on a four-year old girl by **Dr. W. French Anderson and Michael Blaese** on September 14, 1990 at the NIH Clinical Center. She had adenosine deaminase (ADA) deficiency, a genetic disease which made her defenseless against various infections. The corrected white blood cells (isolated from her blood), with normal genes that makes adenosine deaminase were inserted into her body. Dr. Anderson was regarded as **father of gene therapy.** A retrovirus host was used to carry the correct human ADA gene to the cells. But question was how safely and efficiently the correct genes can be transferred into bone marrow cells in animals. The process wasn't harmful, but the number of cells transformed with correct gene was too low to be useful for treatment. Therefore, white blood cells (T cells) was targeted to deliver the correct gene which greatly increases the number of correct genes taken up by the cells. The experiment was so successful that the team began to look for ways to test other delivery system in the people. In **1989,** The team worked with Dr. **Steven Rosenberg** to test the safety and effectiveness of the gene therapy process in cancer patients. The team grew tumor infiltrating lymphocytes (TIL cells) from people with the deadly cancer

Photo: Culver, Anderson, and Blaese with gene therapy patients

(malignant melanoma), and then engineered a virus to put a DNA marker into those cells. These "marked TIL cells" helped the researchers to: in finding which TIL cells is best for cancer treatment; and the engineered virus can be used safely in humans.

In **1993 Andrew Gobea** born with a rare, fatal genetic disease - **severe combined immunodeficiency (SCID)** in which any one of several genes fail to make a protein essential for T and B cell function, was treated with gene therapy approach. Her blood was removed and allele that codes for ADA was obtained and the inserted into a retrovirus. Retroviruses and stem cells (obtained from umbilical cord) were mixed and injected into Andrew's blood system via a vein. For four years T-cells (white blood cells) produced by stem cells, made ADA enzymes and corrected the diseases. After four years repeated treatment was needed.

In 2002 a new gene therapy approach was adapted in aim to correct the gene function. Now repair errors in messenger RNA derived from defective genes were targated. This technique was found to be useful for treatment of the blood disorder 'thalassaemia', cystic fibrosis, and some cancers. (See "Subtle gene therapy tackles blood disorder" NewScientist.com October 11, 2002). Researchers at Case Western Reserve University and Copernicus Therapeutics were able to create tiny liposomes of 25 nanometers that was able to carry therapeutic DNA through pores in the nuclear membrane. (See "DNA nanoballs boost gene therapy" New Scientist.com May 12, 2002). Sickle cell disease was successfully treated in mice with this approach. (See "Murine Gene Therapy" March 18, 2002, issue of The Scientist.) Clinical trials were halted temporarily in 2002, when two of the ten children treated at the trial's Paris center developed a leukemia-like condition but again it resumed after release of regulatory review of the protocol in the United States, the United Kingdom, France, Italy, and Germany. (See also 'Miracle' gene therapy trial halted at New Scientist.com, October 3, 2002).

In 2003 from University of California, Los Angeles research team inserted genes into the brain using liposomes coated in a polymer called polyethylene glycol (PEG). The transfer of genes into the brain was a significant achievement because viral vectors were too big to get across the "blood-brain barrier." This method was found to be useful for treating Parkinson's disease. (See "Undercover genes slip into the brain" New Scientist.com March 20, 2003). In the same year a new stratgies was adapted to treat the pateint with genetic disorder. This stratgies was aimed to silence the gene by using RNA interference (or gene silencing) approach. This stratgies was useful for treating **Huntington's disease**. In this approach short pieces of double-stranded RNA (short interfering RNAs or siRNAs) were designed to match the RNA copy formed from a faulty gene, siRNA degrades RNA of a particular sequence and therefore the abnormal protein product of the gene were not produced. (See "Gene therapy may switch off Huntington's" New Scientist.com March 13, 2003).

In 2006 scientists at the National Institutes of Health (M/H Bethesda, Maryland) success fully treated metastatic melanoma in two patients using killer T cells which was genetically re-targeted to attack the cancer cells. This study was the first demonstration of treating cancer by gene therapy. **In March 2006** diseases of myeloid system were treated by this approach. **In same year** a team of scientists led by **Dr. Luigi Naldini and Dr. Brian Brown** from the San Raffaele Telethon Institute for Gene Therapy (HSR-TIGET) Milan, Italy developed a way to prevent the immune system from rejecting a newly delivered gene since (similar to organ transplantation), gene therapy also faces the problem of immune rejection and thus immune system recognizes the new gene as foreign and rejects the cells carrying it.

To overcome this problem, the HSR-TIGET group utilized (a newly uncovered network of genes regulated by molecules known as) microRNAs which selectively turn off of the gene in cells and thus prevented the gene from being destroyed by immune system. This work was useful for the treatment of hemophilia and other genetic diseases.

In 2007 Moorfields Eye Hospital and University College London's Institute of Ophthalmology announced the world's first gene therapy trial for inherited retinal disease+. **Leber's congenital amaurosis** caused by mutations in the RPE65 gene. The correct gene were delivered using AAV* carrying RPE65 and experiment was successful. The results of the Moorfields/UCL trial were published in New England Journal of Medicine in April 2008.

In September **2009,** *(Nature)* researchers at the University of Washington and University of Florida were able to treat trichromatic vision to squirrel monkeys using gene therapy, a hopeful precursor to a treatment for color blindness in humans.

20.3 TYPES OF GENE THERAPY

Currently there are two approach: 1-Germline therapy; 2-Somatic cell therapy

20.3.1 Germline Therapy

In germ line therapy a fertilized egg is targeted to replace defective gene by correct gene. So that gene get expressed in all the cells of resulting child. Thus a defective mother give birth to healthy offspring. Germ line therapy have a remote prospect and general opinion is strongly negative; such therapy is currently illegal in most of Europe.

20.3.2 Somatic Cell Therapy

Somatic cell entails the transfer of a gene or genes into somatic cells (not germ cells) which directly effects the patient but cannot pass on to next offspring. Ordinary cells are removed (*ex-vivo*) and is transferred back in the body, after transfection. The transfected cell divides further inside body and during subsequent replication gene expresses correctly. Somatic gene therapy has been already in treatments of adenosine deaminase deficiency, in familial hypercholesterolemia, a genetic condition, which affects the livers regulation of cholesterol in the blood.

In lung disease **(cystic fibrosis)** in which DNA were introduced in lung epithelial cell by inhalers and epithelial cell uptakes DNA and expresses correct protein in few weeks. Thus Somatic gene therapy offers the prospect of effective treatment and cure for various fatal disorders. Until now it has only been used experimentally for a small range of genetic disorders; even in these cases treatment is complex, difficult and success uncertain.

20.4 *EX-VIVO* AND *IN VIVO* GENE THERAPY

Gene transfer in somatic cells can be done either *ex-vivo* or *in-vivo.* In the *ex vivo* approach, cells are removed from the patient for transfection, and the therapeutic entity comprises engineered cells are again transferred back to patient. This offers the advantage of more-efficient gene transfer and the possibility of cell propagation to generate higher cell doses. However, it has the notable disadvantages of being largely patient-specific (due to cell immunogenicity) and is costly. The *in vivo*

*AAV, See page 20.10

approach involves direct administration of the gene-transfer vector to patients. In *ex-vivo* approach the cell containing correct gene is mixed with cell containing incorrect gene outside the normal body, the correct cell are selected and amplified and then are returned inside the body in the corrected form, while *in vivo* approach the cell are not corrected of defective person but the correct gene is delivered directly inside the body and the cell itself get chance to make the correction. (See Fig. 20.1).

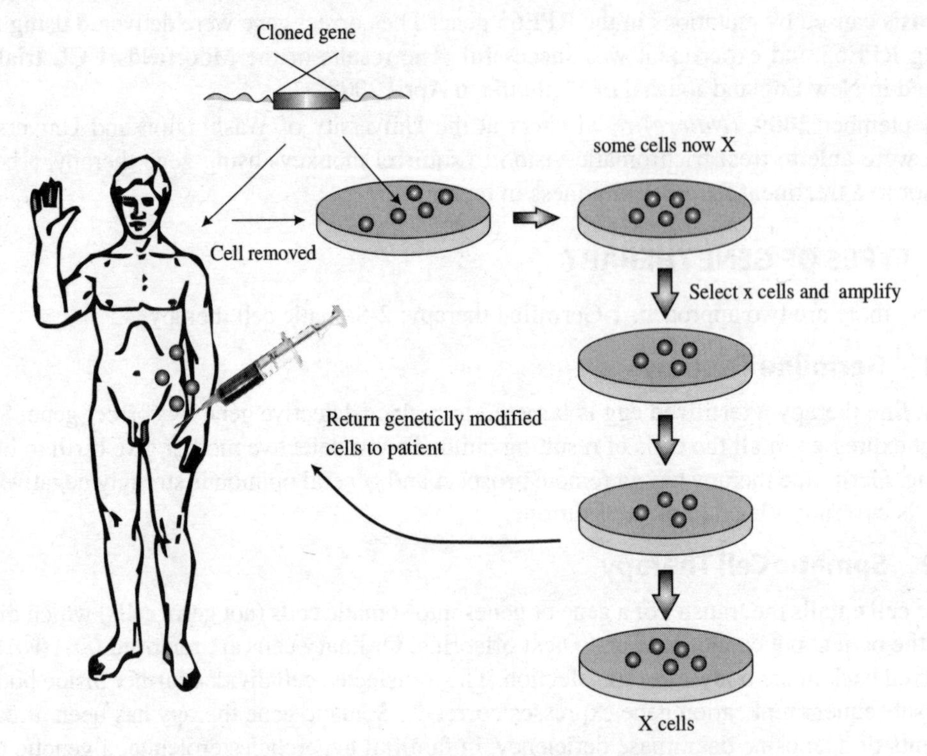

Fig. 20.1 Ex-vivo and in vivo-approach to transfer correct gene

20.5 TECHNICAL ASPECTS OF GENE THERAPY

The most fundamental requirement for successful gene therapy is the effective delivery of a therapeutic gene to the target cell. Once delivered the gene has to enter the nucleus where it will act as a template for production of protein molecule which exerts the **primary therapeutic effect** (cell killing in the case of tumor therapy or cell preservation in the case of neurodegenerative disease). There are several ways to get genes into the cells but the most efficient method is using viral vectors in which their disease causing gene has been disabled, so that they are unable to cause disease. They are engineered in such a way that they pick up and deliver the genes of our choice rather than their own genes. These derivatives of viruses used for gene delivery are known as viral vectors e.g. *retrovirus, adenovirus, adeno-associated virus, lentivirus, herpes virus, AIDS virus etc.* Viral vectors used in gene therapy must be unable to replicate and should have no lytic (ability to rupture a cell) activity. The result of gene therapy is the permanent treatment of disease, hopefully with few or no side effects in the process.

20.6 LIMITATIONS

There are various shortcomings of gene-therapy (somatic)

20.6.1 Short-lived Nature of Gene Therapy

Gene therapy is of very short duration i.e. the patients had to undergo multiple rounds of gene therapy.

20.6.2 Immune Response

As soon as a foreign gene is introduced into human tissues, the immune system attacks and destroys the gene. Furthermore, the immune system's enhanced response to invaders makes it difficult for gene therapy to be repeated in patients.

20.6.3 Problems with Viral Vectors

Viruse vector suspected to present a variety of potential problems to the patient viz toxicity, immune and inflammatory responses. In addition, there is always fear that the viral vector, once get inside the patient may recover its ability to cause disease.

Table 20.1 Disease targeted for gene therapy

Conditions	*Therapy*	*Target cells*
1. Adenosine deaminase deficiency	Adenosine deaminase	Lymphocytes, bone marrow cells
2. Melanoma	Tumor necrosis factor	Tumor-infiltrating lymphocytes, autologous tumor cells
3. Melanoma, glioblastoma, renal cell cancer	IL-2	Autologous tumor cells
4. Hemophilia B	Factor IX	Autologous skin fibroblasts
5. Hypercholesterolemia	LDL receptor	Autologous liver cells
6. Melanoma, colorectal cancer, renal cell cancer	Histocompatibility locus antigen class I-B7 plus beta 2 microglobulin	Tumor cells
7. Glioblastoma, AIDS, ovarian cancer	HSV-TK	Tumor cells, T-cells
8. Cystic fibrosis	Cystic fibrosis transmembrane receptor	Nasal and airway epithelia
9. Breast cancer	Multidrug resistance	Blood CD34+ cells
10. Melanoma	GMCSF	Tumor cells
11. Arthritis	IL-1 receptor agonist	Autologous fibroblasts
12. Amyotrophic lateral sclerosis	Ciliary neurotrophic factor	Encaplulated transduced xenogeneic cells
13. Head and neck squamous carcinoma	p53	Tumor cells
14. Fanconi anemia	Fanconi anemia C	Bone marrow cells

20.6.4 Multigene Disorders

Disorder arises from mutations in a single gene are the best candidates for gene therapy. Unfortunately, some of the most common disorders, such as heart disease, high blood pressure,

Alzheimer's disease, arthritis, and diabetes are caused by the combined effects of thus they many genes (multigene or multifactorial disorders) thus are difficult to treat effectively using gene therapy.

20.6.5 Chance of Inducing Tumor (Insertional Mutagenesis)

If the DNA is integrated at the wrong place in the genome, for example, if it is at site where there is tumor suppressor gene, then it could induce tumor. This has occurred in clinical trials for X-linked severe combined immunodeficiency (X-SCID) patients, in which hematopoietic stem cells were transduced with a corrective transgene using a retrovirus, and this led to the development of T cell leukemia in 3 out of 20 patients.

20.6.6 Religious Concerns

Religious groups may consider the alteration of an individual's genes as tampering or corrupting God's work. In addition deaths of Jesse Gelsinger halted further trial of gene therapy for many years.

20.7 GENE-TRANSFER SYSTEMS (TRANSFECTION VECTOR)

The ideal vector system should have the following characteristics.

1. They should have adequate carrying capacity to carry cDNA and promoter sequence.
2. They should be undetectable by the immune system.
3. They should be safe to the patients.
4. They should have efficiency sufficient to correct the phenotype.
5. They should have long duration of expression and should be safe during re-administered.

A range of physical (*e.g. electoporation, microinjection, particle bombardment, lipofection*) as well as other delivery systems are also under investigation. Each technique has its own advantages and disadvantages (e.g. insert capacity, integration, site of integration, duration and level of transgene expression, ease of administration, tissue specificity, and efficiency of transfection) and therefore, each of the disease might require a different approach. Therfore, gene-delivery system can be devided into viral, non-viral and physical.

1. Viral vector. Viral vectors generally give the most efficient transfection. Their main disadvantages are insert-size limitation, immunogenicity and their synthesis. Many different viruses are being adapted as vectors but the most advanced are

I. Retrovirus (Rv),
II. Adenovirus (Ad)
III. Adeno-associated virus (AAV).
IV. Poxviruses (especially vaccinia) and
V. Herpes simplex virus (HSV).

2. Non viral/Physical method. Non-viral approaches fall into three main categories involving:

I. Naked DNA (electroporation, microinjection, particle bombardment)
II. DNA complexes with liposome
III. Particles comprising DNA condensed with cationic lipid (Lipoplexes and polyplexes)

IV. Oligo approach

Non-viral vectors give less efficient transfection (especially *in vivo*)

3. Hybrid method

I. Dendrimers

II. Chimeraplasty

20.8 VIRAL VECTORS

Viral vectors are mostly derivatives of viruses that infects animal. These viral vectors can be divided into 2 groups:

1. **Free replicating viruses** that multiply within the cell, but do not integrate into the genome of the host.
2. **Integrating viruses,** Retroviruses (HIV and several others) that can integrates into the host genome. Free replicating viruses do not produce stable transgenic cells, because the genetic material does not stay integrated into the host genome. But this is still very useful for transient, (not permanent), introduction of a gene. The integrating viruses (such as HIV) enters the cell, copy their RNA genome into DNA, and then this DNA integrates into the chromosome of the host. The genes carried on the virus can then be expressed, and also stably inherited after cell division. The integration of these retroviruses into the host genome is one reason why these viruses are so dangerous (causing many infections and cancers in vertebrates, including human T-cell leukemia and HIV). They also causes a variety of hematopoetic and neurological conditions, including paralysis, wasting, ataxia, arthritis, dementia, and neuropathy. They vary in size from approximately 8 to 11 kb. (See Fig. 20.10)

20.8.1 Retroviruses

Retroviruses belongs to the family Retroviridae and are composed of a single RNA strand. They uses the enzyme reverse transcriptase to convert RNA into DNA which is integrated into genome DNA. Retroviruses are genetically simple, and have the ability to infect a wide variety of cell types with high efficiency. General structure of retroviruses are shown in Fig. 20.2 which consists of 5′ and 3′ LTRs (long terminal repeats) which enclose promoter/enhancer regions along with striuctural gene, gag (group specific antigen gene, or internal capsid protein gene), pol (reverse transcriptase), and env (envelope protein). A critical element for packaging is the **psi sequence**, which is a cis-acting element. That is, gene must have a psi sequence if it has to be packaged. The protein coding genes are trans acting, since they get diffuse around the cytoplasm. A modified retrovirus might have a *psi* sequence, a gene of interest at a multiple cloning site, and neomycin resistance gene. This would need to be made in a "helper cell" that expressed *gag, pol, and env* in trans from genomic DNA. Modified retroviruses have been used for gene therapy purpose. Retroviruses can transform cells by integrating near to a cellular proto-oncogene and driving inappropriate expression from the LTR, or by disrupting a tumor suppressor gene. This event is termed as **insertion mutagenesis**. Transgene expression can either be driven by the promoter/enhancer in the 5′ LTR, or by alternative viral (e.g. cytomegalovirus, Rous sarcoma virus) or cellular (e.g. beta actin, tyrosine) promoters.

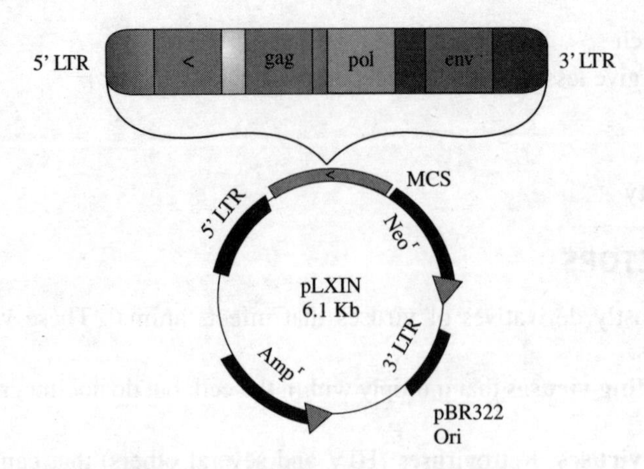

Fig. 20.2 Retrovirus gene structure

20.8.1.1 Murine Leukemia Virus (MLV)

The second viral vector is murine leukemia virus (MLV). When genes are delivered by derivatives of MLV they become stably integrated into the chromosomes of target cell and are maintained for as long as the cell remains alive and also gene activity is easy to control over long periods of time. Many clinical trials have been conducted with these MLV based systems and have been shown to be well tolerated with no adverse side effects. MLV vectors can deliver genes to cells that are in deviding phase

20.8.1.2 Lentiviruses

Lentiviruses are a subgroup with in the general family of Retroviruses but they are distinct from the MLV like viruses, and they are able to infect non-dividing cells. The best studied lentiviruses is HIV. There were number of technical difficulties and first generation vectors could not be used in the clinic as they had potential to generate infectious HIV. Over past two years new HIV based vectors has emerged that are severely disabled containing only the few HIV components that are required for efficient gene delivery to non dividing cells. These were called **as minimal vectors.**

A requirement for retroviral integration and expression of viral genes is that the target cells should be dividing. This limits gene therapy to proliferating cells *in-vivo* or *ex-vivo*, whereby cells are removed from the body, treated to stimulate replication and then transduced with the retroviral vector, before being returned to the patient. When treating cancers in vivo, tumour cells are preferentially targeted. However, *ex vivo* cells can be more efficiently transduced, due to exposure to higher virus titers and growth factors. Furthermore *ex vivo* treated tumour cells will associate with the tumour mass and can direct tumouricidal effects.

20.8.2 Adenovirus

Adenovirus is a group of DNA containing viruses, which causes respiratory disease (ranging from one form of the common cold to pneumonia, croup, and bronchitis), and can also cause gastrointestinal illness, eye infections, cystitis, and rash in humans. The picular feature is their life cycle does not normally involve integration into the host genome, rather they replicate as episomal elements in the nucleus of the host cell and consequently there is no risk of insertional mutagenesis. The detail structure of adenovirus has been shown in Fig. 20.3. Their genome size is approximately 35 kb of which up to 15 kb can be re-placed with foreign DNA. There are four early transcriptional units (El, E2, E3 and E4), which have regulatory functions, and a late transcript, which codes for structural proteins.

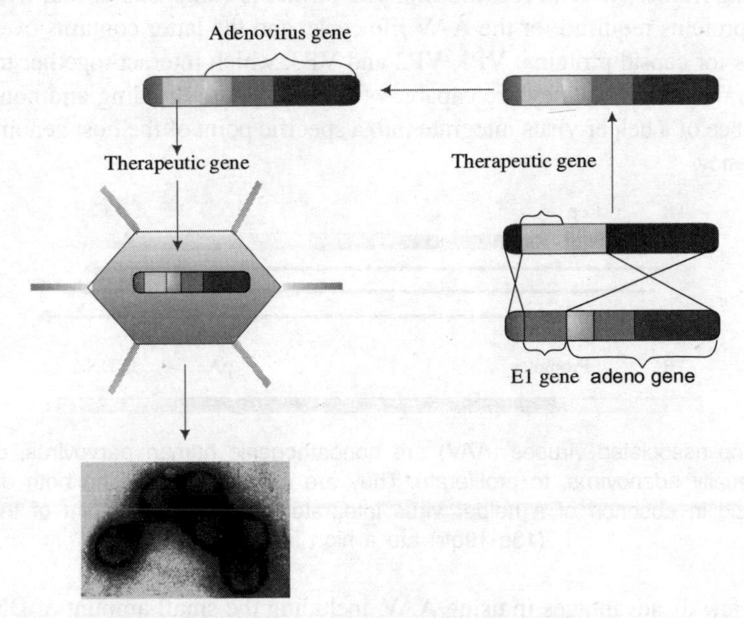

Fig. 20.3 (a) Adenovirus, (b) fusion with transgene containing therapeutic gene and their (c) full expression

Progenitor vectors have either the El or E3 inactivated gene, or E2, a temperature sensitive mutant or an E4 deletion in second generation vectors. The most recent "gutless" vectors contain only the inverted terminal repeats (ITRs) and a packaging sequence around the transgene.

All the necessary viral genes being provided in Trans by a helper virus. Adenoviral vectors are very efficient at transducing target cells *in vitro* and *in vivo*, and can be produced at high titers ($>10^{11}$/ml). Adenovirus is particularly good at infecting epithelial cells. The El genes of adenovirus are required for lytic growth, so a recombinant adenovirus can be prepared (that is **replication defective**) by integrating a gene of interest (a therapeutic gene) into the El locus. Adenoviral vectors can be used to overexpress recombinant proteins in mammalian cells. As a result, the resulting proteins have the relevant posttranslational modifications and folding, which is not possible in prokaryotic systems. Adenoviruses can also be (genetically modified) used in gene therapy to treat **cystic fibrosis, cancer,** and potentially other diseases. Some of the characteristic features of adenoviruses include: They are widespread in nature, and infects birds, and many mammals including humans. There are two genera: Aviadenovirus (avian) and Mast adenovirus (mammalian). They can be used to transfect cells for protein expression and *in vivo* characterization studies. Adenoviral vectors are great to use for transfection because the adenovirus efficiently infects many different cell lines. The virus enters the cell but it does not replicate because of lack of El gene (viral genes which are expressed early i.e. El-minus adenovirus vectors.

20.8.3 AAV (Adino Associated Virus)

Structure of AAV has been depicted in Fig. 20.4. The AAV genome consist of single-stranded deoxyribonucleic acid (ssDNA), either positive- or negative-sensed, which is about 4.7 kilobase. The genome comprises inverted terminal repeats (ITRs) at both ends of the DNA strand, and two open reading frames (ORFs): rep and cap. The former is composed of four overlapping genes encoding Rep proteins required for the AAV life cycle, and the latter contains overlapping nucleotide sequences for capsid proteins: VP1, VP2 and VP3, which interact together to form a capsid of an icosahedral symmetry. They are capable of infecting both dividing and non dividing cells, and in the absence of a helper virus integrate into a specific point of the host genome (19q 13-qter) at a high frequency.

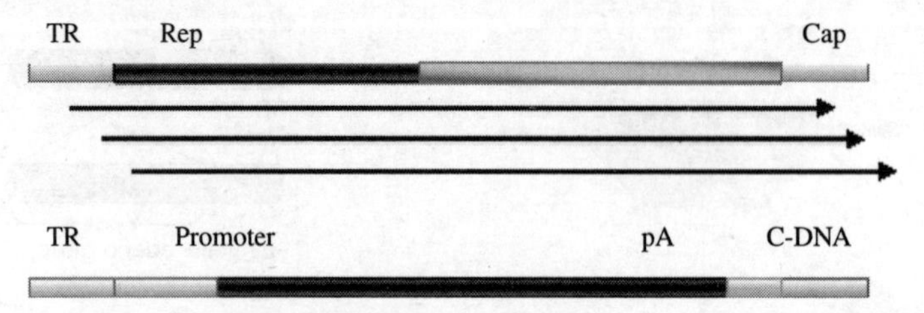

Fig. 20.4 Adeno associated viruses (AAV) are nonpathogenic human parvovirus, dependent on a helper virus, usually adenovirus, to proliferate. They are capable of infecting both dividing and non dividing cells and in absence of a helper virus integrate into a specific point of the host genome (13q-19qtr) ate a high frequency

There are a few disadvantages in using AAV, including the small amount of DNA it only (low capacity). This type of virus is being used, because it is non-pathogenic and in contrast to adenoviruses, AAV do not build an immune response and several trials are going on to treat muscle and eye diseases. However, clinical trials have also been initiated where AAV vectors are used to deliver genes to the brain. This is possible because AAV viruses can infect non-dividing (quiescent) cells, such as neurons in which their genomes are expressed for a long time. To date, AAV vectors have been used for first- and second-phase clinical trials for treatment of cystic fibrosis and first-phase trials for hemophilia. Promising results have been obtained from phase I trials for Parkinson's disease, showing good tolerance of an AAV2 vector in the central nervous system. Other trials have begun, concerning AAV safety for treatment of Canavan disease, muscular dystrophy and late infantile neuronal ceroid lipofuscinosis.

20.8.4 Herpes Viruses

Herpes viruses include herpes simplex virus type-1 (HSV-1), which is common in the general population, but in rare cases can cause **encephalitis**. It is one of the most commonly used vector systems because of its broad host cell range, the ability to transduce neurons, and a capacity to receive large insert. HSV is also being exploited for gene therapy applications. HSV vectors have the advantages of being able to infect non dividing cells (in neuronal cells) that makes HSV an attractive vector for treating neurological disorders such as Parkinson's and Alzheimer's. In addition, the ability of HSV to infect efficiently a number of different cell types, such as muscle and liver, may make it an excellent vector for treating non neurological diseases.

HSV comprise of a linear dsDNA virus of approximately 150 kb, encoding over 70 viral proteins (Fig. 20.5). All herpes virus genomes have a unique long (UL) and a unique short (US) region, bounded by inverted repeats. The repeats allow rearrangements of the unique regions and Herpesvirus genomes exist as a mixture of 4 isomers. Herpes virus genomes has been utilised, in the treatment of some *cancers, retinosis or inflammatory disease*. However, it is a required where sustained gene activity required for many months such as in the treatment of some tumors, neurodegenerative disease and HIV infection.

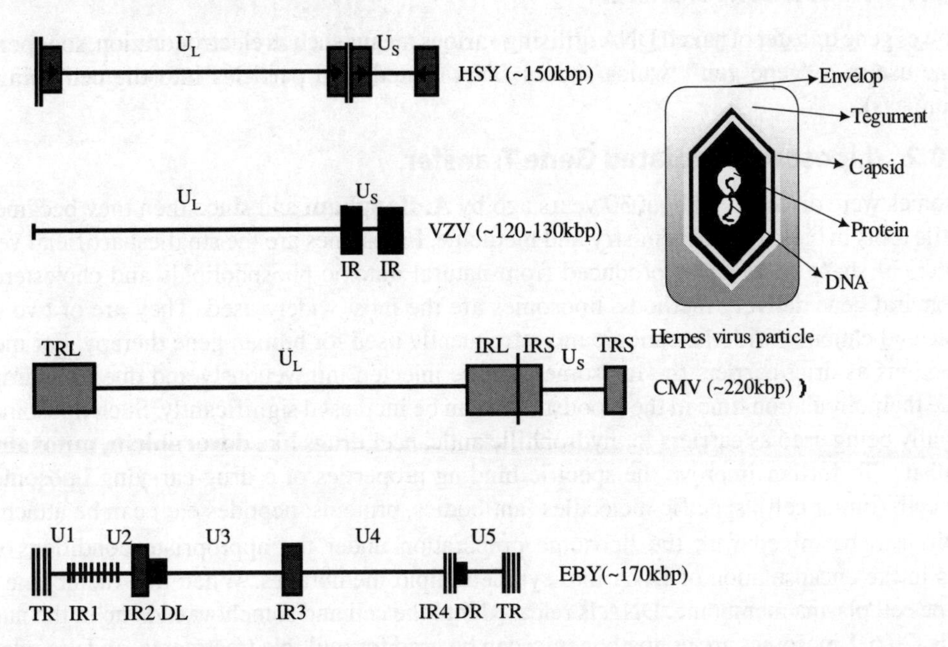

Fig. 20.5 Structure of HERPES genome (L) long (S) short (U) unique (IR) inverted repeat

20.9 PSEUDOTYPING

Viruses in which the envelope proteins have been replaced with envelop of another species are referred to as pseudotyped viruses. The envelope proteins binds to cell-surface molecules of host surface such as heparin sulfate, as well as with the specific protein receptor that either induces entry, promoting structural changes in the viral protein, or localizes the virus in endosomes where acidification of the lumen induces this refolding of the viral coat. In either case, entry into potential host cells requires a favorable interaction between a protein on the surface of the virus and a protein on the surface of the cell.

For the purposes of gene therapy, one might either want to limit or expand the range of cells susceptible to transduction by a gene therapy vector. Many vectors have been developed in which the endogenous viral envelope proteins have been replaced by either envelope proteins from other viruses, or by chimeric proteins. Such chimera would consist of those parts of the viral protein necessary for incorporation into the virion as well as sequences meant to interact with specific host cell proteins. These vectors show great promise for the development of "**magic bullet**" in gene therapies.

20.10 NON-VIRAL METHODS

Non-viral method presents certain advantages over viral methods, such as large scale production and low host immunogenicity. Previously, low levels of transfection and expression of the gene held non-viral methods at a disadvantage; however, recent advances in technology have yielded techniques with improved transfection efficiencies similar to those of viruses.

20.10.1 Naked DNA Transfer

It involves gene transfer of naked DNA utilising various means such as electroporation, sonoporation, and the use of a "gene gun" (which shoots DNA coated gold particles into the cell using high pressure gas).

20.10.2 Liposome Mediated Gene Transfer

Liposomes were discovered about 30 years ago by **A. Bangham** and since then they became very versatile tools in biology, biochemistry and medicine. Liposomes are the smallest artificial vesicles of spherical shape that can be produced from natural untoxic phospholipids and cholesterol. Of the nonviral gene delivery methods, liposomes are the most widely used. They are of two types, anionic and cationic, with the cationic more frequently used for human gene therapy. For medical applications as drug carriers the liposomes can be injected intravenously and due to hydrophilic surface their circulation time in the bloodstream can be increased significantly. Such liposomes are especially being used as carriers for hydrophilic anticancer drugs like **doxorubicin, mitoxantrone** and others. To further improve the specific binding properties of a drug-carrying liposome to a target cell, (tumor cell), specific molecules (antibodies, proteins, peptides etc.) can be attached.

DNA can be mixed with the liposome preparation under the appropriate conditions which results in the encapsulation of DNA into synthetic lipid membranes. When this membrane fuses with the cell plasma membrane, DNA is released into the cell and somehow ends up in the nucleus. See Fig. 20.6. Liposomes are nonpathogenic, can be used for multiple treatments, and are relatively cheap and easy to produce relative to viral vectors, but the efficiency of transfection using the current liposomes is still significantly low compared with that achieved using viral vectors.

To improve transfection, cationic liposomes can be conjugated to defective viral particles, viral proteins, or virally derived peptides that are able to disrupt the lysosome and/or increase DNA transport to the nucleus. Thus, the nonviral vectors are being developed to be more like viruses, whereas the viral vectors are being developed to be more like liposomes.

20.10.2.1 Cationic Liposome

Cationic liposome binds to negatively charged DNA tightly by electrostatic interaction. Several cationic lipid-DNA complexes give fairly efficient transfection (*ex vivo* transfection of endothelial cells in the lungs following injection into the tail vein in mice, and of airway epithelial cells following direct installation into the lungs). Such formulations have been extensively tested for delivery of the cystic fibrosis transmembrane receptor gene following administration into the nose and lungs of cystic fibrosis patients.

The main disadvantages of cationic lipids are formulation instability, inactivation in blood, relatively low transfection efficiency and poor targeting. Some improvement has been achieved recently by using cationic polymers to compact the DNA before mixing it with the lipid component

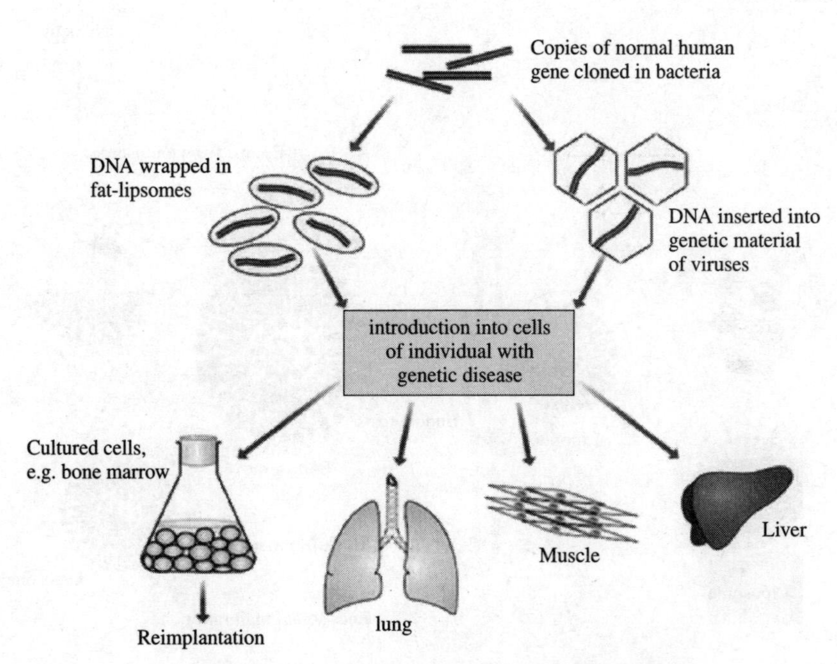

Fig. 20.6 Liposome medicated gene therapy

Transfection efficiency is limited by poor uptake and inefficient trafficking of DNA to the site of gene expression in the nucleus. therfore DNA have to be protected by some polymers.

20.10.2.2 Lipoplexes and Polyplexes

To improve the delivery of the new DNA into the cell, the DNA must be protected from damage. Therefore, lipoplexes and polyplexes were created that have the ability to protect the DNA from undesirable degradation during the transfection process. When the organized structure is complexed with DNA it is called a **lipoplex or genosome.** Cationic lipids, due to their positive charge, were first used to condense negatively charged DNA molecules so as to facilitate the encapsulation of DNA into liposomes. Later it was found that the use of cationic lipids significantly enhanced the stability of lipoplexes. Also as a result of their charge, cationic liposomes interact with the cell membrane. Endocytosis was widely believed as the major route by which cells uptake lipoplexes and likely to be degraded in lysosomes. DNA released from lipoplexes or polyplexes into the cytoplasm may enter the nucleus via two different, non-mutually exclusive, mechanisms. First, transfecting DNA may enter the nucleus during mitosis while the nuclear membrane breaks down. This passive nuclear localization mode is depend on the cell division activity and on the plasmid lifetime in the cytoplasm. The second mechanism postulates an active, energy-dependent nuclear transport into the nucleus. This mode requires the presence of specific sequences in the plasmid that mediate its interaction with proteins such transcription factors. The DNA-protein complex being transported into the nucleus through a nuclear localization signal-mediated transport that requires importins and other nuclear transport mediators. It was also recently postulated that lipoplex-filled endosomes may fuse directly with the nuclear membrane, permitting a direct entry of DNA into the nucleus (see Fig. 20.7)

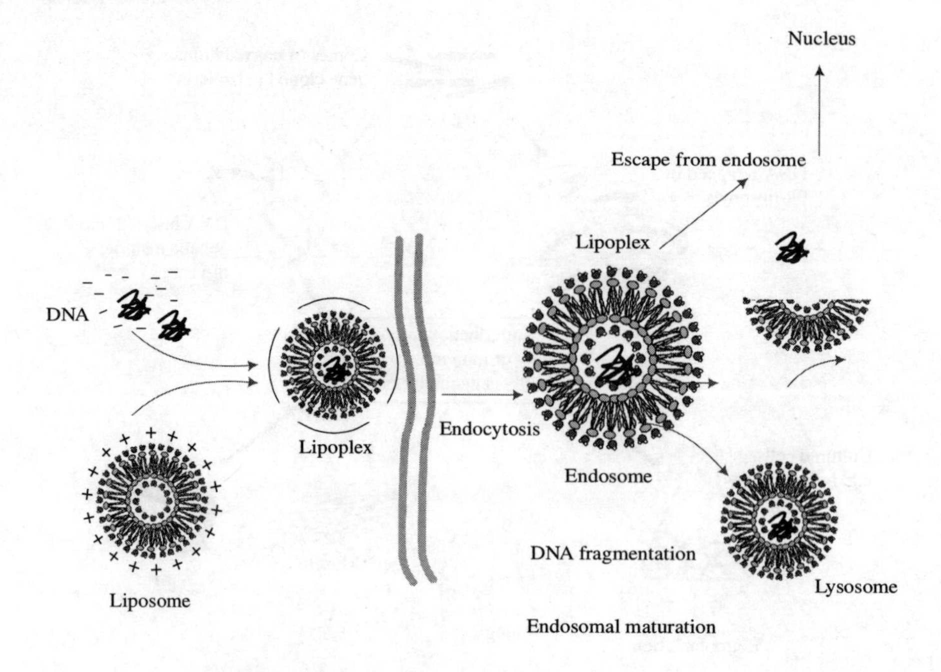

Fig. 20.7 Lipoplex mediated transfection and endocytosis

Endosomes are formed as the results of endocytosis, however, if genes can not be released into cytoplasm by breaking the membrane of endosome, they will be sent to lysosomes where all DNA will be destroyed before they could achieve their functions. It was also found that although cationic lipids themselves could condense and encapsulate DNA into liposomes, the transfection efficiency was very low due to the lack of ability of "endosomal escaping". However, when helper lipids (usually electroneutral lipids, such as DOPE) were added to form lipoplexes, much higher transfection efficiency was observed. Later on, it was concluded out that certain lipids have the ability to destabilize endosomal membranes so as to facilitate the escape of DNA from endosome; therefore those lipids are called **fusogenic lipids**. Although cationic liposomes have been widely used as an alternative for gene delivery vectors, a dose dependent toxicity of cationic lipids were also observed which could limit their therapeutic usages. The most common use of lipoplexes has been in gene transfer into cancer cells, where the supplied genes have targeted tumor suppressor control genes and decrease the activity of oncogenes. Recent studies have shown lipoplexes to be useful in transfecting respiratory epithelial cells, so they may be used for treatment of genetic respiratory diseases such as cystic fibrosis.

Complexes of polymers with DNA are called **polyplexes.** Most polyplexes consist of cationic polymers and their production is regulated by ionic interactions. One large difference between the methods of action of polyplexes and lipoplexes is that polyplexes cannot release their DNA load into the cytoplasm, so co-transfection with endosome-lytic agents such as polyethylenimine, chitosan and trimethylchitosan (to lyse the endosome that is made during endocytosis, the process by which the polyplex enters the cell) such as inactivated adenovirus must occur.

DOPE = Phosphatidyle thanolamine (dioleoyl, PE) Hui et al 1996

20.10.3 Condensed DNA Particles

Many cationic polymers have been used to condense DNA by electrostatic interaction into small particles, with a view to protecting the DNA from degradation and enhancing uptake via endocytosis. The most important polymers are heterogeneous polylysine, defined-length oligopeptides and polyethylene imines. In general, such formulations aggregate in physiological conditions to form entities that are too large to penetrate effectively through solid tissues, and too large to be efficiently injection into the tail vein in mice, with targeted transfection appearing feasible. To date, though, no clinical studies have been undertaken with this approach. Different methods are used to get the alleles into the patient's cells, including: using fat droplets in nose sprays, using cold viruses that are modified to carry the allele i.e. viruses go into the cells and infect them for the direct injection of DNA

20.10.4 Oligonucleotides

Synthetic oligonucleotides in gene therapy is targated to inactivate the specific genes responsible for disease. There are several methods by which this can be achieved. One strategy uses **antisense approach** specific to the target gene. Another appraoch is to use small molecules of RNA called as siRNA that signal the cell to cleave specific unique sequences in the mRNA transcript, and thus disrupting translation capability of the faulty mRNA, and therefore expression of the gene stops (see Fig. 20.8). A further strategy uses double stranded oligodeoxynucleotides as a decoy for the transcription factors that are required to activate the transcription of the target gene. The transcription factors bind to the decoys instead of the promoter of the faulty gene, which reduces the transcription of the target gene, by lowering the gene expression. Additionally, single stranded DNA oligonucleotides have been used to direct a single base change within a mutant gene. The oligonucleotide is designed to anneal with complementarity to the target gene which serves as the template base for repair. This technique is referred to as **oligonucleotide mediated gene repair**, targeted gene repair, or targeted nucleotide alteration. Recently Asthama has been treated using this approach. It is known that **Asthma** is triggered by exposure of allergen and is a chronic pulmonary disease primarily affecting the airways. It is characterized by hyper-responsiveness, airway inflammation, and reversible obstruction. In **allergic asthma**, a common feature appears to be a polarization of T-lymphocyte function, highlighted by a predominance of CD4-positive T cells producing interleukin (IL)-4, IL-5, IL-13, and IL-10. Enhanced secretion of these cytokines results in immunoglobulin (IgE) synthesis, eosinophilic lung inflammation, goblet cell hyperpla

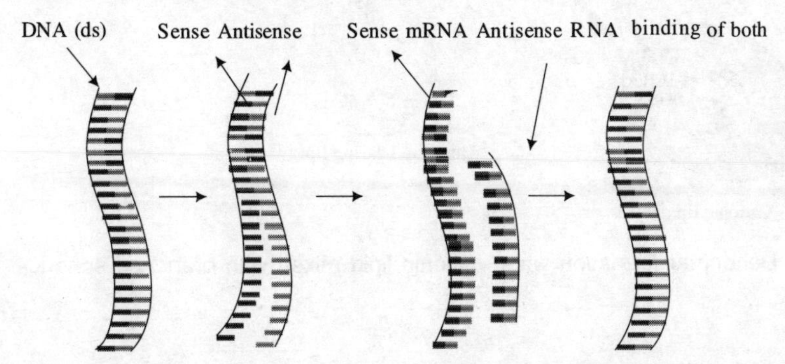

Fig. 20.8 Antisense approach of gene therapy. Prevention of translation by supplying antisense RNA which binds to sense mRNA strand and double stranded RNA cant be translated

sia, and mucus hyperproduction. In association with Th2 immunity and IgE production, mast cell and basophil activation is also increased. The imbalance between Th2 and Th1 immunity results in allergic asthma and gene therapy approach try to restore the balance by interfering with IL-4 or IL-5 and their receptors to prevent the accumulation of eosinophils, and targeting downstream mediators.

Antisense is an attractive option because each mRNA molecule produces large quantities of a single protein. By attaching to a single mRNA strand responsible for producing a single protein, may be a more efficient way of eliminating unwanted proteins. It is more efficient than using antibodies or interfering with their receptor function. In addition, since an antisense drug only binds with its exactly opposite strand, it is highly specific.

20.11 HYBRID METHODS

Every method of gene transfer have some shortcomings, therefore hybrid molecules have been developed that combine two or more techniques. e.g. Virosomes. Virosome formed when liposomes are combined with an inactivated HIV or influenza virus. This has been shown to have more efficient gene transfer in respiratory epithelial cells than either viral or liposomal methods alone. Other methods involve mixing other viral vectors with cationic lipids or hybridising viruses.

20.11.1 Dendrimers

A dendrimer is a highly branched spherical shape macromolecule (see Fig. 20.9). The surface of the particle may be functionalized in many ways and many of the properties of the resulting construct are determined by its surface. In particular it is possible to construct a cationic dendrimer, leads to a temporary association of the nucleic acid with the cationic dendrimer. On reaching its destination the dendrimer-nucleic acid complex is taken into the cell via endocytosis.

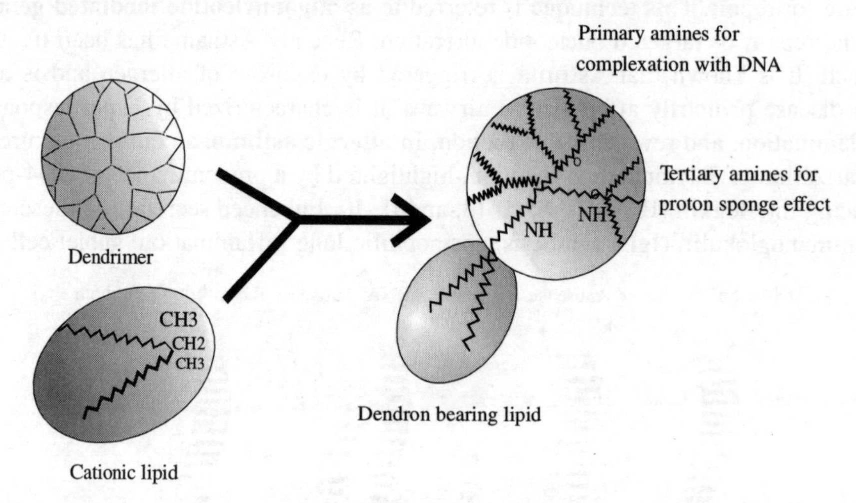

Fig. 20.9 Dendrimer formation when cationic lipid mixed with branched spherical molecule

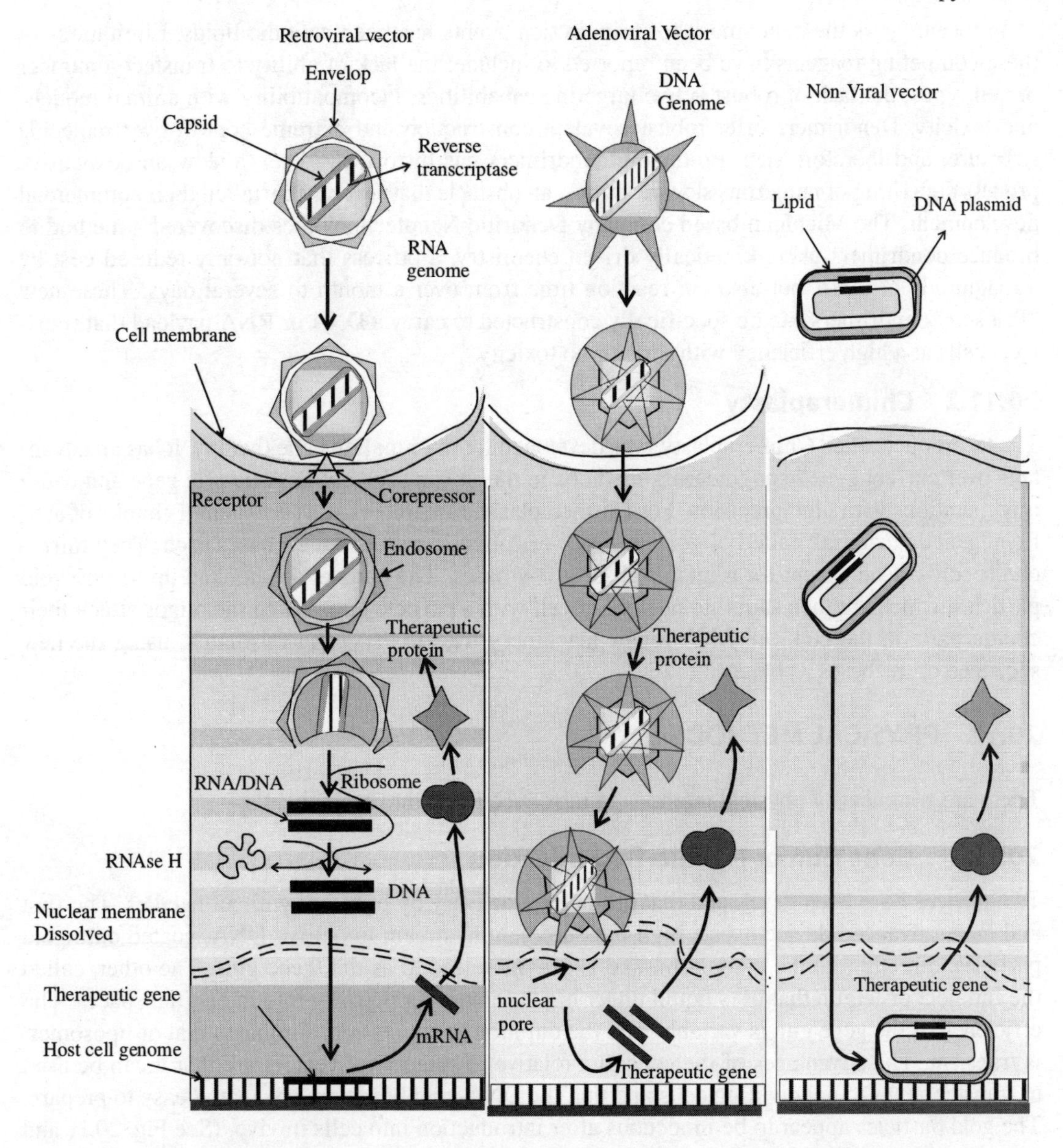

Fig. 20.10 Comparative mechanism of cell expression of viral and nonviral method (Redrawn from nature). Retro viral vectors first converts RNA into DNA by reverse transcriptage and then this DNA is incorporated in the host vector which express correct protein (theraputic) while adenoviruses have DNA as genetic material, therefore they need not to do this and directly are incorporated into the host DNA and expressed the correct protein. In non-viral gene delivery by liposome, the gene enters the cell by fusing with membrane and thus reaches inside the nucleus and get incorporated in the host genome, which along with host genome expressed the correct protein.

In recent years the benchmark for transfection agents has been cationic lipids. Limitations of these competing reagents have been reported to include: the lack of ability to transfect a number of cell types, the lack of robust active targeting capabilities, incompatibility with animal models, and toxicity. Dendrimers offer robust covalent construction and extreme control over molecule structure, and therefore size. Producing dendrimers has historically been a slow and expensive process consisting of numerous slow reactions, an obstacle that severely curtailed their commercial development. The Michigan based company Dendritic Nanotechnologies discovered a method to produce dendrimers using kinetically driven chemistry, a process that not only reduced cost by a magnitude of three, but also cut reaction time from over a month to several days. These new "Priostar" dendrimers can be specifically constructed to carry a DNA or RNA payload that transfects cells at a high efficiency with little or no toxicity.

20.11.2 Chimeraplasty

The technique, called Chimeraplasty, was developed for mammalian gene therapy. It has an advantage over current genetic engineering methods in that it can seek out any specific gene and cause tiny mutations with high precision. For Chimeraplasty, researchers start with small chunks of artificial genetic material, called oligonucleotides or "oligos", with about 25 bases each. They mirror one specific gene except for a mismatch of a few bases. The chunks are hooked up to tiny gold particles, which are then shot into nucleus of cell with a particle gun. When the oligos attach their counterparts in the cell, the DNA repair machinery tries to "fix" the mismatch, using the new sequence of the bases as blueprint.

20.12 PHYSICAL METHODS

There are a number of physical methods to introduce DNA into animal cells.

20.12.1 Gene Gun or Needle-free Injection

Two devices have been developed that allow gene delivery by injection without needles. The first and more advanced device uses a high-pressure helium stream to deliver DNA, coated onto gold particles, directly into the cytoplasm, and is often referred to as the 'gene gun'. The other, called the Intraject or Jetgun that uses liquid under high pressure for delivery into interstitial spaces. The efficiency of the gene gun is variable, and the duration of expression, similar to that of liposomes, is transient. The advantages of the gene gun relative to some viral vectors are that it can be used to transfer genes to nondividing cells and that the DNA-gold beads are cheap and easy to prepare. The gold particles appear to be innocuous after introduction into cells in vivo. (See Fig. 20.11 and 20.12)

The gene gun may have applications in gene transfer to skin, where it can be used to vaccinate against specific viral- or tumor-associated antigens or to deliver genes that can promote wound healing. It may also have uses for *ex vivo* approaches, in which tissue or tumor explants can be "shot" in culture and then reintroduced into a patient. These skin cells (e.g. Langerhans cells, which are antigen-presenting cells that normally capture foreign antigens from invading organisms) present them to naïve T cells to elicit primary immune responses.

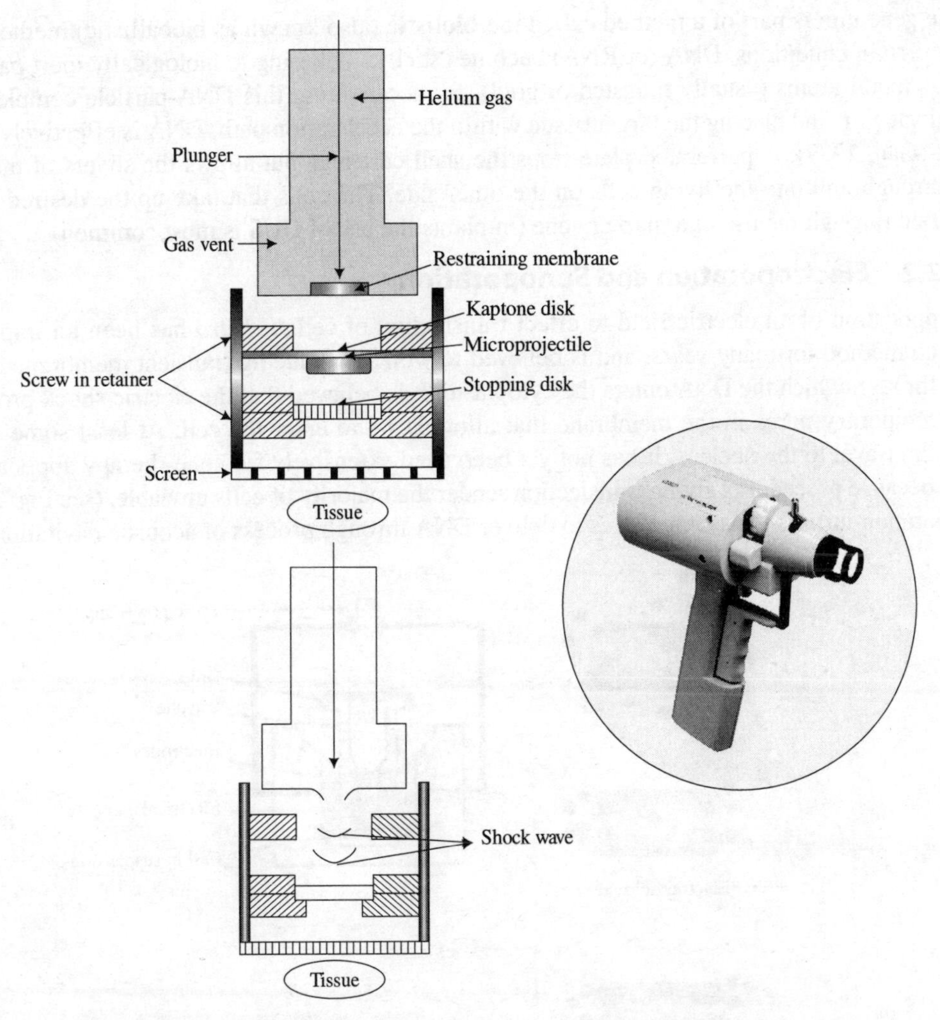

Fig. 20.11 Helium powered gene gun works by pushing DNA coated particles into tissue. Images redrawn from Williams, 1991 Helios Gene Gun schematic (courtesy of BioRad)

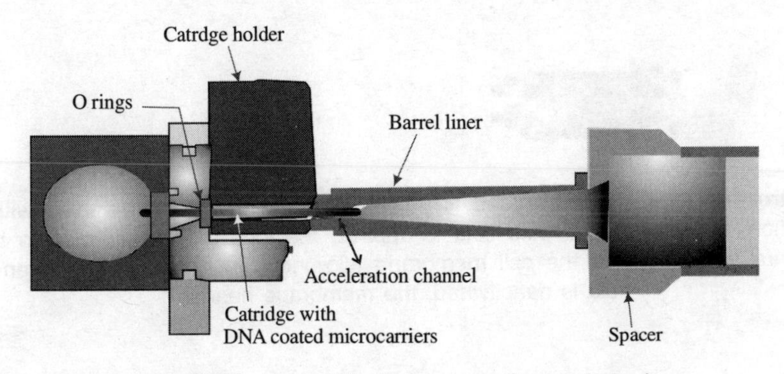

Fig. 20.12 A standard gene gun (internal view)

The gene gun is part of a method called the **biolistic** (also known as bioballistic) method, and under certain conditions, DNA (or RNA) become "sticky," adhering to biologically inert particles such as metal atoms (usually tungsten or gold). By accelerating this DNA-particle complex in a partial vacuum and placing the target tissue within the acceleration path, DNA is effectively introduced (Gan, 1989). A perforated plate stops the shell cartridge but allows the slivers of metal to pass through and into the living cells on the other side. The cells that take up the desired DNA, identified through the use of a marker gene (in plants the use of GUS is most common).

20.12.2 Electroporation and Sonoporation

The application of an electric field to effect transfection of cells *in vitro* has been an important research method for many years, and is believed to work by inducing transient membrane breakdown through which the DNA enters the cytoplasm. It is believed that the electric shock produces small temporary pores in the membrane that allow DNA to enter the cell. At least some of the DNA can travel to the nucleus. It has not yet been used extensively for gene-therapy applications, partly because procedures giving transfection render the majority of cells unviable. (see Fig. 20.13) Sonoporation utilises ultrasonic waves to deliver DNA through process of acoustic cavitation.

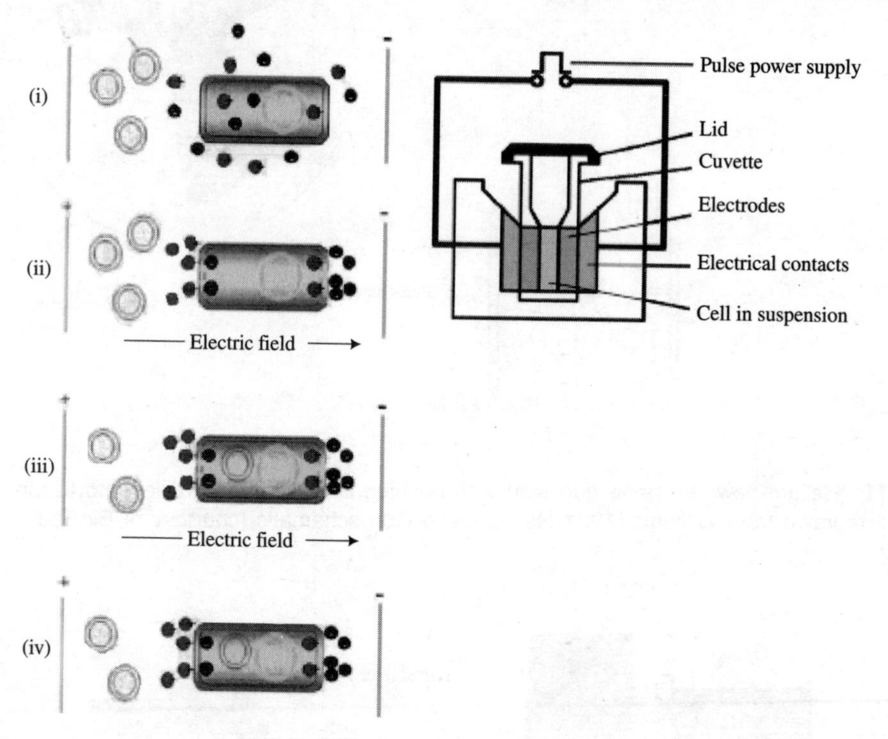

Fig. 20.13 Structure of instruments used for electroporation. (i) Plasmid mixed with host cells for electroporation, (ii) when the electric field is applied, the ions move according to their charge, (iii) Pathways are formed across the cell membrane allowing DNA to enter, (iv) when the electric field is deactivated, the membrane heads.

20.12.3 Calcium Phosphate Addition and Magnetofection

Precipitates of DNA form when DNA is mixed with calcium chloride in form of Calcium phosphate. When these DNA precipitates are added to animal cells growing in culture, the precipitated DNA can be taken up by the cells, again transferred to the nucleus and expressed. Magnetofection utilises magnetic particle to bring DNA complex with monolayer cell.

20.13 COMPARISON OF VECTORS

The ideal vector allows subsequent readministration and avoids immune response. Alternatively, integration into the host genome, preferably in a site specific location, could ensure that the trans gene is not lost during the lifetime of the cell, with greater tissue specificity and regulation of the transgene expression. As yet no one vector has all these requirements, but improvements are being made to all systems, e.g. immunological problems of adenoviruses, production problems of AAV, the short duration of Retroviruses, etc. It is likely that all delivery systems will find a niche, either as dictated by biological needs, e.g. HSV-1 for neuronal applications. Non-viral approaches are less likely to cause an immune response. Thus development of these systems has to focus on improving effectiveness of gene delivery. Viral vector systems have been hampered by problems such as the immune response to the viral proteins and, particularly in the case of AAV, limited packaging capacity. Non-viral vectors have the potential to avoid these problems (for review see Ratko et al., 2003). The non-viral system that has been mostly characterized in clinical trials of gene therapy. The data from the clinical trials suggested that further improvements in effectiveness of the non-viral vector systems are necessary. To improve the delivery of DNA into cells a number of approaches are being explored. For example, a short protein that binds to integrins on the surface of cells can improve the effectiveness of liposome mediated gene delivery by ten fold in cell culture and four fold in animal studies. Small protein sequences called **nuclear localization signals** enhanced liposome mediated transfection by eight-fold in a cultured human cystic fibrosis epithelial cell line although the exact mechanisms have not been adequately determined.

20.14 ROLE OF PROMOTER

Tissue specific promoters, allow a second level of control over transgene expression. The most frequently used promoters are viral in origin, often derived from a different virus than the vector backbone, for example cytomegalovirus promoters have been used in all vector systems. Viral promoters have the advantages of being smaller, stronger and better understood than most human promoter sequences. One of the drawbacks of viral promoters is that a range of commonly used viral promoters were silenced by IFN-gamma and TNF-alpha at the level of mRNA stability. By contrast a constitutive human promoter, **Beta actin,** was less effected. A range of cellular promoters have been developed for specific tissues including the liver (albumin), muscle (myosin light chain and endothelial cells and smooth muscle.

Tumor specific promoters are also being used in cancer therapies, including *tyrosine kinase* for B16 melanoma, DF3/MUC1 for certain breast cancers and fetoprotein for hepatomas. The temporal expression of the transgene construct can be controlled by drug inducible promoters, for example, cAMP response element enhancers in a promoter; cAMP modulating drugs can be used. Alternatively repressor elements can prevent transcription in the presence of the drug. Spatial control of expression has been developed by using ionizing radiation (radiotherapy) in conjunction with the ergl gene promoter.

20.15 REGULATED GENE EXPRESSION

After successful gene delivery, it may be advantageous to regulate transgene expression. Indeed, for many diseases, over expression of the therapeutic gene may be as deleterious as compare to under expression. Thus, gene must have regulated expression. These systems are detailed below.

20.15.1 Rapamycin-Regulated Gene Expression

The ability to bring two proteins together by addition of a drug, potentiates the regulation of a number of biological processes, including transcription. For example, the binding of rapamycin to FK506-binding protein (FKBP) results in its dimerization, which can be reversed by removal of the drug. The dimerized chimeric transcription factor can then bind to a synthetic promoter sequence containing copies of the synthetic DNA-binding sequence. The limitation to this system, is the requirement for delivery of genes in same cells where the "transcription factor" resides. This limitation can be overcome by expressing both genes from a polycistronic message using an internal ribosome entry site.

Table 20.2 Success of different approaches

S.No.	*Vector*	*% of success*
1.	Adenoviruses	26
2.	Retroviruses	24
3.	Naked DNA	17
4.	Poxviruses	6.9
5.	Vaccinia Viruses	6.5
6.	Adeno-associated virus	3.4
7.	Herpes Simplex Virus	3.4
8.	RNA Transfer	1.3
9.	Others	2.4
10.	Unknown	3.1

20.15.2 Cytokine, Steroid, or Heavy Metal-Induced Gene Expression

Certain promoter contains binding sites specific for hormone receptors and thus, they are able to be induced by the addition of steroids. Alternatively, certain promoters such as the metallothionein promoter can be induced by the addition of heavy metals. Although steroid- or heavy metal-inducible promoters have been used extensively in tissue culture experiments, their applications to gene therapy may be limited because of the pleotropic effects of the inducers ***in vivo***. Thus, promoters that can respond to certain local stimuli may be more suited for gene therapeutic applications. For instance, promoters that are induced by specific cytokines, such as interferon-γ or interleukin (IL)-1, may be induced at sites of inflammation. For treatment of inflammatory diseases, promoters induced by pro-inflammatory stimuli may allow a certain level of *in vivo* regulation in response to the local environment.

20.16 APPLICATION AND ADVANTAGES

Treatment or prevention of disease by gene transfer, is regarded as a potential revolution in medicine. This is because gene therapies are aimed at treating or eliminating the causes of disease, whereas most current drugs treats the symptoms. This radical improvement is possible because the gene-based approach which provide superior targeting and prolonged duration of action. Hence, in comparison with other forms of therapy, it permits biological effects that are more subtle and better localized to the most appropriate cells. Ultimately, this will translate into substantial improvements in therapeutic ratio and cure-rate for diseases that are presently untreatable or poorly managed. Gene therapy also has the important advantage of having a broad platform technology, applicable to a wide range of diseases. The first clinical studies involving gene transfer began in 1990 and since then gene therapy has become the focus of a whole new industry.

20.16.1 Sickle Cell Anemia

Sickle cell disease is a genetic disease affecting 80,000 people in the US alone. The Red Blood Cells (RBCs) loses their normal disk shape and become sickle shaped. Due to this abnormality there is hindrance in the movement of these sickled RBC. The movement inside the small blood vessels is restricted and this results in piling, clumping inside the vessels. A team led by Allyson Cole-strauss and Kyonggeum Yoon of Thomas Jefferson University in Philadelphia experimented on cells containing a mutant gene that causes sickle cell anemia. To make their genetic drug they combined normal DNA with RNA of the same gene. When they injected it into the diseased cells, the RNA/DNA particles homed in on the particular stretch of the genome that matched their codes and formed triple stranded DNA, that covered the mutation. The cells DNA repair machinery then apparently replaced the mutation with the normal code thus permanently curing 10 to 20 percent of cells. The researchers still have to demonstrate that this technique works in human cells and in human bodies. Boosting blood cell production does little good for patients, whose blood cells are malformed, such as those of sickle cell anemic. (See Fig. 20.13).

On Dec. 4, 2008, Derek Persons, at St. Jude hospital developed a technique to insert the gene for gamma-globin into blood-forming cells using a harmless viral carrier. The researchers extracted the blood-forming cells, performed the viral gene insertion in a culture dish and then re-introduced the altered blood-forming cells into the body. The hope was that those cells would permanently generate red blood cells containing fetal hemoglobin, alleviating the disease. In the experiments, reported in online issue of the journal Molecular Therapy, the researchers used a strain of mouse with basically the same genetic defect and symptoms as humans with sickle cell disease. The scientists introduced the gene for gamma-globin into the mice's blood-forming cells and then introduced those altered cells into the mice. The investigators found that months after they introduced the altered blood-forming cells, the mice continued to produce gamma-globin in their red blood cells.

20.16.2 Cystic Fibrosis (CF)

CF is caused by a defect in the gene encoding CF transmembrane conductance regulator, which is a chloride channel protein important for normal lung function. In theory, the delivery of the normal CF gene to cells in the lung should result in correction of the defect and a prolonged life span for those having such disorder. Adenoviral vectors have been uemployed for the treatment of CF because of their natural infection behavior to the lungs, but replication-defective adenoviral vectors are apparently unable to infect the appropriate human airway target cells and are responsible for inducing an inflammatory response. Therfore, AAV vectors have been used for treatment of several patients with CF, on the basis of promising experiments. An alternative approach is to use liposome for lung gene transfer, but the efficiency of gene transfer has been low, with only marginal clinical efficacy. Thus, until the appropriate vector system is developed, CF may be difficult to treat by current gene transfer technology because of the refractory nature of the appropriate target cells to viral transduction and the requirement for long term gene expression.

20.16.3 Duchenne Muscular Dystrophy

Duchenne muscular dystrophy (DMD) is an X-linked dominant disease caused by a defect in the gene encoding dystrophin. The pathogenesis of the disease involves necrosis of both skeletal and cardiac muscle and eventually results in death, generally during the teenage years. Although both the defective gene in the disease and the appropriate target cells for therapy are known, successful application of gene therapy to the treatment of DMD is hindered by the great amount of muscle tissue to be targeted. Moreover, the basal lamina in muscle apparently prevents effective infection of the myoblasts by current viral vectors.

20.16.4 Arthritis

Although arthritis is not generally considered to be a genetic disease, it, too, is amenable to treatment by gene therapy. Rheumatoid arthritis is a systemic autoimmune disorder mani-fested by pathophysiological changes in the joints. It currently is treated by systemic ad-ministration of both steroidal and nonsteroidal drugs, which can reduce inflammation, but do not prevent the progression of the joint disease. There are also side effects associated with systemic administration of these antiinflammatory agents.

It has been demonstrated that the pathophysiological changes in the joints can be reversed by intra-articular delivery of specific genes encoding anti-inflammatory and immunosuppressive proteins. In particular, the intra-articular delivery and expression of IL-1 and tumor necrosis factor-a inhibitors resulted in the reversal of disease progression in a rabbit model of antigen-induced arthritis. Recently, the first gene therapy trial for rheumatoid arthritis has been initiated.

In the Phase 1 protocol, synovial cells were removed from a distal joint of the patient during regularly scheduled joint surgery. The cells were genetically modified by retrovirus in culture to express an IL-1 inhibitor and then reinjected into the patient's four knuckle joints scheduled to undergo joint replacement surgery 1 week later. In this initial study, the knuckle joints (that need to be replaced) served as a target for gene transfer. This strategy allows for access to the genetically modified tissue to determine the extent of synovial cell engraftment, transgene expression, and reversal of the arthritic condition.

20.16.5 Lysosomal Storage DiseasesÍ

Gaucher disease is a lysosomal storage disease resulting from mutations in the gene encoding glucocerebrosidase (GC). Although the gene is defective in all cells of people with the condition, it is the macrophages that are affected specifically by the GC defect.

Protein therapy has shown that introduction of modified, macrophage-targeted glucocerebrosidase (GC) (Ceredase®) into macrophages which results in complete amelioration of disease, suggesting that gene transfer can successfully treat the disease. Initial studies in mice have shown that gene transfer to hematopoietic stem cells, followed by reconstitution into lethally Viral Vectors irradiated recipients, results in expression of human GC in all hematopoietic lineages. These promising preclinical studies led to the initiation of three clinical protocols to treat Gaucher disease by retrovirally mediated gene transfer into CD341 cells isolated from peripheral blood. It is thought that a subset of CD341 cells contains hematopoietic stem cells that could partially repopulate the hematopoietic cells.

20.16.6 Curing Hemophilia B In Mice

Researchers at the Salk Institute reported (in the 15 March 1999 issue of the Proceedings of the National Academy of Sciences) work with mice in which genes for clotting factor IX had been "knocked out" and thus were subject to uncontrolled bleeding like human patients with hemophilia B. These mice were injected (also in the hepatic portal vein) with cDNA for factor IX in AAV, vector (the dog gene) with liver-specific promoter and enhancer sequences. The mice proceeded to make factor IX and were no longer susceptible to uncontrolled bleeding. More recently (2005), injection of embryonic stem cells with functioning factor IX genes into the liver of mice without the genes cured them. Injections of an AAV vector containing the cDNA of factor IX have temporarily produced functional levels of factor IX in patients with hemophilia B.

20.16.7 Treating ALS (amyotrophic lateral sclerosis)

ALS (amyotrophic lateral sclerosis) is a human disease in which motor neurons degenerate. (It is often called "Lou Gehrig's disease" after the baseball player who died from it. A similar disease can be created in transgenic mice carrying mutant human genes (for superoxide dismutase) associated with ALS. Researchers at the Salk Institute have slowed up the progression of the disease in these mice by injecting their skeletal muscles with an AAV vector containing the gene for insulin-like growth factor 1 (IGF-1). The vector invaded the muscle cells, moved into the motor neurons attached to them and, through their axons up to the cell bodies. They resulted in reduced destruction of motor neurons, and the mice lived longer than they otherwise would have.

20.16.8 Curing Insulin-Dependent Diabetes Mellitus (IDDM) in Mice and Rats

Researchers in Seoul, Korea reported in the 23 November 2000 issue of Nature that they have used an AAV-type vector to cure mice with inherited IDDM. Rats with IDDM induced by chemical destruction of their insulin-secreting beta cells. The complementary DNA (cDNA) encoding a synthetic version of insulin a promoter that is active only in liver cells and is turned on by the presence of glucose. The DNA encoding a signal sequence (so that the insulin can be secreted) an enhancer to elevate expression of this artificial gene. As a result both groups of animals gained control over their blood sugar level and kept this control for over 8 months. When given glucose, they proceeded to synthesize the synthetic insulin which then brought their blood glucose back down to normal levels.

20.17 GENE THERAPY FOR CANCER TREATMENT

The range of different cancers encountered and the mutations they carry, have led to a variety of strategies for gene therapy namely;

1. Immunopotentiation,
2. Oncogene inactivation,
3. Tumor suppressor gene replacement,
4. Molecular chemotherapy and drug resistance genes.

20.17.1 Immunopotentiation

The aim of immunopotentiation is to enhance the response of the immune system against cancerous cells, and their fast destruction. Immunopotantiation can be done by passive immunotherapy or active immunotherapy. Passive immunotherapy aims to increase the preexisting immune response to the cancer e.g. harvesting tumour infiltrating lymphocytes and treating them to express increased cytokines e.g. IL-2 and TNF-alpha, while active immunotherapy initiates an immune response against an unrecognized or poorly antigenic tumour. The cell population is then expanded *in vitro* and returned to the patient. Tumour cells are often used for active immunotherapy, by genetically modifying them, in order to increase expression of antigen presenting molecules/costimulatory molecules (e.g. B7), by increasing local concentrations of cytokines (e.g. IL-2) or tumour antigens (erbB2 oncoprotein). The cells are then irradiated prior to being returned to the patient, preventing the reintroduction of replication competent tumour cells. These approaches have been termed **cancer vaccines**. Oncogene inactivation uses the same techniques employed for dominantly inherited monogenic diseases.

The oncogene may be targeted at the level of the DNA, RNA transcription or protein product. Oligodeoxynucleotides are short single stranded pyrimidine rich DNA sequences that form a triple helix with purine rich double stranded DNA sequence. Oligodeoxynucleotides are designed in a sequence specific manner to target the promoter regions of oncogenes, (e.g. erb2;). At the RNA level, antisense technique prevents transport and translation of the oncogene mRNA by providing a complementary RNA molecule (e.g. c-myc).

20.17.2 Molecular Chemotherapy

Molecular chemotherapy involves killing of tumor cell by a gene coding for a toxic product. The gene of choice is usually herpes simplex virus thymidine kinase (HSV/TK) which converts the prodrug **ganciclovir** into toxic metabolites, **cytosine deaminase**, which converts 5-fluorocytosine to 5-fluorouracil; and **nitroimidazole reductase**, which converts CB1954 to a cytotoxic alkylating agent as shown in Fig. 20.14. An advantage to this system is that all the transduced cells will be killed, allowing allogenic tumour cells to be prepared in advance. HSV/TK can also be used in other gene therapy protocols, and also has benefit that the treatment to be aborted at any time. This virus-directed enzyme-prodrug therapy is called as the **bystander effect**. This involves diffusion of the activated cytotoxic drug from the cell expressing the enzyme to kill surrounding tumour cells. This might

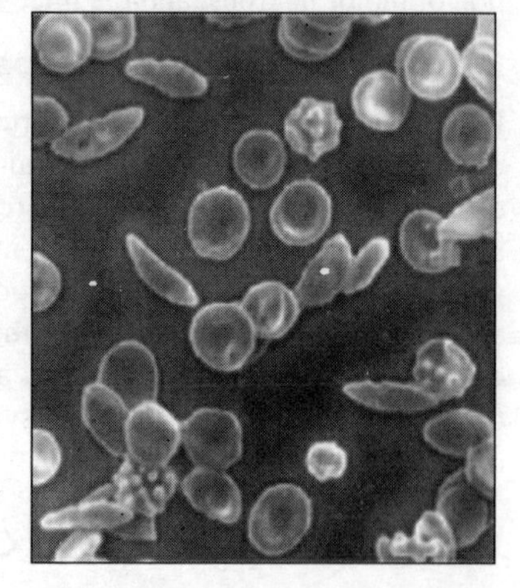

Fig. 20.14 Sickle cell blood cell

be mediated by diffusion through the lipid domain of the cell membrane or through gap junctions. If the activated drug has a short half-life, this further reduces the possibility of escape into the systemic circulation, which could cause side effects. Linkage of the gene to a tumour-specific promoter (such as the promoter for carcino embryonic antigen or -fetoprotein) ensures a higher level of enzyme transcription in tumour cells compared with normal cells.

20.17.3 By Replacing Tumor Suppresor Gene

In about half of lung cancer cases, a gene called p53 has mutated and thus falls to encode a protein that oversees programmed cell death. In the absence of this protein, which helps to curb the growth of damaged or abnormal cells, cancer can gain a foothold. Replacing such defective p53 genes with fresh ones has shown promise against a variety of cancers in animal experiments and studies of a few patients. The researchers injected the tumors with an adenovirus engineered to contain p53 genes. The virus was modified to prevent it from replicating and thus causing the upper respiratory infection that it might otherwise bring about. During the 6 months treatment period patients received one to six monthly injections of the modified virus. The researchers delivered a range of doses -from 1 million to 100 billion viral units to gauge any toxicity of the treatment. 3 of the 28 patients died of cancer before doctors could make a 1-month follow up examination.

20.17.4 For treating X-SCID

Gene therapy trials using retroviral vectors to treat X-linked severe combined immunodeficiency (X-SCID) represent the most successful application of gene therapy to date. More than twenty patients have been treated in France and Britain, with a high rate of immune system reconstitution observed. Similar trials were halted or restricted in the USA when leukemia was reported in patients treated in the French X-SCID gene therapy trial. To date, four children in the French trial and one in the British trial have developed leukemia as a result of insertional mutagenesis by the retroviral vector. All but one of these children responded well to conventional anti-leukemia treatment. Gene therapy trials to treat SCID due to deficiency of the Adenosine Deaminase (ADA) enzyme continue with relative success in the USA, Britain, Italy and Japan. system after infusion into a non ablated, affected individual. Initial results in at least one of the clinical trials have suggested that long-term elevated GC expression levels can be achieved following multiple infusions of retrovirally infected CD341 cells. If this approach proves successful, then a variety of hematopoietic diseases can be treated by gene transfer into CD341 cells, followed by transplantation.

Table 20.3 Method of gene therapy and their effect

Method	*Anti cancer effect*
Compensation of mutation	augmentation of different tumor suppressor gene, inhibition of expression of dominant oncogene
Immunopotantiation	Passive immunotherapy, augmentation by genetic enhancement of cell targeting or cell killing capacity of tumor-tumor infiltrating lymphocyte active immunotherapy, augmentation of immune recognition of cancer cells
Molecular immunotherapy	Selective delivery of toxin or toxin gene to cancer cells

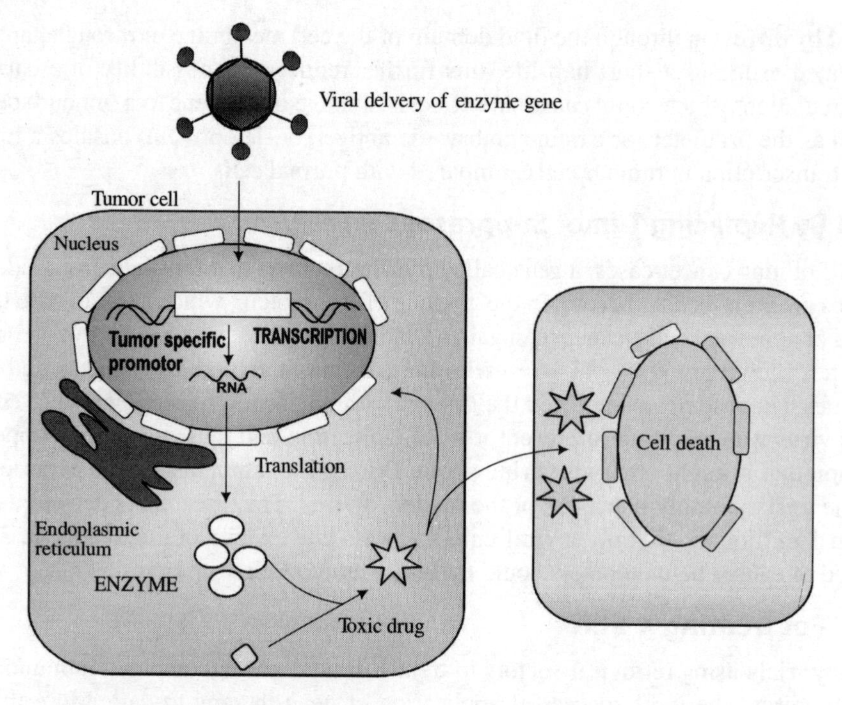

Fig. 20.15 Treatment of cancer by molecular gene therapy in which a gene is delivered by viral vector inside the nucleus which expresses to form enzyme in the cytoplasm by process of translation which convert a prodrug into toxic product

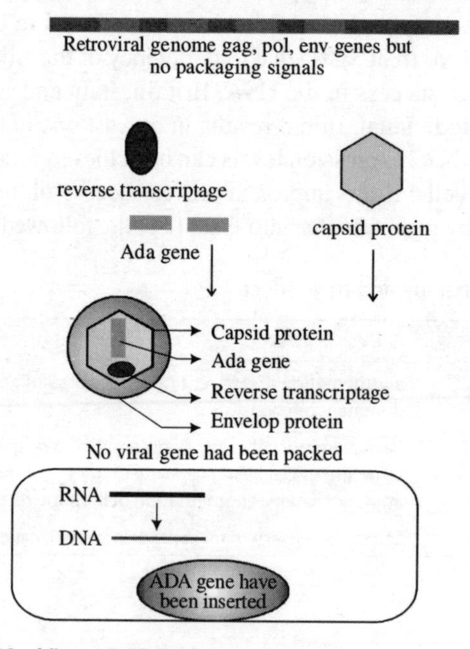

Fig. 20.16 Virus mediated gene delivery for SCID treatment

Table 20.4 Different vectors and their advantages or disadvantage

Vector	*Advantages*	*Disadvantages*
Retrovirus	• High efficiency transduction of appropriately targeted cells. • Chromosomal integration ensures long-term gene expression.	• Requires dividing cells. • Limited size of DNA insert (8 kb). • Potential for insertional mutagenesis.
Adenovirus	• High transduction efficiency. • Broad range of target cells. • Does not require cell division. • Low risk of insertional mutagenesis. • Large insert size (28-32 kb). • Replicates well (10^{10}/cell; titers 10^{13}/ml).	• Transient gene expression. • Immunogenicity (natural pathogen). • Cytopathic effects. • Clear potential for serious problems.
AAV (adeno-associated virus)	• Does not require cell division. • No known disease association. • "Site-specific" integration (70%). • Vector genome has no coding sequences.	• Potential for insertional mutagenesis. • Limited size of DNA insert (5 kb) • Integrates with in a gene.
Nonviral vectors	• No risk of infection. • No limitation on insert size. • Completely synthetic.	• Low efficiency of transduction. • Limited target cell range. • Transient expression.

BRAIN QUEST

1. Describe the strategies followed in gene therapy.
2. Describe the mechanism of Gene therapy.
3. Write short note on diseases that has bean treated by gene therapy.
4. First disease that was treated by gene therapy was __________
5. Describe the methods used to be followed in gene therapy.
6. Why viral vectors are most fascinating vehicle for gene therapy?
7. Define term Immunopotantiation.

TRANSGENIC ANIMALS

21

A transgenic animal is a genetically altered animal for specific characteristics therefore they are also called as genetically modified animal. The improvement in the process technology resulted in more success in production of transgenic animals for many purposes and industrial applications.

21.1 INTRODUCTION

The earliest transgenic animal was prepared by just transferring the desired DNA by injection into a fertilized mouse egg, but the major disadvantages was, it was not possible to control the site of integration of the foreign DNA and also their expressin, therfore they were called as "overexpressors". However, in new molecular genetic techniques that allow a particular gene (in some cases in a specific organ or tissue) to be switched on or off for a short period of time in the adult mouse. This added feature of reversibility provided greater control over the timing and duration of a resulting phenotype, and more closely mimics the effect of pharmacological intervention. It also serves to minimize the physiologic consequences of specific gene modifications in the animal.

21.2 HISTORICAL DEVELOPMENTS

The first gene transfers into mice were successfully executed in 1980. This led to the development of several mutant mice strains including "MutaMouse," the first transgenic animal engineered to detect gene mutations *in vivo.* MutaMouse was created by inserting DNA from the bacteriophage lambda (lgt10) into the genome of the CD2 hybrid mouse. The first transgenic mice for therputic purpose was produced for tPA (tissue plasminogen activator to treat blood clots), in **1987**. In **1997**, the first transgenic cow, **Rosie**, was produced for protein-enriched milk (2.4 grams per litre). This transgenic milk was more nutritionally balanced than natural bovine milk, and was better for babies or the elderly people with special nutritional or digestive needs since it was having the human gene alpha-lactalbumin. The A. I. Virtanen Institute in Finland produced a calf with a gene that makes the substance that promotes the growth of red cells in humans. In **2001,** two scientists at Nexia Biotechnologies in Canada spliced spider genes into the cells of lactating goats. The goats began to manufacture silk in their milk in the form of tiny silk strands. By extracting polymer strands from the milk and weaving them into thread, the scientists were able to create a light, tough, flexible material that could be used in microsutures, and tennis racket strings.

21.3 CLASSIFICATION OF TRAITS

Currently, various type of transgenic animals have been produced for the different purposes, therefore the range of genetic modifications can be divided in three broad areas of input, output, and value-added traits. Examples of each are described on next page.

Table 21.1 Landmarks in transgenic animal production technology in livestock (cattle, goats, sheep and swine)

Year	*Transgenic landmarks*	*Reference/Discoverer*
1985	Transgenic pig and sheep	Hammer et al. (1985)
1986	Embryonic cloning by nuclear transfer in sheep	Willadsen (1986)
1991	Transgenic dairy cattle	Krimpenfort et al. (1991)
1991	Transgenic sheep producing altered milk	Wright et al. (1991)
1992	Transgenic pigs resistant to viral infection	Muller et al. (1992)
1994	Pig expressing inhibitor of human complement system	Fodor et al. (1994)
1997	Somatic cloning by nuclear transfer in sheep (Dolly)	Wilmut et al. (1997)
	Transgenic livestock production as a model of human disease.	Petters et al. (1997)
1998	Transgenic cattle produced by nuclear transfer	Cibelli et al. (1998)
2000	Transgenic sheep produced gene targeting	McCreath et al. (2000)
2001	"Ecologically correct" transgenic pig	Golovan et al. (2001)
2002	Production of biopolymer fiber from transgenic cells	Lazaris et al. (2002)
	Calf with human artificial chromosome	Kuroiwa et al. (2002)
2003	Transgenic cattle producing altered milk proteins compounds	Brophy et al. (2003)
	Complete gene inactivation in pigs	Phelps et al. (2003)
2004	Sequential inactivation of 2 bovine genes	Kuroiwa et al. (2004)
2005	Transgenic cow resistant to bacterial infection (mastitis)	Wall et al. (2005)

21.3.1 Input Traits

An input trait helps livestock and dairy producers in increasing production efficiency and used for animals that have faster, more efficient growth rates, to increased the production of milk, wool, and resistance to diseases caused by viruses and bacteria.

21.3.2 Output Traits

An output trait helps in enhancing the quality of the animal product such as leaner and tendered beef and pork, milk that lacks allergenic proteins, or results in increased amounts of cheese and yogurt.

21.3.3 Value-Added Traits

Producing large amounts of therapeutic proteins in animal milk to treat human diseases or proteins that have industrial value comes under value added traits.

21.4 GENETIC MANUPULATION

21.4.1 Transgenesis Process

Transgenesis process starts with construction of gene containing desired sequence. A complete transgene is called as cassette which includes start (ATG) sequences, a promoter, structural gene, and a terminator gene with poly A tail in sequential order (See Fig. 21.1). The complete process can be summarised in following steps. (1) Isolation of the gene of interest (one strand of DNA or C-DNA). (2) DNA is cut at specific points by restriction enzymes. (3) Ligation of DNA piece with a vector, which may be a virus or part of a bacterial plasmid. (4) Transformation to incorporated DNA into host organism.

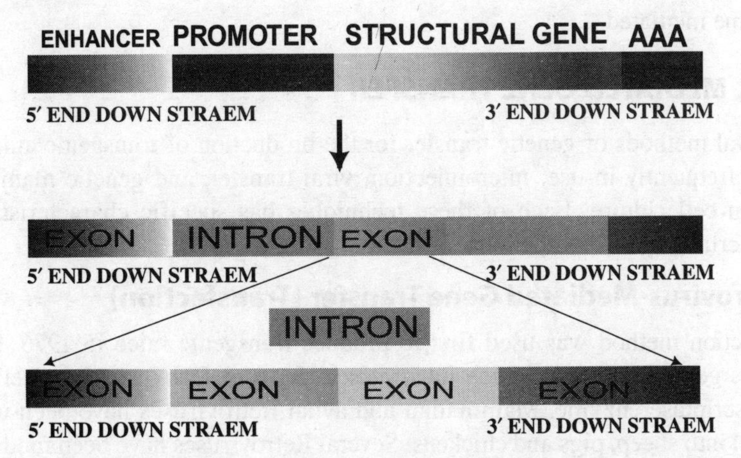

Fig. 21.1 (a) Schematic diagram of DNA regions containing enhancer, promoter, specific segment of DNA with poly A tail at 3 end and ATG sequence at 5 end forming the complete transgene expression cassette. (b) C-DNA containing exons only as compared to genomic DNA

Since the size of eukaryotic genome are very large because of introns, therefore whole genome cant be used for transgene preparation. For transgene preparation only Exons are essential which encode for the protein of interest while introns (intervening sequence) are said to be junk DNA. Therfore cDNA can be utilized to contruct transgene which contains only exons and have smaller size.

21.4.2 Delivery Vehicle

There are a number of different methods for inserting the new genetic material; outlined below.

21.4.2.1 Biological Methods

i. Viral Transfer (Retroviral infection)
ii. Sperm-medicated transfer
iii. Artificial chromosome
iv. Nuclear transplantation

21.4.2.2 Gene Knock Out/In

i. Homologous recombination
ii. Cre lox mediated
iii. Trasposon mediated (SB/Tcl)
iv. Si RNA/antisense

21.4.2.3 Physical Methods

i. Microinjection (of cells oocytes with DNA)
ii. Electrofusion

iii. Particle bombardment (gene gun)
iv. Calcium phosphate mediated
v. Liposome mediated

21.5 VIRAL MEDIATED GENE TRANSFER

There are several methods of genetic transfer for the production of transgenic animals, but three techniques are frequently in use: microinjection, viral transfer, and genetic manipulation of the embryonic stem cell culture. Each of these techniques has specific characteristics which lend themselves to certain research functions.

21.5.1 Retrovirus-Mediated Gene Transfer (Transfection)

Retroviral infection method was used first to produce transgenic mice in 1976. Retrovirus carries RNA as its genetic material which integrates in form of DNA into host cell with the help of reverse transcriptase enzyme. Mammalian and avian Retroviruses have been used to transfer genetic material into sheep, pigs and chickens. Several Retroviruses have been modified to contain a transgene which is then integrated into the target animal's genome as part of the normal viral life cycle. These include *Moloney Leukaemia Virus* (MLV) which causes lymphoid leukaemia in mice, rats, and hamsters; *Rous Sarcoma Virus*, which is associated with cancer formation in humans; and *Avian Leukosis Virus*. Viruses have been exploited which can repeatedly infect different cells (called 'replication competent') or defective viruses (that are able to infect the cells once). When replication competent viruses are used, the transgene are inserted many times in the founder (F1 generation), in the transgenic offspring. Recently lentivirus constructs have been made and used to infect embryonic tissue resulting in the generation of transgenic rats and mice (Rubinson et al. 2003). These vectors have a significant range of cell types that they can infect making them applicable to the generation of transformed somatic or stem cells for use in nuclear transfer. Thus there now exists retroviral vectors that can make the generation of transgenic animals much more efficient. Viral transfer may be done at any stage of embryonic development, however they have limited capacity to carry transgenes (8-10 kilobases) and the offspring produced were chimeric in nature and many inbreeding is required (10 to 20 generations) until homozygous transgenic animals are obtained with the transgene present in every cell.

Limitations

There are serious safety concerns about viral vectors, especially **replication competent viruses**. There is the danger that the retrovirus could recombine with wild viruses and form entirely new pathogens (disease-causing agents). There are also practical drawbacks to viral transfer of genetic material. There is a size limit of DNA which can be inserted, making it unsuitable for larger gene constructs. Viral vectors cannot replicate in early embryo cells so all the GM animals produced are chimeras, where the transgene only appears in some of their cells. The percentage of offspring which inherit the transgene is therefore even lower than normal.

21.5.2 Sperm Mediated Gene Transfer (SMGT)

Lavitrano et al. 1989 uses sperm cells as vectors for introducing foreign DNA into embryos to produce germ line transgenic mice. Incubation of sperm with DNA-containing solution resulted in DNA uptake by the sperm and the subsequent *in vitro* fertilization and embryo transfer resulted in successful production of transgenic mice at high efficiency. Indeed, numerous studies have confirmed that, in several spieces, the sperm can bind transgenes and even carry the genes to the embryos. For this either testes can be transformed or sperm can be transformed by using microinjection technique or by using retroviruses. The results of these approaches have been encouraging and transgenic mice and fish have been produced.

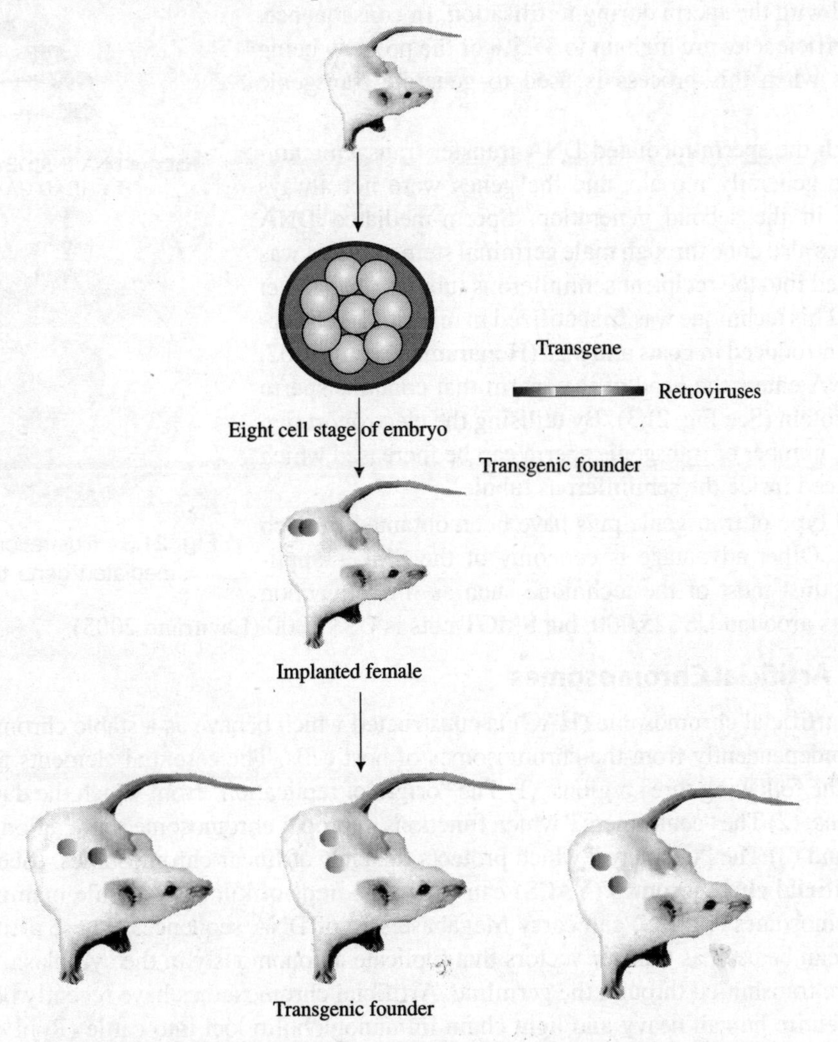

Fig. 21.2 Transgenic mouse by retrovirus gene manipulation method

More recently, **Brinster et al. 1996** examined the potential of spermatogonia cells to carry out the target DNA. Studies shows that spermatogonia could be cultured after transfection, and were able to fertilize the oocytes to produced offspring. Therefore, it could be used as a powerful means to produce transgenics animal with transgenic spermatognia to get desired transgenic offspring. To increase the efficiency of sperm uptake of DNA various approaches are examined. One is to attach the recombinant DNA to the sperm head via an antibody fused to the DNA (Chang et al. 2002). The antibody recognises surface proteins common to sperm from cattle, pigs, sheep, chicken, goats, mice and humans. This approach ensures that the DNA remains associated with the sperm during fertilisation. In consequence, reported efficiencies are high up to 37.5% of the progeny being transgenic when this process is used to generate transgenic pigs.

Through the sperm-mediated DNA transfer transgenic animals were generally mosaic, and the genes were not always expressed in the second generation. Sperm-mediated DNA transfer was also done through male germinal stem cell, that was transplanted into the recipient seminiferous tubules (Nagano et al. 2001). This technique was first utilized in mice and then successfully introduced in goats and pigs (Honaramooz et al. 2002, 2003). DNA enters the head of the sperm that contains sperm binding protein (See Fig. 21.3). By utilising the electroportaion technique, number of transgenic sperm can be increased which can be placed inside the seminiferous tubule.

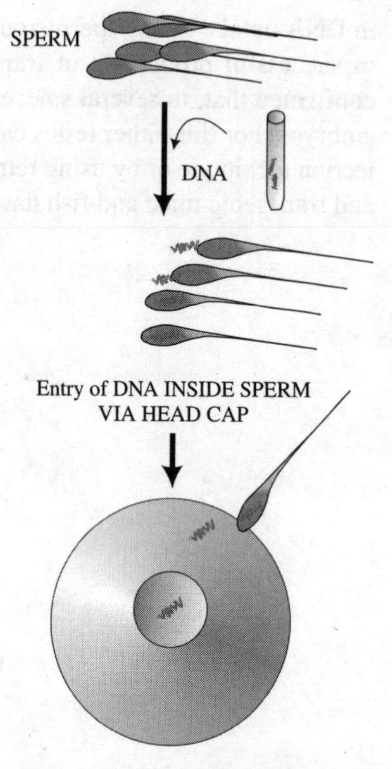

Fig. 21.3 Illustration of sperm mediated gene transfer

Several type of transgenic pigs have been obtained by such procedure. Other advantage is economy of the SMGT applications against most of the technique such as microinjection which costs aroound US$25,000, but SMGT cots is US$1,000 (Lavitrano 2003).

21.5.3 Artificial Chromosomes

A human artificial chromosome (HAC) is constructed which behave as a stable chromosome and replicate independently from the chromosomes of host cells. The essential elements for chromosome are the following three regions: (1) The "origin of replication" from which the duplication of DNA begins, (2) The "centromere," which functions in proper chromosome segregation during cell division, and (3) The "telomere," which protects the ends of linear chromosomes. (See Fig. 21.4). **Yeast artificial chromosomes** (YACS) can carry hundreds of kilobases, while mammalian artificial chromosomes (MACS) can carry Megabase size of DNA sequences. These artificial chromosomes can be used as transfer vectors that replicate autonomously in the cytoplasm of the host cell and are transmitted through the germline. Artificial chromosomes have recently been used to insert the entire human heavy and light chain immunoglobulin loci into cattle (Robl et al. 2003) and resulting in production of human polyclonal antibodies. The main advantage of this technology is the ability to transfer large segments of DNA allowing the generation of large constructs that ensure better control of transgene expression. The disadvantage with this technology is the difficulty in handling such large fragments of DNA and the fact that the chromosomes exist separately to the normal chromosomal complement of the cell.

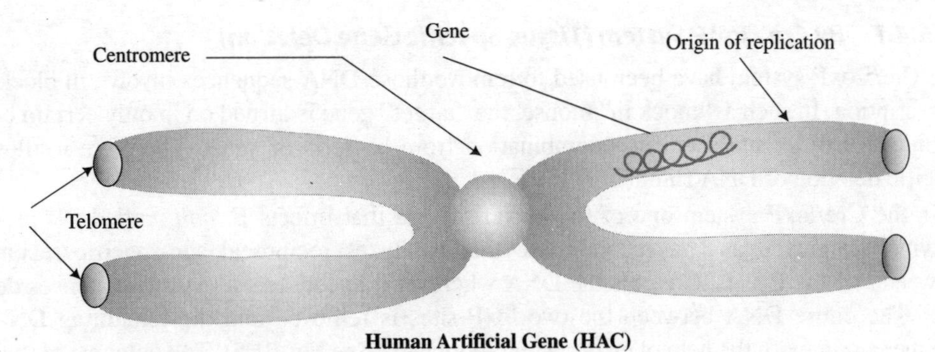

Fig. 21.4 Human artificial chromosome to carry megabase of transgenes

21.5.4 Transposon Mediated Gene Transfer

Transposons are segments of DNA that integrates at many different sites of a chromosome. These DNA sequences code for a transposase enzyme that enables integration of the DNA into the host's chromosome. In vertebrates **TCL**-like transposable elements are found in genomes, though they appear to be inactive (Lohe et al. 1995). Transposons therefore, have been used to generate mutant and transgenic mammals.

Ivics et al. 1997 reconstructed a transposable element from fish based on ***TCL-like transposon***. This transposon has been termed ***sleeping beauty*** (SB) or ***Tol2*** and has been used to integrate DNA into human and mouse embryonic stem cells (Cooper 1998). Sleeping beauty was the first transposon vector developed with the ability to deliver DNA to any vertebrate cell, but it had its own limitations. SB is most effective in mobilizing DNA stretches of up to 2 kb, and does not lose efficacy with loads of 10 kb. The development of SB is particularly significant. This was accomplished by the ten step including repair of the open reading frame, nuclear localization signal, DNA binding domain, and catalytic activity. Because this process required ten major steps, therefore, SB transposase was designated as SB10.

More recently Harris and colleagues had refined this transposon into a single plasmid and they called it ***Prince Charming*** (Harris et al. 2002). The advantage of this transposon is its ease of use in generating site-specific integration of transgenic DNA. This vector system had been used in cell culture showing stability in long-term culture of cells and can be easily applied in the generation of transgenic somatic or stem cell lines.

Regulated and inducible transposase alleles can be utilized to direct transposon-induced mutagenesis at specific sites in the mouse genome to make various model for study of many forms of cancer. The transposon vector itself could be modified, in addition to creating inducible and/or tissue-specific transposase alleles. Mutagenic transposons lacking an internal promoter could be used exclusively to disrupt gene of tumor suppressor genes. Transposons containing insulator/silencing elements could be utilized to make broad alterations to chromatin structure often seen in human tumors.

21.5.4.1 By Cre / lox P System (Tissue Specific Gene Deletion)

The Cre/LoxP system have been used to remove those DNA sequences involve in blocking gene transcription. In such a "knock in" mouse, the "target" gene is turned on in only certain cells. This technique utilizes site specific recombination from bacteria or yeast. Therefore it allows tissue specific deletion of DNA flanked by lox P.

In the Cre/loxP system one of the bacteriophage that infects *E. coli,* called P1, produces an enzyme designated as Cre recombinase that catalyzes reciprocal site-specific recombination between two lox P sites. Cre cuts the DNA wherever it encounters a pair of sequences designated loxP. The entire DNA between the two loxP sites is removed and the remaining DNA ligated together again (with the help of enzyme recombinase). (See Fig. 21.5). The outcome of the Cre-Lox P system depends upon the orientation of the of the Lox -P system. If the repeat is in opposite direction then the exchange inverts the DNA between the two Iox P system or sites; if the repeat in the same (direction) orientation then the intervening sequences is deleted. The repeats are naturally

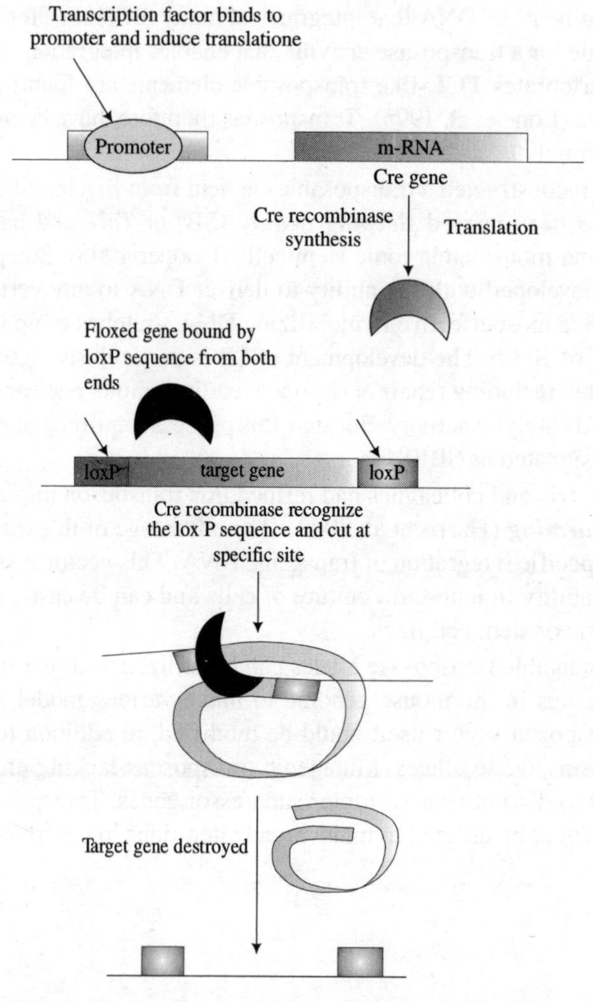

Fig. 21.5 Cre-lox transpogenesis for gene targetting

present in the opposite direction. The ***Cre-Iox P recombination*** system functions when the Iox P site are widely separated for example the two Iox-P site that are essential for circularization of a P1 genome are about 100 kb apart. Specificity of the system is absolute because the Cre recombinase acts exclusively on lox P site.A segment of DNA flanked by Iox P site is called as ***floxed***.Since the DNA between the two lox P site is deleted after the cre recombinase is expressed in double transgenic organism therefore, loss of activity of a gene in a specific tissue can be monitored. The expression of cre is controlled by tissue specific promoter. Cre recombination signal sequence called lox P, can delete genes which are flanked by two loxP sites in the same direction.

The Cre lox P recombination system can also be used to activate a gene in a specific tissue. In this instance, the floxed construct is prepared similar to sequence that prevents transcription. The blocking sequence is present between the promoter and coding sequence of a particular gene. When the cre transgene is expressed in mice with this floxed construct, the DNA sequence that blocks transcription is excised after homologous recombination, thereby enabling the gene to be expressed. Such type of system have been developed (that mimics) for **retina pigmentosa.** For detail patho-physiological study of effects, in human *DiGeorge* syndrome (in which cardiovascular dysfunction occur due to deletion of 22 chromosome).

21.5.4.2 Disadvantages and Advantages

One disadvantage of transgenesis by transposition, could be that multiple copies of a transgene cannot be integrated into one position, and it may be difficult in some cases to achieve very high expression of transgenes. Thus, very low expression is a frequent problem since transgene are often subject to partial methylation and inactivation, as well as other mechanisms of silencing. Transposon mediated gene transfer has advantages in following other applications.

(1) **Introduction of transgenes directly into somatic cells of animals**. Transposons offer the advantage of overcoming the most important barrier to gene transfer, that is, the stable integration of foreign DNA into the host cell chromosome.

(2) **In mouse cancer research**, This allow many genes or gene versions to be tested for oncogenic ability, since it faithfully mimick human tumor development.

21.5.5 Ribonucleic Acid Interference

In 2006 Nobel Prize in Medicine was awarded to **Craig C. Mello and Andrew Z. Fire** for their discovery of RNA interference (gene silencing by formation of double-stranded RNA). RNA interference (RNAi) is a sequence-specific gene silencing process that occurs at the post-transcriptional level. The RNAi machinery is thought to represent an ancient, highly conserved mechanism that defends the genome against viruses and transposons. New evidence shows that this transcribed RNA binds to other RNA molecules like mRNA (to inhibit its translation), or even to DNA (to control gene transcription) or proteins (to alter gene transcription as well). These process are called RNA interference (RNAi). After the transcription RNA forms stem loop like structure (as shown in Fig. 21.6) due to H-bond and such RNA is called as shRNA. ***Dicer*** (yeast RNA polyemrease exonuclease III) recognizes the double-stranded regions of the stem loop and cleave the dsRNA to form a small dsRNA between 20 and 25 nucleotide pairs long strand called as **siRNAs**. The cleaved RNA might then bind a host mRNA, inhibiting its translation. If the complementarity of the mRNA and miRNA is less than ideal, then binding of miRNA to the mRNA may only attenuate the translation of the mRNA. (see Figs. 21.6 and 7)

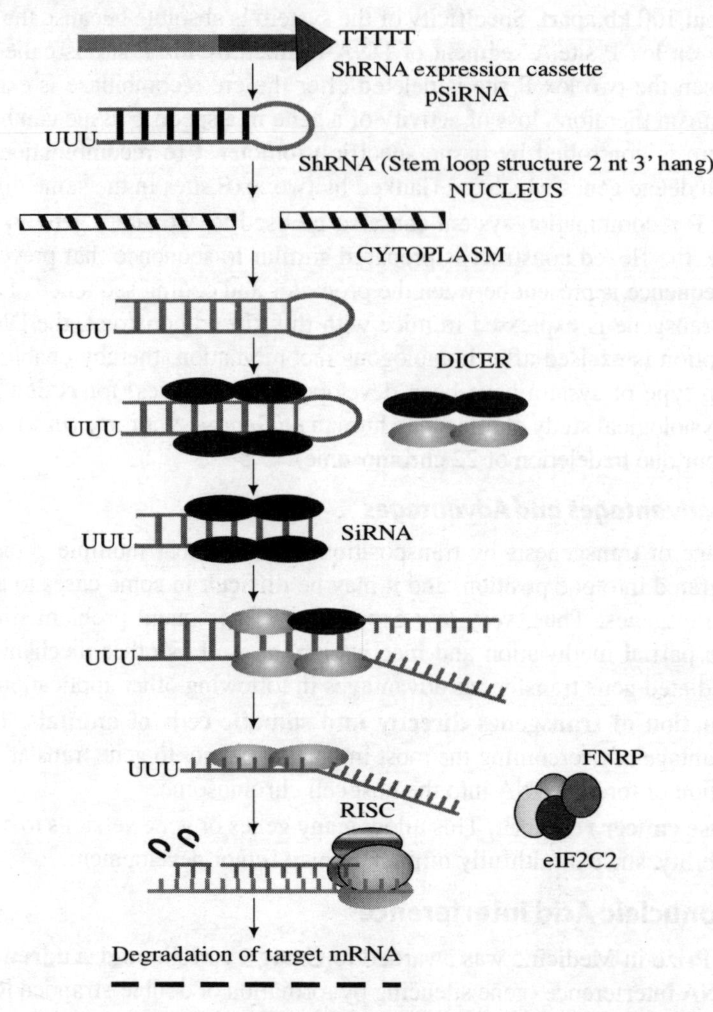

Fig. 21.6 Short hairpin RNAs (shRNA) expressed from RNA polymerase III promoter are processed by dicer (yeast RNAse III family) into small interfering RNA (SiRNA), 19-23 nt RNA duplex with 2-3 symmetric 2-3 nt 3' overhang and 5 phosphate groups. SiRNA are unwounded and incorporated in an RNA induced silencing complex (RISC) which contains Dicer and cellular proteins related to the EIF2C2 translation initiation factors. The antisense strand guides RISC to the target m-RNA for endonucleolytic cleavage

siRNAs consist of 21 to 23 nucleotides, which specifically bind to complementary sequences in target mRNAs and shut down the gene expression. The target mRNAs are degraded by exonucleases and no protein is translated. Target destruction occurs by a complex structure called as **RISC** which consists of RNA and several other proteins. RISC complex cleaves the RISC-siRNA- mRNA complex (i.e. one of the components in the RISC complex is an RNA nuclease.) This mechanism might be linked to defense mechanisms of virally-infected cells by inhibition of viral mRNA translation. Actually RISC cleaves the target mRNA in the middle of the complementary region, ten nucleotides upstream of the nucleotide paired with the 5' end of the guide siRNA. The cleavage reaction does not require ATP. However, multiple rounds of mRNA cleavage requires presence of ATP for release of cleaved mRNA. RISC/miRNP complexes catalyse hydrolysis of the target-RNAphosphodiester linkage, yielding 5' phosphate and 3' hydroxyl termini. (Ref. Review Gunter Meister and Thomas Tuschl, Nature, 2004)

21.5.5.1 Applications

RNAi effect can be achieved locally by delivering siRNAs, shRNA-expressing plasmid DNAs, or viral particles directly into the target tissue, as has been demonstrated for the retina, the brain, and muscle. *Ex vivo* infection of bone marrow cells followed by injection of siRNAs into lethally irradiated mice has also proven effective to induce stable gene knockdown in hematopoietic cells. The RNA interference mechanism can be used to generate **transient or permanent knock downs** for specific genes. The RNA interference mechanism can be used to generate **transient or permanent knock downs** for specific genes. For transient gene knockdown, synthetic siRNAs can be transfected into cells or embryonic stages. Further, gene deletion may be productive for manipulating expression in **metabolic pathways** controlled by negative regulator genes. (e.g. myostatin in muscle), and has been used to **delete the Prion gene** in an experimental sheep. One approach that has been proved to be useful is the use of chimeraplasts to generate single base pair alterations in targeted genes. This may be useful in introducing a **single nucleotide polymorphism (SNP)** such as that associated with the Callipyge locus. For transient gene knockdown, ***synthetic siRNAs*** can be transfected into cells or at embryonic stages. For stable gene repression, the siRNA sequences must be incorporated into a gene construct. The conjunction of siRNA and lentiviral vector technology have been used for gene transfer efficiency with highly **specific gene knockdown for livestock**. RNAi has also been used to **study loss of function** for a variety of genes in several organisms including various plants, *Caenorhabditis elegans* and *Drosophila*, and permits loss-of-function genetic screens and rapid tests for genetic interactions in mammalian cells (Hannon, 2002; Williams et al. 2003). RNAi has been used to **identify gene products essential for cell growth** to cause subtype and species-specific knockdown of various protein kinase C (PKC) isoforms in both human and rat cells (Irie et al. 2002), and to specifically target degradation of an oncogene product (Wilda et al. 2002). RNAi has also been used to specifically target and prevent viral infections by HIV-1 and HCV in cell culture (Park et al. 2002) and intact animals (McCaffrey et al. 2002). These observations open the field for further studies toward novel gene therapy approaches for anti-cancer or anti-viral treatments using siRNAs or shRNAs.

21.5.6 Homologous Recombination (Gene Targeting)

Homologous recombination is used generally to delete a gene. This method may have a limited use in applications for transgenic livestock as usually researchers seek to add genes in not remove them.

A knockout mouse can be prepared by using homologous recombination that replaces one allele in two to more generations of selective breeding. Knock out technique is also referred to as gene targeting. Genes are knocked out or disrupted during homologous recombination mechanism, when transgenes are inserted into specific genetic locations often using somatic or germ cells (serve as recipients). This technique allows scientists to target specific genes for inactivation and mutagenesis and is particularly useful in promoting an understanding of the molecular mechanisms associated with toxicity.

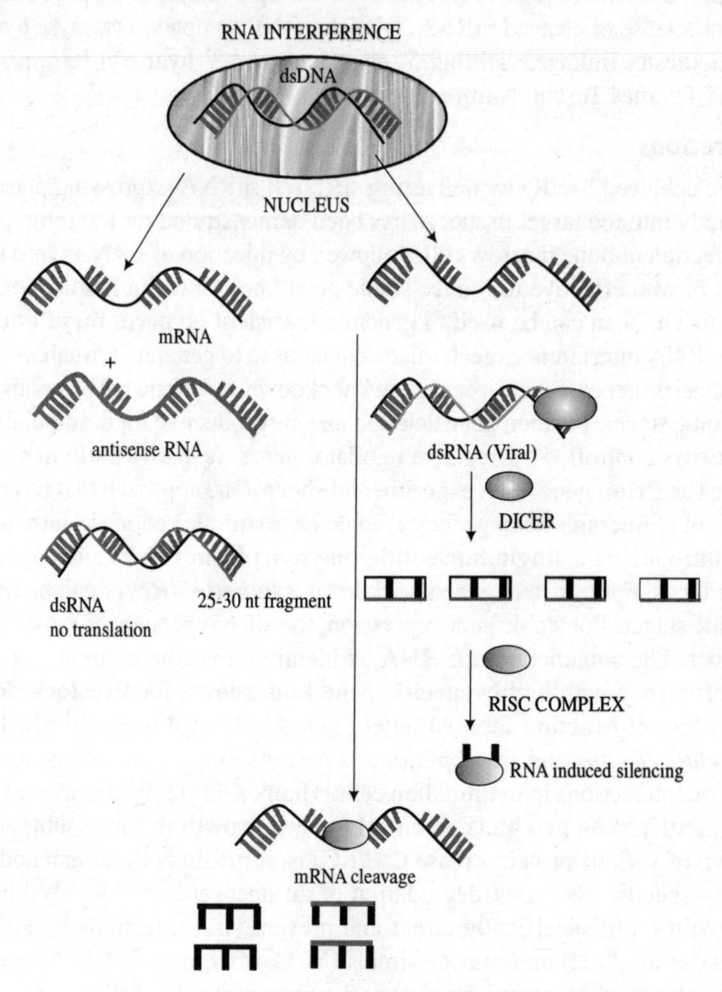

Fig. 21.7 Application of Si-RNA in target destruction

21.5.6.1 *Knock In*

Knock-in mice is used to study the exogenous expression of a protein. A knock in mouse is generated by targeted insertion of the transgene at a selected locus (Refer to Fig. 21.5). The insert is flanked by DNA from a non-critical locus, and homologous recombination allows the transgene to be targeted to the specific, non-critical integration site. In this way, a researcher has complete control of the genetic environment (surrounding the overexpression cassette and it is likely that the DNA did not incorporate itself into multiple locations). Site-specific knock-in result in a more consistent level of expression of the transgene from generation to generation because it is known that the overexpression cassette is present as a single copy. Also, because a targeted transgene is not interfering with a critical locus, the researcher can be more certain that any resulting phenotype is due to the exogenous expression of the protein. Although the generation of a knock in mouse does avoid many of the problems of a traditional transgenic mouse, this procedure requires more time to assemble the vector and to identify ES cells that have undergone homologous recombination.

21.5.6.2 Selection of Gene Targeted Cells

The two most common selectable genes used in gene targeting are the neomycin resistance (neo^r gene, which allows cells to survive in the presence of the antibiotic neomycin, G 418, and the thymidine kinase (TK) gene from the herpes virus. Cells with this gene die in the presence of the antiviral agent Gancyclovir. The neo^r and TK^+ genes are generally used together for maximum selection.

Positive-Negative Selection

After transfection of ES cells in culture with vector containing desired DNA, some cells will have: 1-DNA integrated at non target sites; 2-DNA integrated at correct site or 3-No integration at all. To enrich and identify for DNA integration at the correct site, a procedure called positive negative selection is used. This strategy uses positive selection for cells that have DNA integrate any where in their genomes, and negative selection against any DNA sequence that is integrated at non-target sites.Positive-negative selection is done for cells containing transgene as: Two blocks of DNA sequence that are homologous to separate regions of the target site (genel and gene 2) for example A DNA sequence coding for neomycin resistance (neo^r) and other genes for thymidine kinase (tk) seperated from HSV gene. The arrangement of these sequence is key to success of positive-negative selection procedure. (1) TK gene and neo^r must be between gene 1 and gene 2. (2) HSV -tk gene must be outside gene 1 and gene 2. If integration occurs at non-target site (i.e. not at gene 1 and gene 2), either one or both HSV-tk genes has a high probability of being integrated along with other sequences. (See Fig. 21.8). If integration event is due to gene recombination by a double cross over at target site, HSV-tk genes will be excluded and TK gene and neo^r gene will be incorporated into genome. When such cells grow in the presence of neomycin, all the cells lacking neo^r gene will be killed (Positives election). When cells grown in presence of Gancyclovir (GCV) as well, the cells that express tk will be killed because tk converts GCV into a toxic compound that kills cells (Negative selection). The cells that survive this duel selection scheme will have DNA integrated at the target site.

21.6 ANIMAL CLONING

Animal Cloning is the process by which an entire organism is reproduced from a single cell taken from the parent organism. This means that cloned animal is an exact duplicate in every way of its parent. This is the process where the nuclear material from an oocyte or a female sex cell is removed and a somatic cell's or body cell's nucleus is inserted in to the egg cell. The newly formed zygote has the potential to divide into a blastocyst, and, if implanted, the zygote will form containing genetic copy of the organism from somatic cell's nucleus.

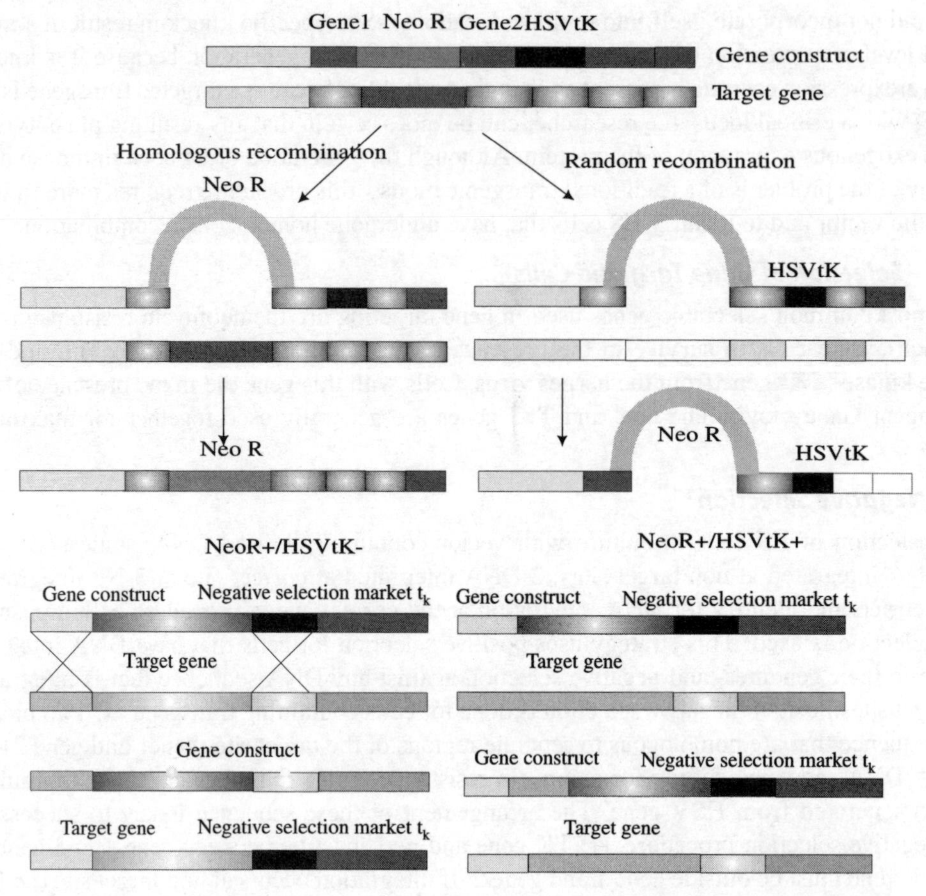

Fig. 21.8 Selection based on homologous and non homologous recombination

21.6.1 History

Back in the 20s, the German experimental embryologist **Hans Spemann** showed that after five divisions in the fertilized egg, nuclei retained the potential to program the complete development of an adult. Using strands of baby hair, fertilized amphibian (newt) eggs were tied so that they were constricted into two halves with the nucleus confined to one half and a narrow bridge of cytoplasm connecting the two halves. Spemann found that only the half portion of the tied egg containing the zygote nucleus would divide by mitosis and if a nucleus would cross into the other half and it, too, would begin dividing. Therfore, as long as both the portion of the egg have some of the cytoplasmic region called as the gray crescent, the second half also develops into perfectly formed embryo. Even at the 32-cell stage of development, the nuclei had not lost the potential to program the complete development of the organism. Spemann's result suggested, that cloning is possible and genetically-identical nuclei are able to produce genetically identical individuals. Further the development of micromanipulators has made it possible to remove and add the nuclei from the cells. (see Fig. 21.9)

Scientists have attempted to clone animals. The first fairly successful result in animal cloning was obtained in tadpoles from frog embryonic cells. This was done by the process of nuclear transfer. The tadpoles so created did not survive to grow into mature frogs, but it was a major breakthrough. After this, using the process of nuclear transfer on embryonic cells, scientists managed to produce clones of mammals. Again the cloned animals did not live very long. The first successful instance of animal cloning was that of Dolly Sheep, who not only lived but went on to reproduce herself naturally. Dolly was created by **Ian Wilmut** and his team at the Roslyn Institute in Edinburgh, Scotland, in 1997. Unlike previous instances, she was not created out of a developing embryonic cell, but from a developed mammary gland cell taken from a full-grown sheep. Since then scientists have been successful in producing a variety of other animals like rats, cats, horses, bullocks, pigs, deer, etc.

21.6.2 Cloning by Embryo Splitting

Animals can be cloned by embryo splitting or nuclear transfer from one cell to other cell. Embryo splitting involves bisecting the multicellular embryo at an early stage of development to generate "twins". This type of cloning occurs naturally and has also been performed in the laboratory with a number of animal species. (Figs. 21.9 and 21.10)

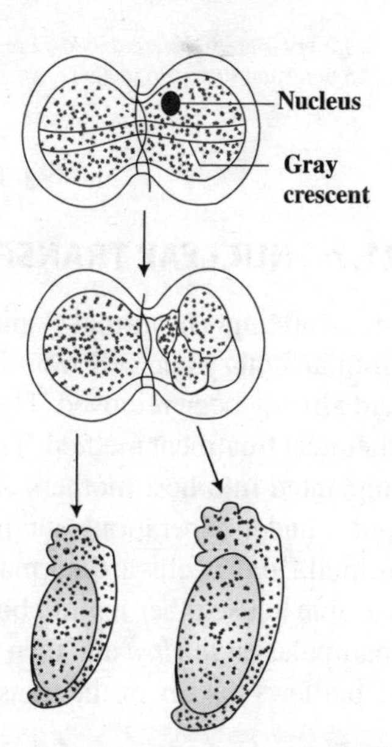

Fig. 21.9 Clone obtained by spemmens binding to cell and separating two portions of eggs

21.6.3 Cloning by Nuclear Transfer

Cloning can also be achieved by nuclear transfer where the genetic material from one cell is placed into a "recipient" (unfertilized egg) where recipient genetic material is removed by a process called enucleation. The first mammals were cloned via nuclear transfer during the early 1980s, almost 30 years after the initial successful experiments with frogs . Numerous mammalian species have been cloned from cells of preimplantation embryos: namely mice, rats, rabbits, pigs, goats, sheep, cattle and even two rhesus monkeys, NETI and DETTO. Tracy (1990-1997) was a transgenic ewe that had been genetically modified by the Roslin Institute, near Edinburgh, Scotland, so that her milk produced a human protein called alpha antitrypsin, a potential treatment for the disease cystic fibrosis. The Roslin Institute is one of the world's leading centres for animal research. Dolly was part of a project to reproduce reliably animals with the genes of Tracy. First animal patented was 'oncomouse', a transgenic mouse genetically engineered to develop cancer.

A new technique for nuclear transfer has been developed in the meantime. A fusion apparatus developed by Zimmermann allows short electric pulses (20-40 ms, 100 V) used for fusion of cellular membranes. Willadsen (1986) had investigated the ability for further development of blastomeres obtained from 8-cell to 16-cell embryos and more than 40 percent of the nuclei transferred into enucleated oocytes developed into blastocysts by the electrofusion technique. Several of these transferred blastocysts developed into healthy lambs, including a pair of monozygous twins. Some blastomeres derived from the 16-cell stage also possessed the capacity for full development.

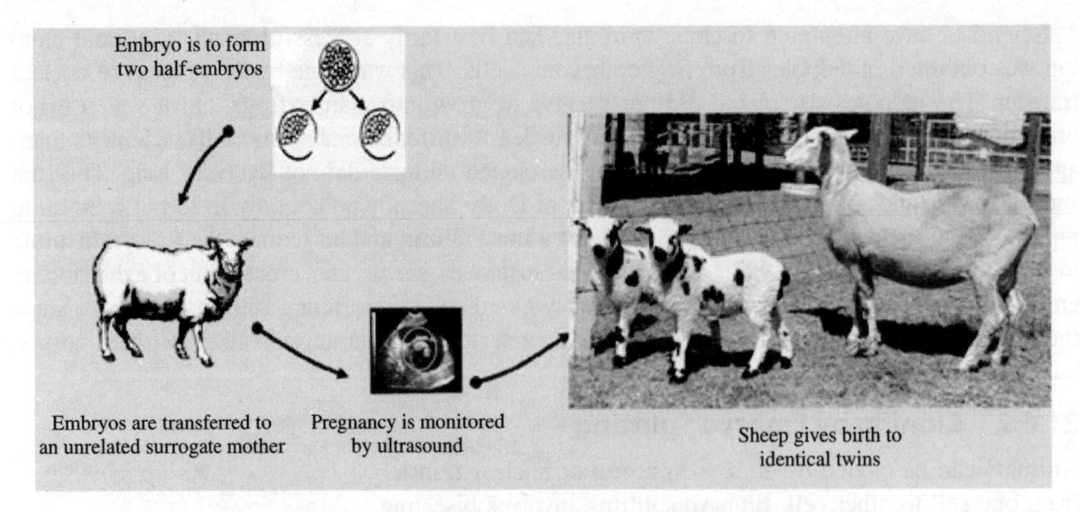
Embryo is to form
two half-embryos
Embryos are transferred to
an unrelated surrogate mother
Pregnancy is monitored
by ultrasound
Sheep gives birth to
identical twins

Fig. 21.10 Cloning by embryo splitting method

21.7 NUCLEAR TRANSFER TECHNOLOGY

Initial attempts in animal cloning were done using developing embryonic cells. The DNA extracted from an embryonic cell were implanted into an unfertilized egg, from which the existing nucleus had already been removed. The process of fertilization was simulated by electric shock or by some chemical treatment method. The cells that developed from this artificially induced union were then implanted into host mothers. The reconstructed eggs are then implanted into a surrogate mother and founder generations obtained are cross hybrid up to 10-20 generation to get final transgenic animal.Donor cells are normally obtained by culturing embryonic cells, but modern techniques are able to use other mature body (somatic) cells. These cells can be cultured, subjected to genetic manipulation *in vitro* and then produce viable animals by means of nuclear transfer. This technique also allows the sex of the transgenic animal to be predetermined.

21.7.1 Induction of Superovulation for Cloning

Large numbers of eggs can be produced by super ovulation by using hormone impregnated in sponge inserted in the vagina 14-20 days prior to insemination.

21.7.2 Insemination Co-culture and Implantation

It is the process by which sperm are placed into the reproductive tract of a female for the fertilization.Following insemination, the fertilised eggs are surgically removed and genetically modified. The cultured microinjected embryos are then surgically implanted into surrogate mothers which have been 'synchronised' with the donor ewes by a similar hormone injections.

21.7.3 New Techniques for Nuclear Transfer

A New micromanipulation techniques have been developed by **McGrath and Solter** (1983) to inject karyoblasts through the zona pellucida into the perivitteline space (see Fig. 21.11, Fig. 21.12).

Karyoblasts are nuclei with an intact nuclear membrane, which are also surrounded by a small fraction of cytoplasm. Karyoblasts obtained from fertilized oocytes can be fused with enucleated cells by means of inactivated Sendai viruses. This technical improvement prevents damage of the cellular membrane during the nuclear transfer. The survival rate of embryo was higher than 90% and more than 90 percent of the manipulated embryos have developed into morula and blastocyst stages *in vitro*. Upon transfer 16% of these embryos developed into the normal mice. The fusion of karyoblasts and enucleated oocytes by Sendai virus has gradually been replaced by Electrofusion which produces short electric pulses (20-40 ms, 100 V) used for fusion of cellular membranes.

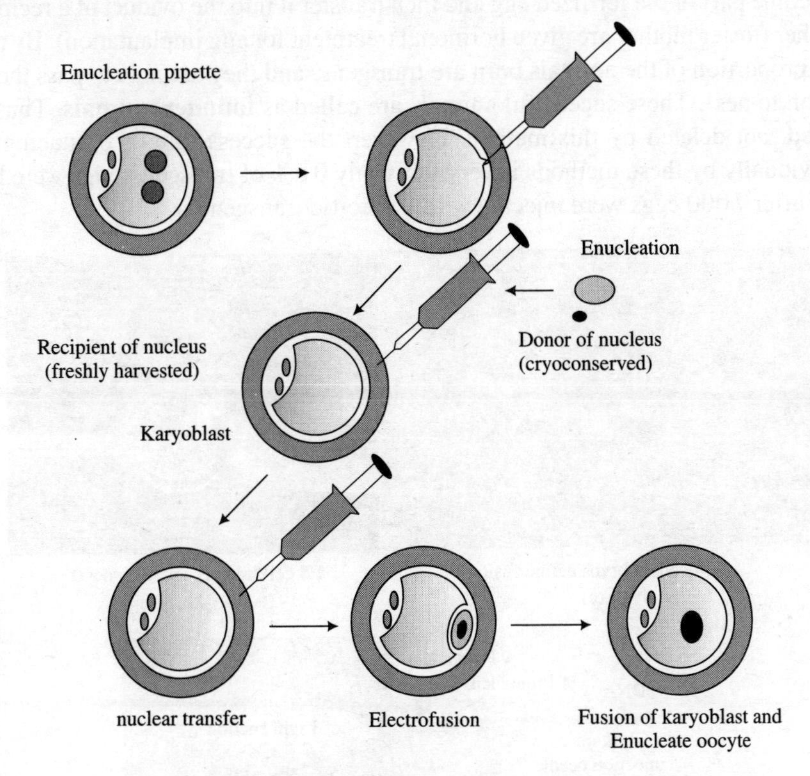

Fig. 21.11 Nuclear transfer method

21.7.4 Microinjection

The first successful production of transgenic mice using pronuclear microinjection was reported in 1980 (Gordon et al., 1980). This method involves the direct microinjection of a chosen gene construct (a single gene or a combination of genes) from another member of the same species or from a different species, into pronucleus of a fertilized ovum. It was one of the first methods proved to be very effective in mammals (Gordon and Ruddle, 1981). The introduced DNA may lead to the over- or under-expression of certain genes or to the expression of genes entirely new to the animal species. The insertion of DNA is, however, a random process, and there is a high probability that the introduced gene will not insert itself into a site on the host DNA that will permit its expression.

In this method a microtube (see Fig. 21.12) is used to hold the unfertilized egg and tiny amount of solution containing many copies of the transgene are injected with the help of microinjetion. The manipulated fertilized ovum is then transferred into the oviduct of a recipient female, (or foster mother) that has been induced to act as a recipient by mating with a vasectomized male. *A major advantage of this method is its applicability to a wide variety of species.* In this method eggs are first harvested from super-ovulated animals and fertilised. Note that fertlised egg should not be in process of zygote formation i.e. two nucleus must be separate in one egg cell. In such cell any one cell of embryo is targeted for DNA injection with pronucleus. Now fertlization takes place and transgene become part of the fertlized egg and then transfer it into the oviduct of a recipient female or foster mother (foster mother are given hormonal treatment for egg implantation). By this method only a small proportion of the animals born are transgenic and they are able to pass the gene from one generation to next. These successful animals are called as **founder animals**. Thus genes can only be added, not deleted by this method. However, the success rate of producing transgenic animals individually by these methods is very low (only 0.6% of transgenic pigs were born with a desired gene after 7,000 eggs were injected with a specific transgene).

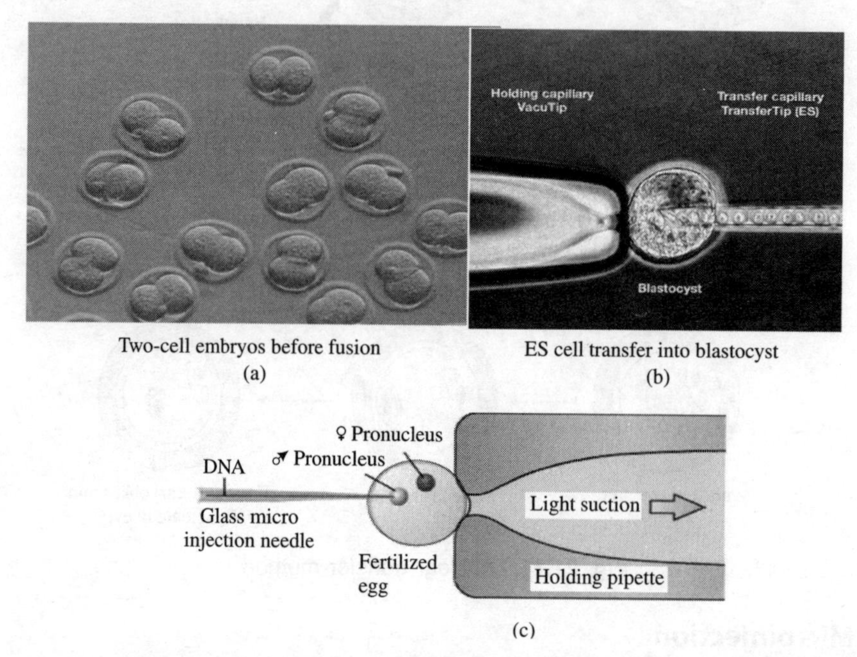

Fig. 21.12 Microinjection method for insertion of transgenic DNA into pronucleous of fertilised egg. The injection pipette is used to thrust through the zona pellucida, cell membrane, cytoplasm and nuclear membrane in a single smooth motion. Even if the membrane appears to have been pierced, the only reliable indication of success is the swelling of the pronucleus (volume = approximately 1 picoliter). The pipette is removed smoothly, and the sperm is injected into egg

Sequence of events in the generation of a transgenic animal by pronuclear microinjection involves (Fig. 21.12 and Fig. 21.15)

1. The double-stranded DNA components of the transgene are combined enzymatically to yield a transgene expression cassette.
2. Transgene cassettes are inserted into plasmid vectors and cloned.
3. Transgene-bearing plasmids are transfected into cultured eukaryotic cells to evaluate expression of the transgene.
4. Plasmid-free transgene fragments are introduced directly into embryonic pronuclei.
5. Manipulated embryos are placed in the reproductive tract of a pseudopregnant recipient.
6. The genomic DNA of live-born pups are analyzed for the presence of the transgene DNA sequence.

   Recently, the microinjection technique has also been applied to down regulate gene expression in mice and rats (for cardiovascular model system). Microinjection techniques are not suitable for poultry because it is extremely difficult to gain access to the fertilised egg when it is still at the single cell stage, which is necessary for any genetic modification to be integrated into all the cells of the developing embryo. Most success with GM poultry has been achieved via viral transfection. However, the Roslin institute has reported producing transgenic birds containing one of two marker genes by microinjection of zygotes (single cell embryos).

21.7.5 Limitations

Microinjection technique has serious limitations, such as:

1. Impossibility to produce knockouts by homologous recombination;
2. Inefficiency in generating embryos in which the injected DNA was stably integrated into the host genome (Nottle et al. 2001); and
3. Production of chimeric transgenic embryos resulting in a mosaic animal with some cells containing the transgene and others not (Keefer 2004);
4. Unpredictability of the site of transgene integration in the host genome and the resulting variation in transgene expression because of the position effect (Clark et al. 1994).

21.8 TRANSGENIC MICE

It is assumed that over 80% of mouse gene (function the same as those in) is similar to humans. Mice also have a short reproduction cycle and their embryos are amenable to manipulation. Mice is therefore, an ideal organism for human model in the study of diseases. Currently over 95% of transgenic animals used in biomedical research are mice. Other transgenic animals include rats, pigs and sheep.

21.8.1 Transgenic Mice From Embryonic Stem (ES) Cell Manipulation

Embryonic stem cell has been manipulated *in vitro* for transgenic production. The DNA construct containing desired gene, vector DNA, promoter and enhancer sequences are cocultured with ES cells for incorporation into the inner cell mass of the blastocysts. Embryonic transfer is then conducted into foster mother resulting in the production of a chimeric animal. (See Figs. 21.13 and 21.14). Inbreeding of such chimeric animal leads to developmentt of homozygous animals.

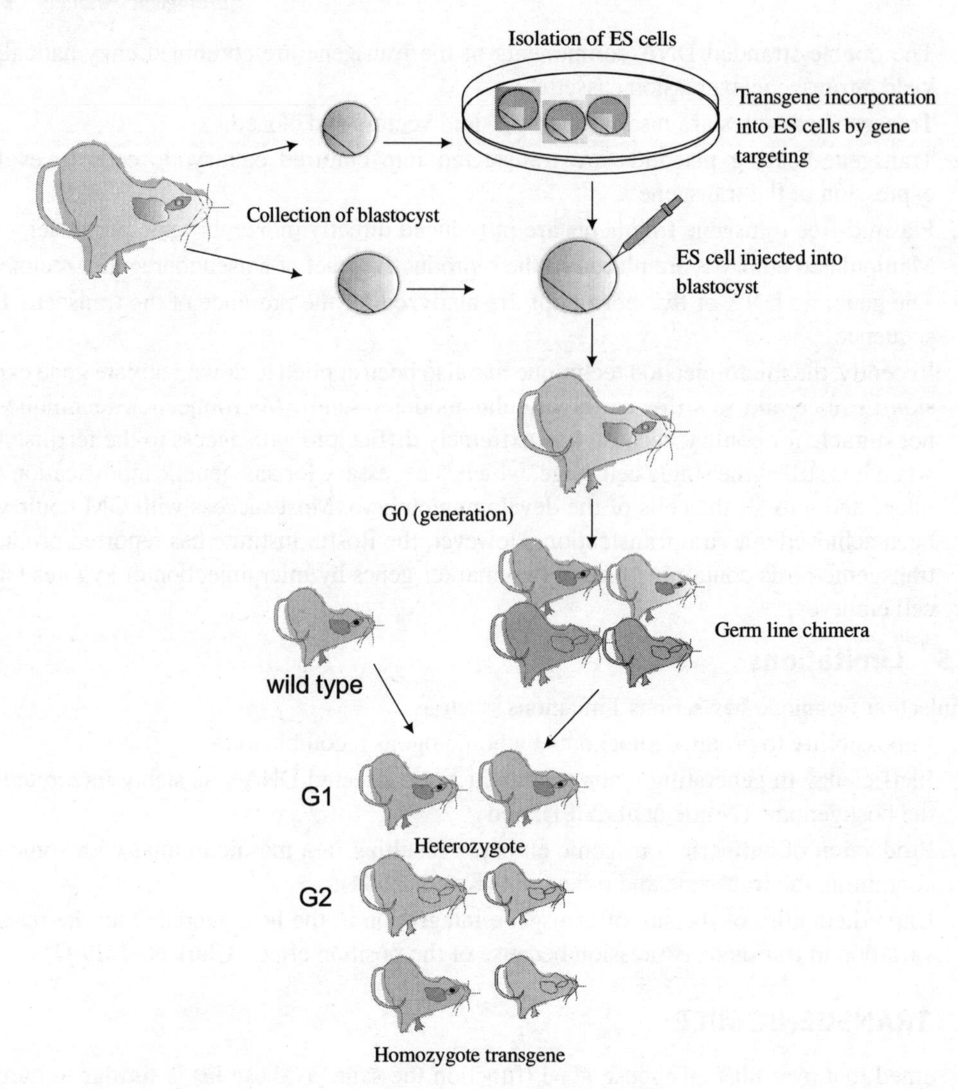

Fig. 21.13 Embryonic stem cell method to get transgenic mice

21.9 TRANSGENIC SHEEP

21.9.1 Creation of Dolly

Method of somatic cloning method has been utilized for creating transgenic sheep, Dolly (See Fig. 21.14). In this method somatic cells from the mammary gland of an adult sheep that have white faces are grown in culture. An egg isolated at 8 cell stage are made enucleated by sucking the nucleus from the egg. After five days, the nutrient level in the culture are reduced which resulted in cease of cell division and cells enter into G_0 phase of the cell cycle. Donor cells (udder cell) and recipient cells (enucleated cells) are co-cultured and are exposed to high pulses of electricity to cause their respective plasma membranes to fuse; and to stimulate the resulting cell to begin mitosis (by mimicking the stimulus of fertilization). The cells are then allowed to grow up to morula stage (solid mass of cells) or even into a blastocyst (6 days). Several of these cells are now implanted into the uterus of surrogate mother to develop female sheep.

21.9.2 Embryonic Stem (ES) Method of Creating Transgenic Animal

Stem cells are undifferentiated cells that have the potential to differentiate into any type of cell (somatic and germ cells) and therefore to give rise to a complete organism. ES cells can be obtained from embryos (blastocysts, see Figs. 21.14 and 21.15) which can be transformed with the desired DNA. Transgenic DNA get inserted by homologous recombination into the stem (ES) cells. These cells are then incorporated into an embryo at the blastocyst stage of development. The resultant is production of a chimeric offsprings. The offsprings have mixed coloring. Fertile chimeras that carries the construct in germ cells is then selectively bred to establish the DNA modification permanently in a new mouse line.

This technique is of particular importance for the study of the genetic control of developmental processes. This technique works particularly well in mice. It has the advantage of allowing precise. targeting of defined mutations in the gene via homologous recombination. Targeted expression is generally controlled by a special segment of DNA called as the promoter sequence. Molecular biologists in the last several years have identified a number of genetic elements in DNA that "insulate" genes from neighboring DNA. These elements, were identified as ***scaffolding elements,*** shown to have potential for improving the activity and regulation of transgenes. When added to transgenic mice, they cause a higher percentage of the founding transgenic mice to have active transgenes. Studies also suggests that it is possible for genes to be regulated more efficiently by their own regulatory DNA.

21.10 METHOD TO GET STABLE TRANSGENIC ANIMAL

All of the methods described above have been used to produce transgenic mice. Variations in these techniques are used for creating smaller organisms including parasites for bio control of possums and other pests while larger transgenic animals are more difficult to create and livestock have only been created using microinjection or nuclear transfer. For any of these techniques the success rate in terms of live birth of animals containing the transgene is extremely low. Providing that the genetic manipulation does not lead to abortion, the result is a first generation (F1) of animals that need to be tested for the expression of the transgene. Depending on the technique used, the F1 generation may result in chimeras. When the transgene has integrated into the germ cells, the so-called germ line chimeras are then inbred for 10 to 20 generations until homozygous transgenic animals are obtained and the transgene is present in every cell. At this stage embryos carrying the transgene can be frozen and stored for subsequent implantation. (See Fig. 21.12)

21.11 DETECTION OF TRANSGENIC ANIMALS

Proof of the stable integration and expression of a gene construct is obtained in three steps:

21.11.1 Southern Blot Analysis

Before the development of PCR, Southern blotting was the method of choice for transgene detection. It involves the isolation of DNA from selected animals, digestion of DNA with restriction enzymes,

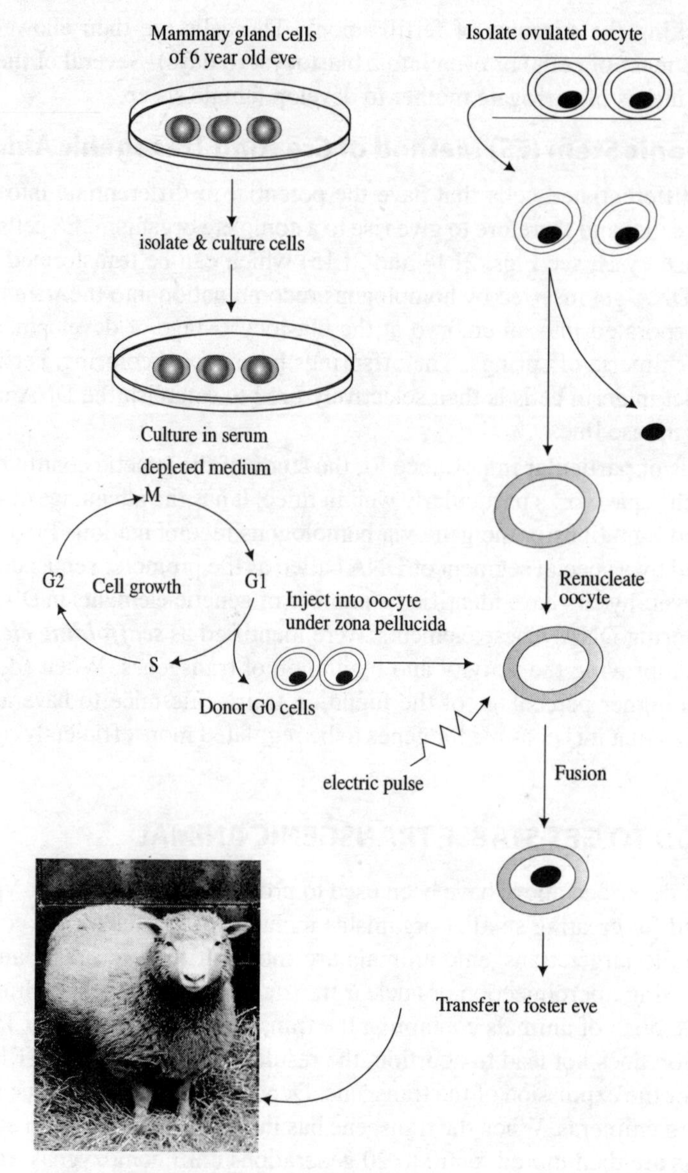

Fig. 21.14 Method of creation of Dolly

electrophoresis on agarose gels to separate DNA by size, denaturation of the DNA and transfer of the DNA to a membrane. This membrane (either nitrocellulose or more recently nylon) is then hybridised with a specific probe (complementary to the transgenic sequence of interest). The probe is usually radioactively, colorimetrically or chemiluminescently labelled such that once hybridised to the transgene DNA, the position of the transgene on the membrane can be visualised.

Southern blot analysis can be applied to: a. Demonstrating presence or absence of a transgene in the sample; b. Determining whether chromosomal integration has occurred; c. Identifying the position of integration relative to other samples; d. Determining whether homologous recombination (directed integration) occurred; e. Determining copy number of transgenes.

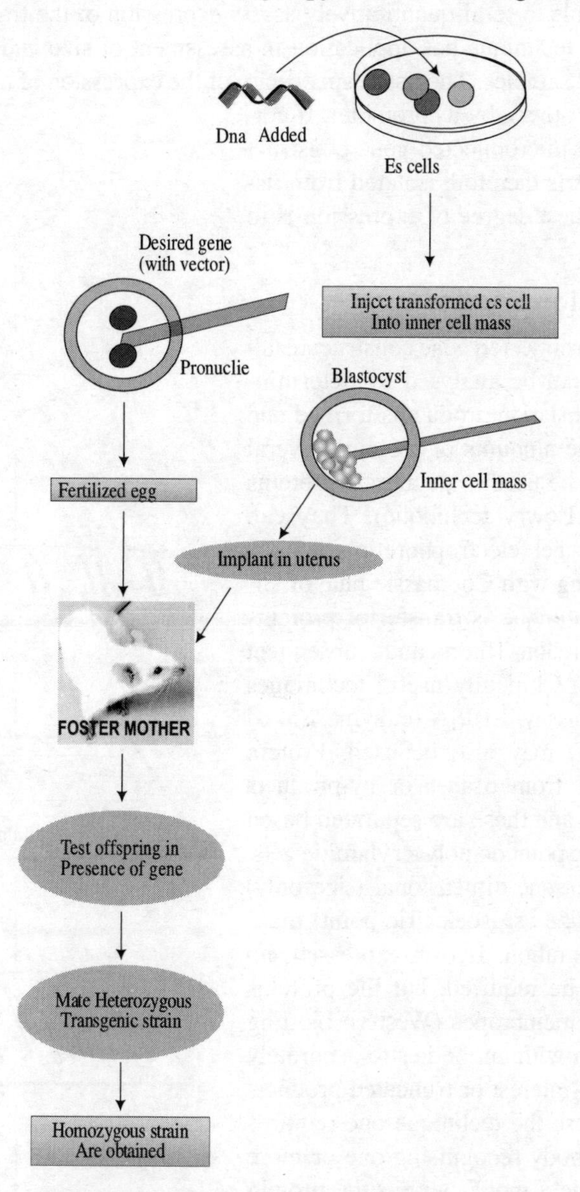

Fig. 21.15 Transgenic mice utilizing ES cells

21.11.2 Northern Blot Analysis

Similar to Southern blotting in many respects (and hence the name), Northern blotting involves the isolation of RNA from selected animals or tissues, electrophoresis to separate transcripts by size and transfer to nylon membranes. This membrane is then hybridised with a specific probe (complementary to the expected expressed transgene transcript). The probe is usually radioactively, colorimetrically or chemiluminescently labelled such that once hybridised to the transgene DNA, the size and quantity of transcript can be visualised. By accurately loading similar amount of RNA per sample, it is possible to semi-quantitatively assess expression of the transgene in different tissues or animals. This technique has application in assessment of size and expression patterns of transcripts in different samples. The tissue specificity of the expression of a gene or gene construct is determined, among others, by its promoter. If correct transcription of a microinjected gene construct is to be analysed RNA is therefore isolated from tissues in which the highest degree of expression is to be expected.

21.11.3 Translation Studies

Transcription of a microinjected gene construct result in translation which can be analysed by determining first whether a translation product is formed and if so by measuring the amounts of protein. Several colorimetric tests can be used to quantitate proteins (Bradford technique, Lowry technique). They can also be analysed by gel electrophoretic methods and subsequent staining with Coomassie blue or silver stain. Another technique is transfer of proteins to nitrocellulose or nylon filters and subsequent Western blot analysis. Clinically useful techniques such as radioimmunoassay, ELISA (enzyme linked immunosorbant assay) may also be used. Protein mixtures are isolated from tissues or by-products (e.g. milk) of interest, and these are separated based upon size or isoelectric point on polyacrylamide gels. This may be done in a one dimensional (size only) or two dimensional (size vs. isoelectric point) manner for increased separation. If over-expressed, no further analysis may be required, but the proteins can be transferred to membranes (Western blotting Fig. 21.16) and probed with antibodies to accurately identify the protein of interest or truncated products (non-functional). To use the technique one requires a highly specific antibody recognising one or more characteristic amino acid motifs within the protein of interest.*

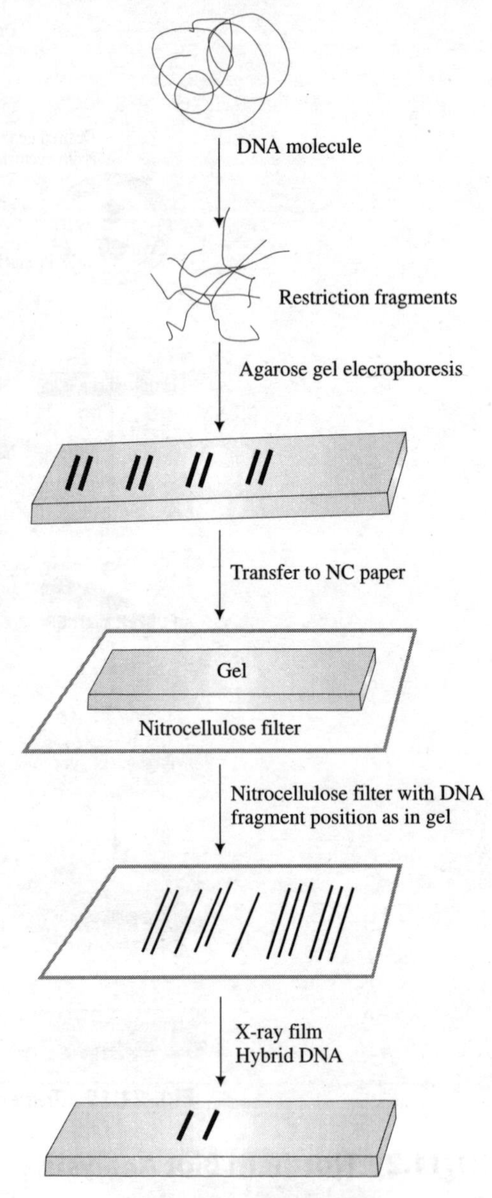

Fig. 21.16 Southern Blot Analysis (Southern, 1975)

This technique finds application in: Characterisation of protein expression in various animal tissues or tissue-derived products; Determination of molecular weight and isoelectric point of the

*Note for many proteins microarray or RT-PCR technique is also helpful and applied nowadays

product; Assessment of product solubility; Determination of quaternary structure (product complexation); Determination of the intracellular stability of the product (i.e. is the protein being rapidly degraded, as demonstrated by a range of smaller than expected products), Qualitative assessment of expression levels.

Table 21.1 Success rate of microinjection in different species. Figures in parentheses are percent efficiency compared to original number of ova injected. (After Hammer et al., 1985)

Animal species	*Number of ova injected*	*Number of offspring*	*Number of transgenic offspring*
Rabbit	1907	218 (11.4%)	28 (1.5%)
Sheep	1032	073 (7.1%)	01 (0.1%)
Pig	2035	192 (9.4%)	20 (1.0%)

21.11.4 Cloning and Verification of the Integrity of the Transgene

The transgene should contain unique markers so that its presence can be easily detected in DNA samples and so that its expression can be assayed and distinguished from endogenous gene expression. Sequencing of junction fragments should be carried out in order to confirm that the transgene has a functional promoter, initiation codon, and polyadenylation signal.

21.11.5 Establishment of a Screening Method

A PCR assays should be established to rapidly identify transgenic animals. A second assay that will detect an endogenous mouse gene, such as beta-globin, or an endogenous rat gene, such as prolactin, is required in order to demonstrate that the DNA preparations are amenable to PCR.

21.11.6 Establishment of an Expression Assay

It's important to show that transgene is expressed. RNA expression can be detected by in situ hybridization or RNAse protection assay with RNA probes. Alternatively, an RT-PCR approach can be used. RT-PCR allows the generation of stable complementary DNA (cDNA) from RNA. The technique consists of two parts, synthesis of cDNA from RNA by reverse transcription (RT) and amplification of a specific cDNA by polymerase chain reaction (PCR). When applied to multiple samples it can be used in a comparative semi-quantitative fashion, or quantitatively using competitive RT-PCR. Absolute quantification, using competitive RT-PCR, measures the amount (e.g., 5.3×10^5 copies) of a specific mRNA sequence in a sample. Dilutions of a synthetic RNA (identical in sequence, but slightly shorter than the endogenous target) are added to sample RNA replicates and are co-amplified with the endogenous target. The PCR product from the endogenous transcript is then compared to the concentration curve created by the synthetic "competitor RNA." This technique has application in semi-quantitative or quantitative assessment of transgene expression in different animals or tissues. RT-PCR will remain a useful technique, but its use will decrease as the newer Q-PCR technologies become more accessible.

21.11.7 Screening of Potential Founders

A genetic founder, denoted "F_0," is a first-generation transgenic animal that develops directly from a microinjected embryo and carries the transgene stably integrated in its genome. Transgene sequences can be analyzed by biochemical and molecular methods to confirm that they are complete and intact. If they are, then the animals are raised to maturity and bred to non-transgenic mates. Their offspring are tested to confirm that the transgene is inherited at the expected frequency.

21.11.7.1 Localising an Expressed Protein

This is the technique for localising an expressed protein within cells and tissues, based on the affinity of specific antibodies. It is used to confirm the tissue specificity of protein expression, or to exclude the inappropriate accumulation of transgene products within cells and tissues. A further value of this technique would be observation of any structural malformation that might result from over-expression of a transgene, though in this case the technique would utilise antibodies to proteins other than the transgenic protein.

21.11.7.2 Protein Functionality

It is axiomatic that many transgene products are functional proteins, and so confirmation of the function *in vitro* or *in vivo* is a key to understanding the impact to the animal of transgenesis. If the introduced transgenes encode an enzyme, it should be possible to demonstrate novel enzyme activity or in the case of "gene knockout" transgenesis, the loss of enzyme activity. Alternatively, the introduced transgene might encode a structural protein and hence one should be able to demonstrate altered cell or tissue morphology or physiological function. An interesting example is the properties of the macrophages in transgenic pigs expressing human decay accelerating factor (hDAF). These were found to be resistant to lysis in the presence of human serum, demonstrating that hDAF is functional and confers resistance to human antibodies and complement (Lavitrano et al. 2002)

Proof of functionality of transgene protein products demonstrates success of the experiment, and suggests that the transgene is stable. Continued assessment of function throughout the animal's life, and in future generations can also be seen as assessment of transgene stability over time. If the transgene is not stable, one would expect reduced or even eliminated function of the transgene product.

21.11.8 Breeding and Analysis of Transgenic Rodents

The final stage in the process is to study animals carrying the transgene.Typically, the transgenic founder animals are bred to mice C57BL/6 (of defined genetic background).

21.12 GENE EXPRESSION REGULATION

21.12.1 Promoters

Promoter is a type of consensus sequence (conserved sequence) which is recognized by RNA polymerase enzyme. The promoter decides when and where the expression of the gene should occur. If the gene has to be expressed in the mammary gland the casein promoter is often utilized, as the casein is one of the important milk proteins. There are two type of promoter strong and weak promoter. Strong promoter has a high rate of transcription and thus large amount of product is formed while weak promoter has, low rate of transcription thus results in low amount of product.

21.12.2 Regulatory Elements

Beside this, there are other regulatory elements such as **enhancer and silencer** that control the overall productivity, enhancer increases, the speed while silencer decreases the amount of product formed. Inducible genes do this job when it get appropriate signal by addition of some chemical present in the growth medium. In contrast repressible gene which is switched off by addition of regulatory chemical. [In gene regulation promoter is central to amount of product formed.] This property can be utilized for the controlled production of recombinant product. There are many promoter which is utilised in development of the expression vector. (According to need fusion protein.)

21.13 PROMOTERS

Metallothionein promoters were among the first eukaryotic promoters used in fish. Metallothioneins are proteins that bind heavy metals in cells, in particular cadmium, copper, zinc and mercury (Maclean and Penman 1990). They supply zinc to zinc-requiring enzymes within the cell, and in detoxification. Metallothioneins are inducible, and their synthesis is up-regulated in the presence of heavy metals. Thus, the promoter sequences of the metallothionein genes may be used to regulate transgene expression in experimental animals, in response to heavy metal administration. However, two other promoters were found promising: the β–actin promoter and the antifreeze protein (AFP) promoter. β-actin and its promoter are evolutionarily conserved sequences, and the promoter can be used to drive ubiquitous tissue expression.

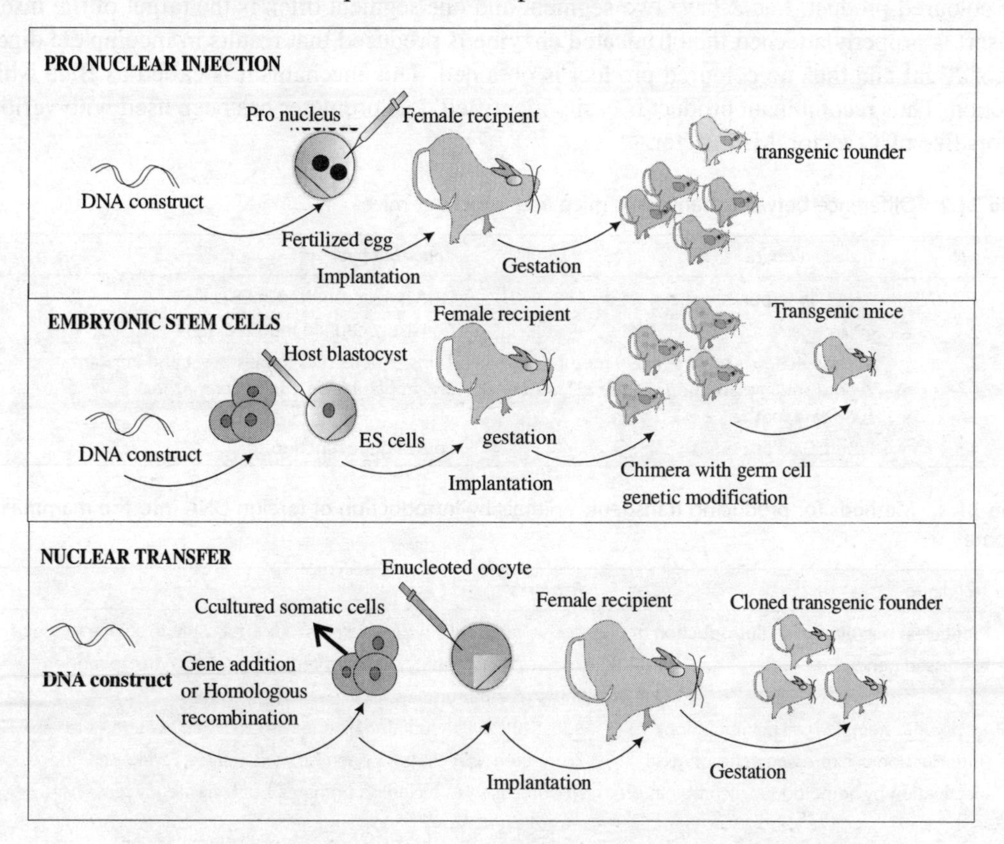

Fig. 21.17 Comparison of all methods of making transgenic animal

The *rat β–actin promoter* was the first to be used successfully in a construct, but later the carp and tilapia sequences were identified. The ocean pout Macrozoarces americanus AFP promoter has also shown promise in Canadian trials with pacific salmon. In general, it is likely that species-specific promoters will give the best expression. For instance trials with the *tilapia β–actin promoter* fused to the β–galactosidase coding sequence (as an experimental reporter of gene expression) have shown significantly higher expression in transgenic tilapia than the carp β–actin promoter (Hwang et al. 2003).

Promoter selection is important both in respect to amount of product and in proper separation of product from the other protein present in the cells cytoplasm. This separation is important because sometime a fusion product is obtained (two product simultaneously) and that should be separated from the fused protein to obtained their full functionality. For large size of protein eukaryotic origin gene are expressed in two separate expression system according to size and amount keeping in mind of their function and purity.

21.13.1 lac Promoter

It is the segment of DNA sequence that control transcription of lac Z gene. Lac Z product is an enzyme β-galactosidase that cleaves colourless substrate X-gal. X-gal is digested, or hy-drolyses to give coloured product. Lac Z have two segment and one segment often is the target of the insert. If insert is properly attached then truncated enzyme is produced that results in incomplete digestion of X-gal and thus no coloured product is obtained. This mechanism is called as Blue white selection. Thus recombinant product is easily identified. Lac promoter has been used with various vectors like pUC vector, M13 vector.

Table 21.2 Difference between transgenic mice and knockout mice

S.N.	*Transgenic mice*	*Knock out mice*
1.	DNA is incorporated in zygote	DNA is in corporated in ES cell
2.	DNA construct	Construct contain mutated gene
3.	Microinjection is main delivery method into zygote and implantation is done in foster or surrogate mother.	Transfer of ES cells to blastocyst and imp lantation and implantation into foster mother
4.	Result: Gain of a gene	Loss of gene function.

Table 21.3 Methods for producing transgenic animals by introduction of foreign DNA into the mammalian genome

Technique	*Remarks*
Pronuclear microinjection (introduction of expressed gene)	Technical simplicity; low success rate; applicable to a wide range of species; most widely used; unpredictable effects due to random transgene integration.
Embryonic stem (ES) cell manipulation (introduction of expressed gene, or gene inactivation by homologous recombination)	substitution of a functional gene with an inactive gene; germ-line competent ES cells have been isolated in mice; ES-like cells identified in other species, including primates, but totipotency remains to be established

Table Contd...

Table Contd...

Cre-lox technique	preferred method with more control over resulting phenotype; time-consuming
Viral vectors	complex; largely restricted to avian species
Cytoplasmic injection	less efficient than direct pronuclear microinjection
Primordial germ cells	chimaeric animals result
Nuclear transfer	large potential for genetically modifying livestock
Spermatogonial manipulation	transplantation

Table 21.4 Consequences of techniques used in transgenesis

Technique	*Remarks*
Gene addition	achieved by all methods
Gene knockout	targeted inactivation of host gene by embryonic stem cell manipulation
Random insertion of mutations	achieved by all methods
Inhibition of gene expression	for example, prevention of translation by hybridisation of antisense RNA with mRNA

21.13.2 Tmp Promoter

This promoter is induced by 3-β indol acrylic acid and occur upstream of coding gene repressed by tryptophan

21.13.3 Tac Promoter

This is the hybrid promoter formed by the trp and lac promoter and is induced by the IPTG.

21.13.4 Pi Promoter

It is the strong promoter recognized by *E. coli* RNA polymerase. Promoter is repressed by CI genes that code a repressor protein and is expressed at low temperature. This fact can be utilized for preparation of shuttle vector. These expression vectors have hybrid promoter and thus they can be controlled in different condition. For example PI at low temperature expresses while *E. coli* expresses at high temperature.

21.13.5 Gal-Promoter

These promoters are induced by galactosidase substrate. Cloned gene is placed under con-trol of gal promoter in yeast. This method provides straight forward system for regulatory expression of cloned foreign gene. Yeast can give higher yield of recombinant product but one problem they have they are unable to glycosylated the product.

21.13.6 Aox Promoter

Aox denotes for alcohol oxidase promoter isolated from *S. cervisiae* yeast expression vector system (*Pichia pastoris*) and is induced by methanol. They are advantageous in the sense that they can effectively glycosylated the protein.

21.13.7 Hsp 70 Promoter

They are called as heat shock protein and is utilized for expression of eukaryotic gene. They are induced by the temperature. Usually they are active at higher temperature.

21.13.8 Metallothionine Promoter

Metallothionine promoter is mouse promoter and is switched on by addition of zinc salt to culture medium and is utilized for expression of various mammalian genes during transgenic expression.

21.13.9 Heterologous Inducible Promoter

Heterologous inducible promoters are advantageous because foreign gene is induced and no other endogenous protein or mammalian gene can be expressed. Certain gene is maintained at very high copy number and utilizes selection system like DHFR dihydrofolate gene in methotrexate.

21.14 SPECIAL VECTORS FOR EXPRESSION

Vector is an important part of gene expression. It has promoter terminator and RBS site (ribosome binding site). These are the flanking nucleotide segment that is recognized by many transcription factor, for e.g. promoter is recognized by sigma subunit of RNA polymerase that starts the reaction. Termination mark the end of gene and is recognized by rho protein. RBS is short sequence of nucleotide in m-RNA and is recognized by ribosome for attachment. This also gives signal for attachment of initiator t-RNA because initiation codon is present few nucleotides downstream of this site.

The capability of a gene delivery vector to transport large transgene sequences (e.g. >10 kb) is essential for the delivery of genes together with their associated regulatory elements. However, the ability of different viral vectors to carry such large nucleic acids varies sig-nificantly. For example, AAV vectors can only accept inserts of up to 4.7 kb. Most retroviral, adenoviral and lentiviral vectors can accept inserts of up to ~8 kb. Vaccinia and retroviral, adenoviral and lentiviral vectors can accept inserts of up to ~8 kb. Vaccinia and gutless AdV can accept inserts of up to 25 kb, and vaccinia vectors can accept inserts of up to ~45 kb. In contrast, the capacity of non-viral vectors is generally much larger than that of viral vectors, with the ability to transfect DNA sequences of >100 kb. On the other hand, non-viral transfection reagents also have DNA size limitations mainly related to difficulties in forming and transfecting DNA/lipid complexes containing large plasmids. Nevertheless, the overall greater insert capacity afforded by non-viral vectors gives them the advantage in this category. Non-viral gene delivery methods have advanced significantly in providing more efficient transfection in cell lines cultured in vitro, they still lack adequate efficiency in many primary cells in vitro and in vivo. Of course, different transfection efficiencies are required for different applications. Transfection efficiencies of non-viral vectors vary significantly for *in vitro* transfections, ranging anywhere from <1% to >98%. In contrast, viral vectors are almost always able to achieve greater gene transfer rates in vivo than non-viral vectors, and they typically transduce >80-90% of cells in vitro. Thus, in terms of overall transfection efficiency, current viral vectors are preferred over current nonviral vectors in most cases.

The gene of higher organism have same system (promoter, structure, regulator) but all they have different nucleotide sequence. Therefore, eukaryotic protein cannot recognize prokaryotic promoter and prokaryotic protein cannot recognize eukaryotic promoter. Therefore, foreign DNA is inserted close to ***E. coli* expression system**. Therefore, foreign DNA will be replicated along with the *E. coli* promoter because these promoters are easily recognized by the *E. coli* replicating and translating enzymes. The promoter is most important part of expression vector.

All these problems can be overcome by using large capacity vector like yeast expression system. Yeast is useful not only in ability to express foreign protein correctly but also provides suitable environment for studying protein -protein interaction. Large capacity cloning expression vector carry strong promoter like SV40 useful in genome analysis, further these construct along with them may be transfected in animal cell.Like retrovirus (adenovirus) origin of replication are maintained episomally by mammalian cells expression vector such as SV40 that also carry enhancer as inducible promoter (heat shock promoter or a modified *E. coli* lac promoter).

21.15 ADVANTAGES AND LIMITATIONS

Somatic-cell nuclear transfer (SCNT) has been used to produce seemingly-healthy cows, mice, rats, goats, pigs, rabbits, cats, mule, horse, and dog. Knock out mice allow investigators in determining the role of a particular gene by observing the phenotype of individuals that lack the gene completely. The proportion of adult cell nuclei to develop into live offspring after transfer into an enucleated egg is very low. High rates of abortion have been observed at various stages of pregnancy after placement of the eggs containing the adult cell nuclei into recipient animals.

Various abnormalities have been observed in cloned cows and mice after birth and this has been found to be somewhat dependent on the type of tissue that originated the nuclei used to make the clone.The reasons for the low efficiency of cloning by nuclear transfer are currently under investigation but it is thought that it may be related to insufficient nuclear reprogramming as the cloned nuclei goes from directing the production of an adult somatic cell to directing the production of a whole new embryo.

21.16 APPLICATIONS

In 1985, the first transgenic farm mammal was produced, (a sheep) called "Tracy". Tracy was created for production of human protein **alpha-1-antitrypsin**. The protein, when missing in humans, can lead to a rare form of emphysema. Many more animal clones have been generated in the mean time. For example, cloned cows appeared in 1999 and now there are cloned pigs that have been modified to reduce transplant rejection of pig organs in humans. Cloned pets (cats and dogs) have been created too. There are even cloned mules. Harvard scientists had designed a genetically engineered mouse, called OncoMouse® or the Harvard mouse, carrying a gene that promotes the development of various human cancers.

Table 21.5 Promoter (Tissue specific)

Promoter Gene	*Expression In*
(Beta)-actin promoter	Many tissues of the transgenic animal
Simian virus 40 T antigen promoter	Many tissues of the transgenic animal
Adipocyte P2 promoter	Fat cells
Myosin light-chain promoter	Muscle
Amylase promoter	Acinar pancreas
Insulin promoter	Islets of Langerhans beta cells
Beta-lactoglobin promoter	Mammary glands

Scientists has engineered the overexpression of the human mitochondrial transporter protein, "uncoupling protein-3" (UCP-3), in skeletal muscle in mice. In this model, the transgenic mice were found to eat more than wild-type littermates, yet remain leaner and lighter. The mice also exhibit lower glucose and insulin levels and an increased glucose clearance rate, leading to the hypothesis that compounds that regulate expression of UCP-3 might be of use in treating obesity.

Mice fed heavy metals are 2-3 times larger

a

Transgenic mice fed with heavy metals grows rapidly 2-3 times larger than normal one.

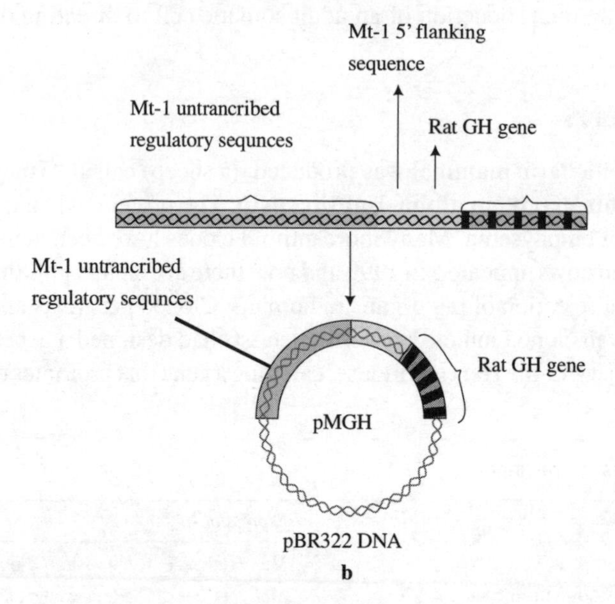

b

Fig. 21.18 Transgenic mice for hormone production

BRAIN QUEST

1. Describe the property of different promoter utilized for specific transgenic production.
2. Describe the difference between gene knock out and gene knock in.

3. Why smaller microorganism is not suitable for production of animal cell products.
4. Describe the procedure of cloning in reference to "Dolly".
5. Describe the different method of transgenic production.
6. Describe the SiRNA approach of controlled gene expression.

APPLICATION OF TRANSGENIC ANIMALS

22

In this Chapter we will study the different application of transgenic animals viz as disease model, as farming reactors and also in improving food quality. Cloning of Dolly, then Polly, Molly and Tracy created a history in the world and till now there are various dynamic changes in methods and quality of these transgenic animals.

22.1 HISTORICAL

1986 Willadsen	Embryonic cloning by nuclear transfer in sheep.
1991 Krimpenfort *et al.*	Transgenic dairy cattle.
1991 Wright *et al.*	Transgenic sheep producing for altered milk compositions.
1992 Muller *et al.*	Transgenic pigs resistant to viral infection
1994 Fodor *et al.*	Pig expressing inhibitor of human complement system by
1997 Wilmut *et al.*	Somatic cloning by nuclear transfer in sheep (Dolly) and
1997 Petters *et al.*	In same year produces transgenic livestock as a model of human disease
1998 Cibelli *et al.*	Transgenic cattle produced by nuclear transfer.
2000 McCreath *et al.*	Transgenic sheep produced by gene targeting method.
2001 Golovan *et al.*	"Ecologically correct" transgenic pig.
2002 Lazaris *et al.*	Production of biopolymer fiber from transgenic cells by and in same year **Kuroiwa *et al*** calf with human artificial chromosome was produced.
2003 Brophy *et al.*	Transgenic cattle producing altered milk proteins compounds.
2003 Phelps *et al.*	Complete gene inactivation in pigs.
2004 Kuroiwa *et al.*	Sequential inactivation of 2 bovine genes.

22.2 INTRODUCTION

A transgenic animal is one which has been genetically altered to have specific characteristics by means of transgenesis (transferring DNA into the animal or altering DNA already in the animal). The application of gene transfer techniques allows the development of new production system for pharmaceutically important proteins; for improving disease resistance such as anti-thrombin III (to treat intravascular coagulation), collagen (to treat burns and bone fractures), fibrinogen (used for burns and after surgery), human fertility hormones, human hemoglobin, human serum

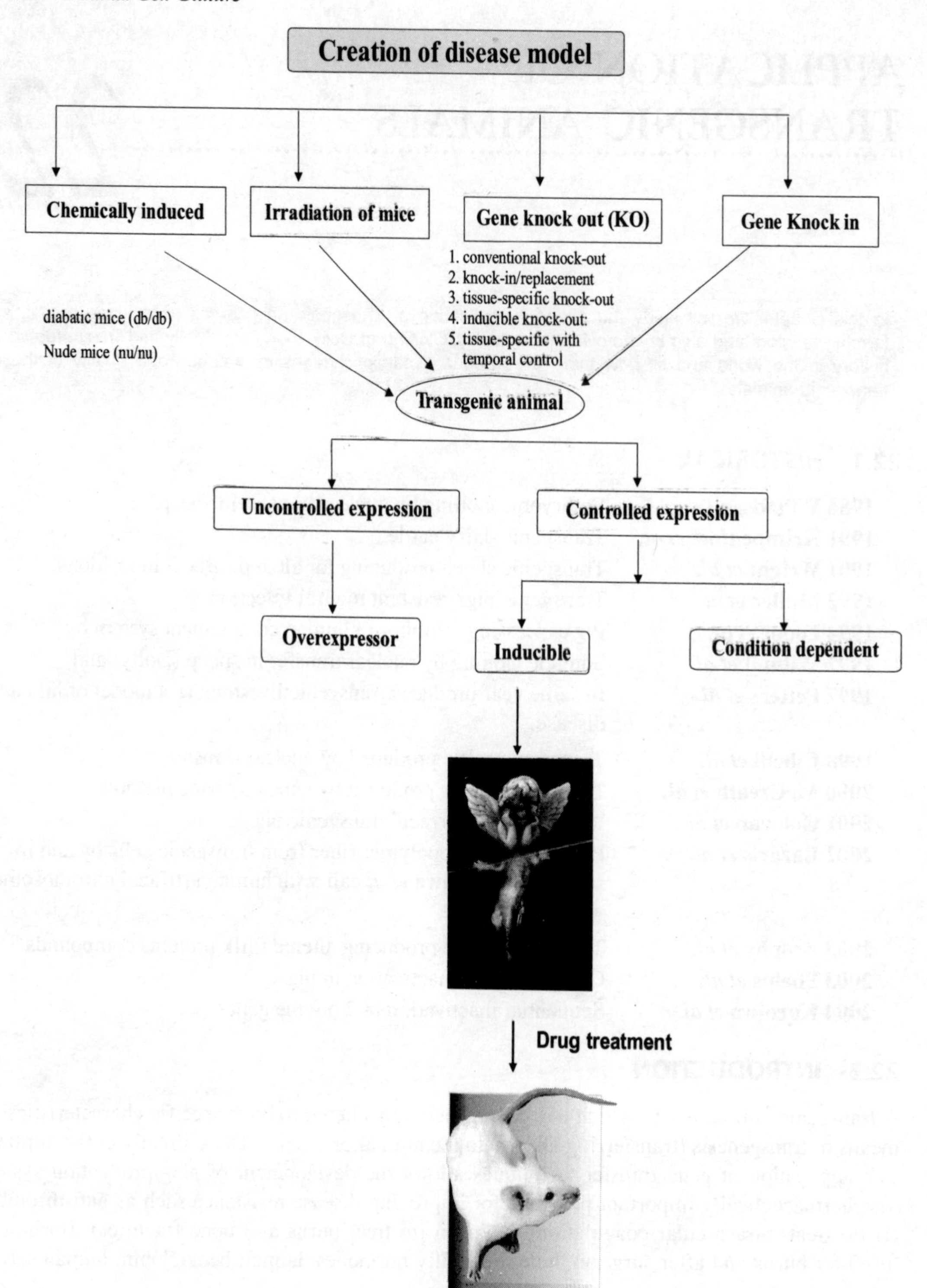

Fig. 22.1 Summary of method of creation of transgenic animal

albumin (for surgery, trauma, and burns), lactoferrin (found in mother milk), tissue plasminogen activator, and particular monoclonal antibodies (including one that is effective against a particular colon cancer) performance of the quality of animal products by modifying metabolic pathways, and hormone status. Several different proteins have since been successfully expressed in the milk of different livestock species and some of them are subsequently in clinical trials. It may be noted that unit cost of proteins is much cheaper when transgenic animals are being used as bioreactors for production of Proteins. It has been estimated that use of animals is 5 – 10 times more economical in operations costs whereas 2 – 3 times cheaper in startup costs as compared to cell culture production methods. Other emerging application involve selective improvement of specific traits and as a sources of organs transplantation. Although, both the above applications hold great potential, however lot of basic research and streamlining of regulatory approval needs to be undertaken before such commercial applications can be optimally realized.

In addition transgenic animals are used by researchers to understand, stages and symptoms of a disease, and also to screen potential therapies or drugs. Transgenic model are also being used as responsive test animal for detection of toxicants in both genotoxicity and carcinogenicity testing for exposure to the test compound. However, extensive care needs to be taken while drawing inferences when natural/complex products are being tested.

More than 50 companies are actively involved in various activities related to transgenic animals such as commercial supply for animal research models, research for production of therapeutic proteins and xenotransplantation. Regarding biopharmaceutical products, several companies have successfully achieved secretion of therapeutic proteins in milk of farm animals at commercially feasible levels of 1 gm/litre. Some of these products are undergoing clinical trials including anti thrombin III from transgenic goats (Phase III), α-1-antitrypsin from sheep (Phase II) and β-glucosidase from transgenic rabbits (Phase 1). In addition to therapeutics, there is great interest in production of monoclonal antibodies in transgenic animals. As compared to global developments, transgenic animal research is still at infancy stage in India. R&D work is being undertaken in limited number of institutions under the aegis of Department of Biotechnology (DBT), Govt. of India. National Institute of Immunology (NII), Delhi. For the first time in India, has successfully established 17 transgenic lines of mice for targeting several important genes for production of biopharmaceuticals and research on human diseases. Centre for Cellular and Molecular Biology (CCMB), Hyderabad has developed a simple transgenic fly Drosophila system for screening and validating a class of drugs targeted against certain types of cancers. A national facility on "Transgenic and gene knock out animals" has also being created.

Table 22.1 Transgenic modification and their applications in cattle

Introduced Modification	*Application*	*Species*	*Refernces*
Insulin like growth factor	Meat	Pig	Pursel et al 1999
Human & Porcine growth, hormone releasing factor	Meat	Pig	Draghia et al 1999
Human growth hormone releasing factor	Meat	Sheep	Rexroad et al 1989
Inducible myostatin (Knock out)	Muscle growth	Mouse	Grobet et al 2003
Sex specific disruption of myostatin	Superior beef bulls	Mouse	Pirottin et al 2005

MILK			
Bovine-α-lactalbumin	Increased milk yield	Pig	Wheeler& Bleck 2001
Bovine-β & α-casien	Milk	Cattle	Brophy et al 2003
FIBER			
Ovine insulin growth factor-1	Wool	Sheep	Demak et al 1996
Ovine growth hormone	Wool	Sheep	Adams & Briegelk et al 2002
Ovine keratine intermediate filament to check shrinkage of wool	Wool	Sheep	Bawden et al 1998
Bacterial serine transacetylase & O-acetyl serine sulphhydrylase	Wool	Sheep	Ward et al 2000
Bacterial isocytrate lyase and malate synthase	Increased glucose supply	Sheep	Ward et al 2000
Human glucose transporter protein 1 & rat hexokinase II	Improved glucose utilization	Fish	Krasnov et al 1999
DISEASES/FOOD SAFETY			
S. simulans lysostaphin	Mastitis resistance	Cattle	Wall et al 2005
Human lysozyme	Food spoilage	Goat	Maga et al 2006

Table 22.2 Gene targeted for development of diseases

Disease Mode	*Gene*	*Technique*
Cystic Fibrosis	CFTR	Gene-Targeted
Ather osclerosis	Apo E, apo (a), Apo A-II	Gene-targeted, Transgentic
anti-Atherosclerosis Gene Therapy	Apo AI, Ape E, LDLR	Transgenic
B-Thalassemia	β -globin	Gene-Targeted
Sickle Cell Anemia	βs (and variants)	Transgenic
Inflammatory Bowel Disease	Interleukine-2, Interleukin-10 and T-cell Receptor, β ; MHC II	Gene-Targeted
Severe Combined Immunodeficiency Disease	Rag-1, Rag-2	Gene-Targeted
Muscluar dystrophy Gene Theraphy	Dystrophin	Transgenic
Alzheimers disease	β -amyloid	Transgenic
Amyotrophic lateral sclerosis (ALS)	Neurofilament heavy chain	Transgenic
Insulin Dependent Diabetes Mellitus	Interferon-	Transgenic
Cancer	Many oncogenes and tumor supressor genes	Transgenic and Gene-Targeted

22.3 MEDICINAL APPLICATIONS

22.3.1 Models of Human Disease Processes

One of the most important applications of transgenic animals was the development of new animal models for human disease. Gene targeting technique was exploited by scientists to create models for human disease. The genetic setup of an animal was modified in such a way that it developed a disease equivalent to human disease. Hundreds of such transgenic rodent lines have been produced by introducing genetic sequences into the genome such as viral transactivating genes and activated oncogenes implicated in specific pathologies. Transgenic rodent models have been characterized for several human diseases including cardio-vascular disease (Walsh et al., 1990), cancer (Sinn et al., 1987), autoimmune disease (Hammer et al., 1990), AIDS (Vogel et al., 1988), sickle cell anemia (Ryan et al., 1990), muscular dystrophy, Lou Gehring's disease, and neurological disease. Transgenic animals enable scientists to understand the role of genes in specific diseases. By either introducing or inactivating particular genes, researchers for the first time discovered the root causes of diseases associated with gene defects. For example, GSK scientists engineered the overexpression of the human mitochondrial transporter protein, "uncoupling protein-3" (UCP-3), in skeletal muscle in mice. In this model, the transgenic mice were found to eat more than wild-type littermates, yet remain leaner and lighter. The mice also exhibit lower glucose and insulin levels and an increased glucose clearance rate, leading to the hypothesis that compounds that regulate expression of UCP-3 might be of use in treating obesity.

22.3.1.1 Model for study of Cardiovascular Disease

Transgenic mice have been developed to study the cause of cardiovascular disease, with the human gene apolipoprotein which was thought to increase the risk of atherosclerosis due to buildup of low density lipoprotein. Other transgenic models was also created for atherosclerosis which express the plasma enzyme, cholesterol transfer protein and apolipoproteins B100 and E3.

22.3.1.2 Model for Study of Cancer

A mouse (disease) model was developed called as 'oncomouse' to study mechanism of cancer. Transgenic mice for p53 allele, shows that it is crucial in checking the uncontrolled growth characteristic of cancer. Thus such mice were susceptible to many types of cancer, most frequently lymphoma (Harvey et al, 1993). The type of tumor is also depends on the promoter placed near to the c-myc gene in the construct e.g.. The mammary tumor virus (MTV) promotor increases the incidence of **breast adenocarcinomas**. In addition, resulted immunoglobulin heavy-chain enhancer (IgH), when inserted along with the c-myc, resulted in development of high incidence of **lymphoblastic lymphomas**. in mice.

The first "oncomouse" was created in 1984 by replacing the normal 'myc' gene with a virus tumor promoter/myc fusion transgene. Both the mice and their offspring developed **carcinomas**. This mouse was made at the Harvard Medical School in Boston for DuPont (Stewart et al, 1984) and was the world's first patented animal. Experiments on oncomouse may lead to preventing and curing multiple forms of cancer.

22.3.1.3 Model for study of Lesch Nyan Syndrome

Lesch Nyan syndrome is single gene disorder, which is characterized by mental retardation and distressing behavioral (abnormalities such as compulsive self-mutilation are observed). Mutation in *phosphoribosyltransferase* gene resulted in severe metabolic and neurobiological symptoms in humans while Mice with the same mutation have no distinct phenotype due to different purine metabolism.

22.3.1.4 Model for Study of Blindness

A transgenic pig has been produced as a model to study human **retinitis pigmentosa**, which causes progressive loss of vision followed by blindness. It will help in future for treating in better way for diseases.

22.3.1.5 Model for Alzheimers/Huntingtons disease

Another important disease model developed was Alzheimer's mouse created in 1995 by a joint effort at Worcester Polytechnic Institute and Transgenic Sciences, Inc. This mouse line overexpressed a gene and thus (Games et al, 1995) overproduced a protein that forms amyloid plaques in the brain, where fibers developed tangles that blocked and degrade neurons, and displayed the symptoms of Alzheimer's disease (Duff et al, 1996). A vaccine was tested on this line of mice that almost entirely prevented the creation of amyloid plaques, and even reduced the damage of the plaques already developed in older mice (Schenk et al, 1999). This was the first Alzheimer's vaccine. This vaccine moved to human clinical trials in 2000 (Jones 2000), and was cancelled in 2001 due to brain inflammation in a minority of patients; however a second generation vaccine by the same company is already in Phase II human clinical trials with no deleterious side effects observed. So far, a vaccine has not yet been FDA approved for the widespread treatment of Alzheimer's, but thanks to the mouse model, researchers are on the right track of preventing and curing Alzheimer's entirely.

In 2004, Crowther et al created a transgenic *Drosophila melanogaster* for Huntington's disease (HD) which is progressive, late onset neurodegenerative disorders caused by a polyglutamine expansion and the dominant neuropathological symptoms was widespread with neuronal loss affecting mainly the caudate nucleus, putamen, and frontal lobes. However, degeneration appeared throughout the brain in another transgenic mouse model of HD that was transgenic for exon 1 of the human HD gene, containing highly expanded CAG repeats, under the control of the HD promoter. Therefore, it was concluded that expansion of cytosine-adenine-guanine (CAG translated into glutamine) trinucleotide repeats in the first exon of the human Huntington (HTT) gene was responsible for HD. Thus Mutant HTT expressed expanded polyglutamine (polyQ) was in the brain and peripheral tissues, but causes selective neurodegeneration that is most prominent in the striatum and cortex of the brain. Although rodent models of HD have been developed, these models do not satisfactorily parallel the brain changes and behavioral features observed in HD patients. Because of the close physiological, neurological and genetic similarities between humans and higher primates, (monkeys) can serve as very useful models for understanding human physiology and diseases.

22.3.1.6 Models for Parkinson's Disease

Parkinson disease is a progressive, degenerative neurologic disease which is characterized by a tremor that is maximal at rest, retropulsion (i.e. a tendency to fall backwards), rigidity, stooped posture, slowness of voluntary movements, and a masklike facial expression. Parkinson's disease involves the formation of neuronal inclusions resulting in premature death. There are two inherited forms of Parkinson disease which are due to the mutations in the alpha-synuclein and parkin genes. In Drosophila the expression of human alpha-synuclein results in loss of motor control, development of neuronal inclusions and degeneration of dopaminergic neucrons, making it an excellent model to study Parkinson disease. To investigate the role of Parkin, Haywood and his associate successfully produced transgenic Drosophila.

22.3.1.7 Models for in Toxicology Study

Several of the currently existing transgenic models (e.g., MutaMouse, OncoMouse) have an application in the assessment of carcinogenesis and mutagenesis. Gene p53 (+/-) animal model in mice have been constructed using a targeted knockout cassette of the tumor suppressor gene p53 which is known to control cell cycling and the apoptosis sequence. This animal model was valuable in the testing of compounds for their potential carcinogenicity because lack of the gene allows genetic damage to remain in cells through mitotic cycle. This model has been found most susceptible for study of **genotoxic carcinogens**. Most tumors recovered that have lost the normal allele therefore identifying the logical mechanism of tumor induction e.g. creation of Tg:AC transgenic model for SV40 T antigen have been done and FDA has accepted this model as an alternative to the mouse oncogenicity test in some cases. In Fig. 22.2 a method of assay of carcinogenic agents have been described in which mouse DNA after exposure is isolated and phase assay is done on sensitive bacteria and also Tg : Ac model helps in testing the effect of carcinogens.

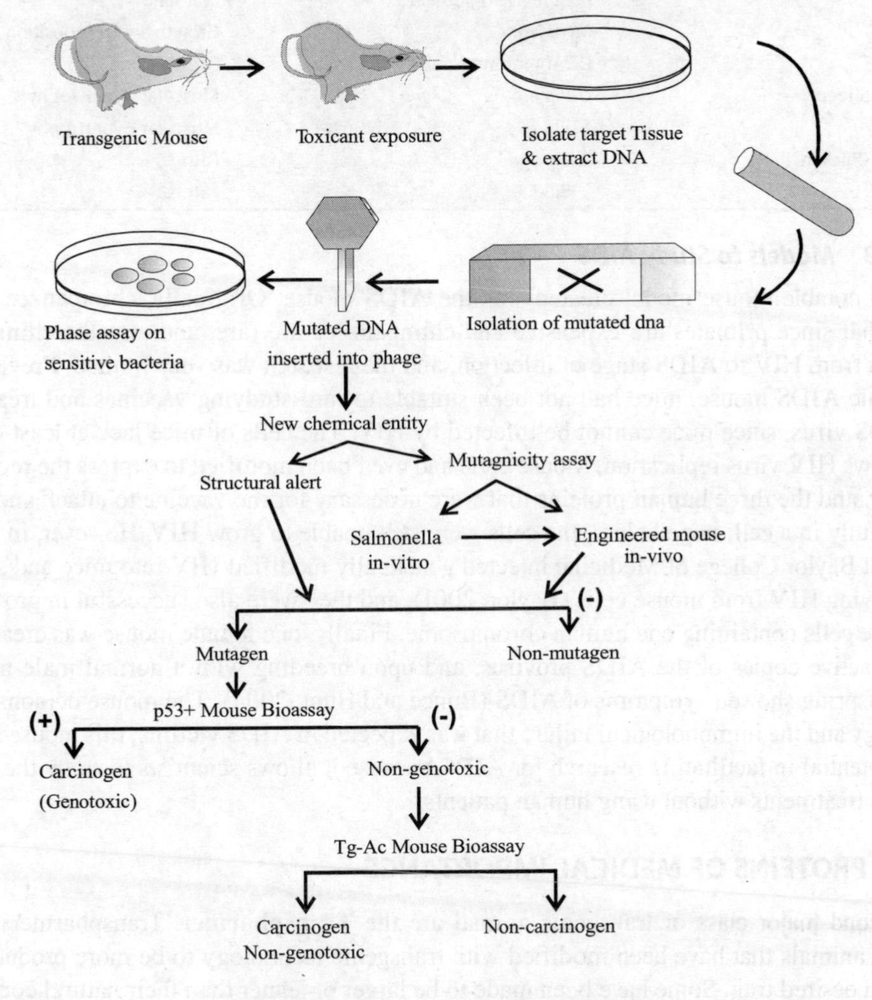

Fig. 22.2 Mice model for carcinogenesis

Table 22.3 Candidate genes associated with meat quality in form animals

Animal	*Candidate genes*	*Traits*
Pig	HAL	Meat Quality/Stress
	MC4R	Growth and Fatness
	RN, PRKAG3	Meat Quality
	AFABP/FABP4	Intramuscular Fat
	HFABP/FABP3	Intramuscular Fat
	CAST	Tender Mess
	IGF2	Growth and Fatness
Cattle	CAST	Meat Tenderness
	Leptin/Thyroglobulin	Marbling
	Myostatin	Growth and composition
	DGAT1 Intramuscular	Fat/Marbling
Sheep	Callipyge	Muscular Hypertrophy
	GDF8	Muscular Hypertrophy
Chicken	EX-FABP	Fatness
	L-FABP	Fatness

22.3.1.8 Models to Study AIDS

Another notable mouse model created was the AIDS mouse. Originally, chimpanzee was only animal but since primates are expensive and chimpanzees are rare, and also the chimps do not progress from HIV to AIDS stage of infection, and the research was very limited. Previous to the transgenic AIDS mouse, mice had not been suitable toward studying vaccines and treatment for the AIDS virus, since mice cannot be infected by HIV. The cells of mice lack at least one factor that allows HIV virus replication. Mouse cells had even been modified to express the receptor, co-receptor, and the three human proteins that were necessary for the vaccine to attack and replicate successfully in a cell; nonetheless, the cells were still unable to grow HIV. However, in 2001, scientists at Baylor College of Medicine injected genetically modified HIV into mice and succeeded in retrieving HIV from mouse cells (Baylor, 2001), and they were also successful in growing HIV in mouse cells containing one human chromosome. Finally, one female mouse was created which carried active copies of the AIDS provirus, and upon breeding with a normal male mouse one third offspring showed symptoms of AIDS (Bunce and Hunt, 2004). The mouse demonstrated the pathology and the immunological failure that was expected in AIDS victims; this mouse model has great potential in facilitating research for AIDS because it allows scientists to study the virus and possible treatments without using human patients.

22.4 PROTEINS OF MEDICAL IMPORTANCE

The second major class of transgenic animal are the **Transpharmer**. Transpharmers are agricultural animals that have been modified with transgenic technology to be more productive or to exhibit a desired trait. Some have been made to be larger or leaner than their natural counterparts, and some transpharmer animals secretes hormones in their milk. The large amounts of proteins produced by the animals are also advantageous because of the post-translational modifications to the proteins, (cleavage and glycosylation), which is time consuming and difficult to perform in laboratory (Houdebine 1997).

Milk-producing transgenic animals are especially useful for medicines. Transgenic animal enables to produce certain pharmaceutical protein in milk, urine, blood, sperm, or eggs e.g. monoclonal antibodies. These proteins serve in treatments for cystic fibrosis, hemophilia; osteoporosis, arthritis, malaria, and HIV. In theory, large quantities of the human protein could be produced in the animal's milk after subsequent purification for use in medical therapies. (see Table 22.4-5).

Table 22.4 List of targeted protein production in livestock

Protein	*Purpose*	*Species*
Lysostaphin	Resistance to mastitis	Goats
Bovine α-lactoglobulin	Improved milk production	Pigs
Insulin like growth factor	Increased growth rate	Pigs
Monoclonal antibodies	Resistance to gastroenteritis	Pigs
E.coli Phytase	Phytase utilization	Pigs
Envelop from avian luekosis	Increased diseases resistance	Chiken
Cercopin peptide	Increased diseases resistance	Catfish

Table 22.5 Farming products currently in development

Animal	*Drug/protein*	*Use*
Sheep	alpha1 anti trypsin	Deficiency Leads To Emphysema
Sheep	CFTR	Treatment of cystic fibrosis
Sheep	Tissue plasminogen activator	Treatment of thrombosis
Sheep	Factor VIII, IX	Treatment of hemophilia
Sheep	Fibrinogen	Treatment of wound healing
Pig,	Tissue plasminogen activator	Treatment of thrombosis
Pig	Factor VIII, IX	Treatment of hemophilia
Goat	Human protein C	Treatment of thrombosis
Goat	Antithrombin 3	Treatment of thrombosis
Goat	Glutamic acid decarboxylase	Treatment of type 1 diabetes
Goat	Pro542	Treatment of HIV
Cow	Alpha-lactalbumin	Anti-infection
Cow	Factor VIII	Treatment of hemophilia
Cow	Fibrinogen	Wound healing
Cow	Collagen I, collagen II	Tissue repair, treatment of rheumatoid arthritis
Cow	Lactoferrin	Treatment of GI tract infection, treatment of infectious arthritis
Cow	Human serum albumin	Maintains blood volume
Chicken, cow, goat	Monoclonal antibodies	Other vaccine production
Pigs	Swine cattle	For better growth, growth hormone
Cattle	Extra copy of caseien, removal of alpha lactoglobuline	Altered milk composition
Goat	Spider gene	Biosteel production in milk
Swine	Phytase gene	Reduced phosphorus in pig
	CD 55 FACTOR CD 59	Xenotransplantation

One of the first notable transpharmer was sheep, (created in 1997) to produce **human clotting factor IX in milk** which play an essential role in blood coagulation, and its deficiency results in hemophilia B. This disease is currently treated with Factor IX derived mainly from human plasma. Recombinant FIX produced in the milk would provide an alternative source at lower cost, and free of the potential infectious risks associated with products derived from human blood" (Schnieke, et al. 1997).

In another example, a transpharmer sheep was developed to produces the human protein alpha-1-antitrypsin, (AAT), in milk that was vital to patients with a protein deficiency that leads to dangerous lung problems. PPL Therapeutics of Edinburgh, Scotland, (that deveploped the Dolly sheep), created a animal producing a cheaper and more efficient hormone. The protein was limited in supply and costs was each patient $80,000 each year; but with the transpharmer sheep producing the hormone, the price was more reasonable (White 1999). It is important that due to production of foreign animal proteins in the milk another have no side effect in the transpharmer animal.

Transpharmer animals produced was Herman Bull. Herman was engineered for lactoferrin and it was expected that female offspring produce the protein in their milk and may alleviate the need for human babies to drink formula or mother's milk unnaturally low in protein. Herman Bull fathered eight calves in 1994 (Biotech Notes 1994) with expected result in offspring. Studies transgenic cattle show that the lactoferrin production was high in the milk and "few hundred such animals could supply thousands of kilograms of the milk annually, (Van Berkel et al, 2002).

In addition to cow and sheep, transpharmer goats have been created with advantage over cows that goats were not susceptible to **"mad cow"** disease and yet still could produce abundant milk. Two of the major companies involved in transhparmer goat creation are Genzyme Transgenics Corporation based in Massachusetts, and Nexia Biotechnologies based in Quebec, Canada. Genzyme is focusing on developing goat-pharmed medicines and tools for medical research, while Nexia is pursuing the more industrial applications of goat pharming (Gaffney 2003). Genzyme Transgenics Corporation, in early 1999, announced in Nature about transpharmer **goat** named **Grace**. "The goat's milk was positive for the **human protein ATIII**, (a protein normally found in human plasma that regulates blood clotting)" (GTC Biotherapeutics 1999). Researchers used this protein in a treatment of heart disease and stroke and its Pre-clinical trials was successful and later approved by the FDA. The market for ATIII is $200 million annually, and this amount of the transgenic protein can be produced by a herd of less than 100 goats (Gaffney 2003). While the goats may prove to be medically and pharmaceutically useful, they also have a strong potential for profit for example recently developed "enviro pig", had a **phytase** gene placed in its salivary glands to allow better utilization of phosphorus in foodstuffs.

The last group of Transpharmer animals produced was transgenic rats, mice and some rabbits. The first Transpharmer animal produced was a mouse in 1987 for **α-lactoglobulin in their milk** (Houdebine, 1997). Mice are useful for transpharming because they are relatively for cheap and reproduce swiftly. However, small animal it was difficult to milk, and the process is relatively time consuming and lacks efficiency. Nonetheless, they are suitable for use when only a small amount of protein is needed (Houdebine, 1997). For example, when protein is needed for studying its biochemical properties, (as in crystallization). The protein in the milk is enough (only a few milliliters) and often provides milligrams of protein needed for research (Houdebine 1997). Often the Transpharmer mice and rabbits are milked using a vacuum and oxytocin to stimulate the mammary gland to release milk, and the milk can be collected daily while the animal is lactating. For a rabbit, the amount of milk collected per nipple is between 10 and 20 ml (Houdebine, 1997), so although the animals are small, the amount of protein can be a merit for production of recombinant proteins at a commercial level.

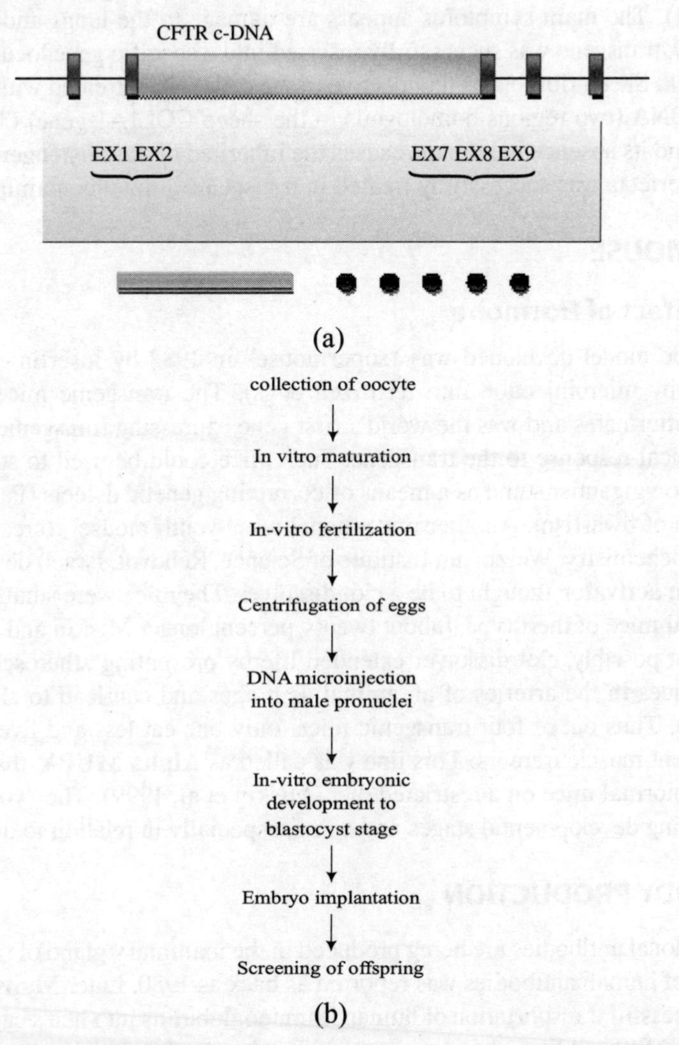

Fig. 22.3 (a) Transgene for preparing c-DNA of cystic fibrosis. For preparing transgenic animal C-DNA is placed within beta casein gene so that CFTR can be secreted into milk with membrane bound fat globule and CFTR can be extracted from fat globule. (b) Method of production of transgenic cow

The human lung is constantly challenged by the dust, spores and bacteria. One such important protein is AAT, against enzyme elastase. Elastase can harm the elastin in the lungs which maintains the elasticity of lungs by releasing a protein called **α-1 antitrypsin' 'AAT',** (which now referred to as 'α-1 proteinase inhibitor') has been successfully expressed in sheep, which bind to the elastase and prevent the damage lungs.

Pateint with low level of alpha 1-antitrypsin resulted in the disease Alpha1-Antitrypsin Deficiency (A1AD or Alpha1). The main symptoms appears are damage to the lungs and sometimes to the liver. In July 2000, transgene was successfully inserted into a specific gene locus was alpha1-antitrypsin, in the milk. Sheep fibroblasts (connective tissue cells) when treated with a vector containing segments of DNA (two regions homologous to the sheep COL1A1 gene) (This gene encodes Type 1 collagen and its absence in humans causes the inherited disease **osteogenesis imperfecta**). Osteogenesis imperfecta was successfully treated in transgenic animal containing COL1A1 gene.

22.5 SUPERMOUSE

For Study of Effect of Hormone

Another transgenic model developed was "supermouse" in 1982 by inserting the gene of a rat growth hormone by microinjection into fertilized eggs). The transgenic mice grew noticeably larger than their littermates and was the world's first gene expressing transgenic animals, with an obvious phenotypical response to the transgene. Such mice could be used to study the effects of growth hormone, on gigantism, and as a means of correcting genetic defects (Palmiter et al, 1982) such as correction of dwarfism. Another mice model was "youth mouse", (created in 1997 at the Department of Biochemistry, Weizmann Institute of Science, Rehovot, Israel) developed for **urokinase plasminogen activator** thought to be a clot dissolver. The mice were smaller, with life much longer than normal mice of their type, [about twenty percent longer Miskin and Masos, (1997)]. It was explained that possibly, clot dissolver extended life by preventing atherosclerosis, (a process that develops plaques in the arteries of an animal as it ages and can lead to clots, hemorrhages, and heart attacks). Thus out of four transgenic mice, only one eat less and lived longer, but also displayed infrequent muscle tremors. This line was called as **Alpha MUPA**, that shows the same characteristics as normal mice on a restricted diet (Miskin et al, 1999). The "youth mouse" could be useful in studying developmental stages and aging, especially in relation to diets.

22.6 ANTIBODY PRODUCTION

Numerous monoclonal antibodies are being produced in the mammary gland of transgenic goats. A preliminary trial of human antibodies was reported as back as 1980. Later **Mansoor Mohammed et al** reported successful transportation of human immunoglobulins into hen's egg. To achieve this, stably transfected DT40 cell lines secreting recombinant human IgG3 and IgA (rhIgG3 and rhIgA) were injected into laying hens. Within the egg yolk the DT40 cells colonized the host and secreted rhIgG3 and rhIgA. Small amount of deposition of rhIgA was also observed in the egg white. This demonstrated the possibility of human immunoglobulins and other foreign proteins production in the chicken's egg. Thus transgenic hens is a competitive manufacturing biosystem to produce human biopharmaceuticals. An interesting development was the generation of **trans-chromosomal animals** e.g. A human artificial chromosome was prepared containing the complete sequences of the human immunoglobulin heavy and light chain loci and introduced into bovine fibroblasts. Trans chromosomal bovine offspring expressed human immunoglobulin in their blood. This system could be a significant step in the production of human therapeutic polyclonal antibodies.

Table 22.6 Animal genes transferred (promoter or enhancer/structural transgenes)

Goat	A variant of tPA gene (LAtPA)
Sheep	mMT/hGH, mMT/TK, mMT/bGH, mMT/hGRF, oBLG/hFIX, oBLG/alpha1AT, oMT/oGH
Rabbit	mMT/hGH, hMT/hGH, rbEu/rb,c-myc
Pig	mMT/hGH, mMT/bGH, hMT/pGH, MLV/rGH, MLV/rGH, bPRL/bGH
Fish	hGH, mMT/hGh, mMT/bGal, cd-crystallin SV/hygro, AFP
Cow	BPV, lactoferrin
Chicken	ALV, REV

22.7 GENETIC MODELS

22.7.1 Genetic Models for Developmental Changes

Frequently used model in genetic research are transgenic fruit flies (*Drosophila melanogaster*) to study the effects of genetic changes on development. In addition, transgenic mouse models was also developed for human genetic disorders (*rag-1* and *rag-2* genes to study) related to immune system.

The largest use of genetic models is in the identification of chemically induced mutations, that elucidates the mechanisms responsible for tissue specific and developmentally regulated gene and expression involve signal transduction pathway and hormonal factor that modulates the activity of genes. The detail of successful protein have been described in Table 22.7. Expression of the enzyme telomerase, which is primarily responsible for the formation and rebuilding of telomeres, was reported to suppressed in most somatic tissues postnatally. Recent studies revealed that telomere length can be established (early in pre-implantation development) by a specific genetic programme that correlates with telomerase activity.

Table 22.7 Current status of livestock

Area of application	Early experimental phase	Experimental phase	Advanced experimental stage	Practical use
BIOMEDICINE				
Gene pharming				
Antibody production				
Xenotransplantation				
Solid organ				
Cell/Tissue				

Table Contd..

Blood replacement
Disease rsistance

Agriculture
Carcass composition
Leaner meat
Harmful Lipid
Lactation performance
Environemntal Effect
Wool production
Disease resistance
Fertility

22.7.2 Cystic Fibrosis

Cystic fibrosis (CF) is an autosomal recessive disease that affects 1 in 2000 births and is caused by genetic defect in transmembrane conductance regulator (CFTR) gene as a result ion and water imbalance occurs in several secretory tissues. The net effect is the accumulation of thick secretions in the lung, in the exocrine pancreas, in the hepatobiliary system, and in reproductive epithelia. These abnormal secretions lead to the characteristic pathology viz pulmonary obstructive disease infections, pancreatic insufficiency, infantile meconium ileus, and biliary stones and obstruction. The CFTR gene was cloned (in 1989) and was found to code for a cyclic AMP (cAMP) activated chloride channel present in the apical membrane of certain secretory for epithelial cells. In contrast to this the gene-targeted mice for CFTR gene display varying degrees of disease. The mutant CF mouse was having *ion permeability defects* in airway, respiratory epithelia and the characteristic histopathology of CF appeared in respiratory, hepatobiliary, and male reproductive epithelia, with little dysfunction in these organs. Since then CF mice has been exploited for unlimited supply of live tissue, and to study the underlying pathophysiology of the defect. Most humans, however, found to have gross morbidity in these organs with severe pulmonary infections, which were not observed in the CF mice. The differences in cystic fibrosis between mice and humans may be a consequence of species variation in the CFTR gene product and relative chloride channel that may adequately compensate for the CFTR mutation in mice. Despite these differences, the animals were useful in the study of the disease and the development of treatment strategies. For example, the evaluation of drugs such as the secretagogue amiloride could be studied in the CF mice. Transgenic mice carrying normal CFTR gene into the respiratory and gastrointestinal epithelia of effected mice, were safely reversed. Such study shows that even low levels of CFTR transgene expression can correct the ion transport deficiency in the lungs and intestines of CF mice, and now serving as the foundation for human trials of CF gene therapy.

22.8 AGRICULTURAL APPLICATIONS

22.8.1 Disease Resistance

Scientists till date had produced only few disease-resistant animals, (influenza-resistant pigs) but a very limited number of genes are currently known to be responsible for resistance to diseases in farm animals. Therefore, a better understanding on immune response specially on the major histocompatibility complexes, which will led to the better understanding of the genetic basis of disease resistance or susceptibility. Manipulation of MHC gene in farm animals can have a beneficial effect on the disease resistance property in farm animals. The mouse Mx1 protein is known to act against influenza viruses. Therefore, transgene with mouse Mx1 cDNA controlled by the human metallothionein IIA promoter (hMTIIA::Mx), along with the SV40 early enhancer/promoter region (SV40::Mx) into pigs show high-level synthesis of Mx1. There were also attempts to produce knock out pigs for the intestinal receptor for the *E. coli* K88.

Most of the bovine farm animals get affected by the inflammatory reaction of mammary gland (called as mastitis) because of attack of various bacteria (*Staphylococcus aureus S. uberis, E. coli, Streptococcus dysgalactiae, and S. galactiae*). The main bacteria found to be responsible is *S. aureus*. Only a few hydrolases like **lysostaphin** (that is potent peptidoglycan hydrolases) secreted by *S. simulans* has a bactericidal effect against other staphylococcus bacteria and can be used for making transgene. The levels of these anti-microbial peptides (lysozyme and lactoferrin) are many times higher in human milk than in bovine milk. Thus transgenic expression of lysozyme and lact of errino gene in mice reduced the frequency of mammary gland infections and cattle shown to have high resistance against diseases. Lycostaphin has been shown to confer specific resistance against mastitis caused by *Staphylococcus aureus*. A recent report indicates that transgenic technology could be used to produce cows that express a lycostaphin gene construct in the mammary gland, thus making them **mastitis-resistant.**

Other prevalent disease is mad cow caused due to mutation in PrPc gene also called as BSE **bovine spongiform encephalopathy**. *Brucellosis* is important diseases worldwide caused by Bruecella bacteria. The introduction of NRAMP1 variant gene had promising prospects in development of brucellosis resistance. In absence of this antigen pigs reported to developed resistance against infections associated with *E.coli*. Scientists are investigating and transgenic swine trypanosomes, nematodes, bacterial viral infections and other genetic diseases.

22.8.2 Milk Production

Milk protein genes have been cloned from a variety of mammals utilising the promoter elements of one or more species in the mammary glands of mice, sheep, goats, cattle, rabbits and pigs to enhance the quality of milk. There are three main modification have been done in the dairy industry for (1) 'humanizing' cows' milk (primarily to enhance the properties of infant formula); (2) Increasing the proportion of valuable protein component; and (3) Reducing lactose to increase potential markets for milk. Milk contains mainly casein (80% of total) encoded by S1- αS2- β and κ-casein gene (which is important in cheese production). Recently genetically modified bovine had been produced by adding additional copies of gene CSN2 and CSN3 for α and β casein. That resulted in 20% increase in α-casein and 2 fold increase in β-casein. This had improved the cheese production. The presence of 10% to 20% of the altered casein in milk produced by a transgenic cow could increase proteolysis (e.g., protein break down) of cheese. Thus, increasing the number of copies of the gene casein, reduces the size of the micelles and made it more susceptible to the digestion with chymosin. Female bovine fetal fibroblasts has been engineered to express additional copies of transgene for two types of casein: bovine β-casein and α-casein. Milk from the cloned animals was enriched for β- and α-casein, resulting in a 30 % increase in the total milk casein or a 13 % increase in total milk protein.

In most of the part of the world 80% people are intolerant to lactose. Therefore, to **improve/ reduce lactos**e in the milk α transgenic swine for alpha lactoalbumin deficient gene has been created as a result about 0.4 - 0.9 g of bovine α– lactalbumin per liter were reported in transgenic sows. In transgenic animals addition of bovine α-lactalbumin resulted in the increased concentration of a lactalbumin in milk by 50%. Further, it has been shown that piglet feeding on such milk had higher growth rate. Apart from α-lactalbumin, the presence of some additional gene also affected animal growth such as IGF - I, EGF, TGF - β and Lactoferrin. These proteins are extremely important for the better health conditions of neonate, as they are responsible for the development and maturation of the gut; the immune system and the endocrine organs. Lysozyme is another important protein offering better protection to the developing neonates and such protein can be expressed in the milk of transgenic swine.

'Humanization' of milk is possible by the addition of genes encoding for human **lactoferrin and Lysozyme** in aim to increase the antibacterial properties in infant formula. Increased lactoferrin is expected to improve iron absorption in infants. (Lysozyme is present in human milk at 3,000 times the concentration found in cows' milk, while lactoferrin is present at 8-80 times).

22.8.3 Increased Meat Production

Meat quality traits is very difficult to define, and it can be classified into three categories: taste, processing and nutrition. In domestic species (pig, sheep, rabbit) growth hormone was enhanced by attaching the metallothionein promoter. Subsequently transgenic swine and cattle were prepared for **c-ski oncogene** which targets skeletal muscle. In other experiment transgenic mice and sheep were created that separately express transgene encoding growth hormone-releasing factor (GRF) or insulin-like growth factor I (IGF-I).

Saeki et al. (2004) generated transgenic pigs that carried the fatty acid de-saturation gene for a 12 fatty acid desaturase from spinach. Levels of linoleic acid (18:2n-6) in adipocytes of pigs was 10 times higher than those from wild type pigs along with better acid ratio, unsaturated, fatty acid, Tenderness (muscle fiber), Water -holding capacity, Color, Oxidative stability and uniformity and also Flavor. Japanese researcher reported to have genetically modified pigs with a gene from a spinach plant, FAD2, which produced an enzyme involved in fat metabolism having less fat and were 'healthier' to eat.

22.8.3.1 Myostatin Suppression for Increasing Muscle Mass

Another potential approach to increase muscle mass in pigs was modification of the myostatin's gene responsible for muscle decrease. Certain breeds of cattle, such as the **Belgian Blue**, have enlarged muscles, primarily due to a marked increase in the number of muscle fibers. Recently, mice with targeted deletion of a growth differentiation factor, GDF-8, (a member of the TGF superfamily), were shown to increase a two to threefold increase in muscle mass (McPherron et al. 1997). **GDF-8, or Myostatin** has subsequently been shown to be the same gene associated with muscling in cattle (Smith et al., 1997). Meta Morphix, Inc. is pursuing vaccine development approaches for improving production efficiencies in poultry, swine, cattle and fish.

22.8.3.2 *Superpig*

"Superpig," thought to have characteristics like grow bigger and faster, thus may prove to be a more efficient food source due to higher levels of growth factors (Miller et al, 1989) e.g. the famous Beltsville pig (made in Beltsville, Maryland under the supervision of the US Department of Agriculture). Unfortunately, the Beltsville pigs had many health problems, such as arthritis (Connor 1999) ulcers, heart problems, lameness, kidney disease, and pneumonia (Animal Aid, 2006). The pigs were euthanized, and biologists imposed a voluntary moratorium on performing any further studies on mammals involving growth hormone.

22.8.3.3 *Superfish*

Another attempt to get more efficient food source was creation of 'superfish'. One species of these fish, the tilapia, was engineered by microinjection to overexpress own growth hormone. This animal was not transgenic, but it was genetically engineered. It showed accelerated growth, but it reached an adult size no larger than normal tilapia (Martinez et al, 1996). Similar techniques have been used on salmon (Devlin et al, 2001). The transgenic salmon produces the growth hormone continuously, instead of turning it off depending on the season (Biotechnology, 2006). The eggs of a species of usually slow-growing trout were microinjected with the gene of a salmon that grew quickly after many generations of selective breeding (Devlin et al, 2001).

There is considerable opposition to the creation and farming of these "superfish". But the transgenic fish look like a much more likely source of food than any transgenic animal species.

22.9 INDUSTRIAL APPLICATIONS

22.9.1 Material Fabrication

It is not possible to 'farm' spiders because they are aggressively territorial and it was very difficult to produce the silk successfully in bacteria because of its structure. Nexia Biotechnologies Inc has engineered goats to produce spiders' silk in their milk, (BioSteel®). The protein was the silk used in spiders' webs and is one of the strongest materials in the world. Spiders' silk has remarkable properties and will undoubtedly be a very useful material because of its extreme strength, flexibility and light weight.

In 2001, two scientists at Nexia Biotechnologies in Canada spliced spider genes into the cells of lactating goats for silk production in the milk. By extracting polymer strands from the milk and weaving them into thread, the scientists could create a light, tough, flexible material could used in such applications as military uniforms, medical microsutures, and tennis racket strings.

22.9.2 Transgenic Silk Worms (Insects)

Transgenic silk worms have been produced for gene human type III procollagen in cocoons. This study demonstrated the viability of transgenic silkworms (insects) as a tool for producing useful proteins in bulk. Tomita et al 2003, used a c-DNA to encode human type III procollagen protein and mini-chain of C-propeptide having a deleted fibroin light chain (L-chain) with an enhanced green fluorescent protein (EGFP). The resulting cDNA was ligated downstream of the fibroin L-chain promoter and inserted it into a **piggyBac vector**. Silkworm eggs were then injected with these vectors, producing worms displaying EGFP fluorescence in their silk glands. This study demonstrates the viability of transgenic silkworms (insects) as a tool for producing useful proteins in bulk.

22.10 IMPROVEMENT IN FIBER PRODUCTION (WOOL)

Currently wool production meet with several limitations such as limitation of glucose supply, limited hydrolysis of cellulose in rumen, shrinkage in wool due to limited supply of cysteine therefore various strategies have been tried for improving wool quality. Two types of transgenic sheep are being developed for increased wool production. One method was introduction of bacterial genes into sheep to produce the protein cysteine, often the limiting factor in the rate of wool production. Other approach was to improve **'metabolic repair'**. Two bacterial genes viz cysE (serine acetyltransferase) and cysM (O-acetylserine- sulphudrolase) and the gene isolated from *E.coli* encoding SAT and OAS (acetyl transferase, and O-acetyl serine sulphydrylase) were fused with metallothionein (MT-la) promoter isolated from long terminal repeat of Rous sarcoma virus (RSVLTR). The next approach was to introduce an insulin-like growth factor into the sheep to increase glucose supply. Mouse ultra-high-sulfur keratin promoters were linked to an ovine insulin-like growth factor 1 (IGF1) to improve shrinkage problem in wool. Glucose supply have been increased by introducing two gene of glyoxalate cycle, *Cys E and Cys M* encoding enzyme SAT and OAS isolated from bacteria in TCA cycle. Glucose produced in this way not only reduces the use of amino acid but also supplied the energy needed for wool growth. Introduction of growth hormone gene does not improved wool production. Insertion of new genes into rumen bacteria belongs to genera bacteroides, *Rumnococcuus and butyrivibro* helped in coversion of cellulose and hemicelluloses into sugar, so that simultaneously plants proteins can be hydrolyzed into amino acids. These amino acid pools can be recycled back to resynthesis of bacterial proteins. The sheep can also use amino acid reserve for its keratin protein.

The main problem was the shrinkage in the wool felting in wool industry but improved wool now had the property of fine texture, resilience and durability since wool is comprised of between 50 to 100 different proteins representing some 14 different protein families. The major proteins of wool fiber was keratins, which carry sulphur-containing amino acid, cysteine and Mammals can synthesize cysteine from methionine (the other sulphur-containing amino acid) but are incapable of synthesizing either of these amino acids using dietary inorganic sulphur. As a result, cysteine/methionine are limiting for wool growth in many dietary situations. Supplementing diet with purified sulphur containing amino acids or formulating diets that contain higher natural sources of these amino acids is the obvious method of improving

Fig. 22.4 Sheep for wool production

the rate and efficiency of wool growth. The plant fodder have been genetically engineered to produce proteins with high amount of specific amino acids, which are required for keratin wool and vicillin rich in cysteine amino acids, which is required for keratin wool protein. These two genes were expressed in tobacco leaves for fodder to animals. However, these works were not so promising and further work was required to improve wool production and its quality.

22.11 GROWTH AND CARCASS COMPOSITION

Human growth hormone has been successfully introduced into sheep in order to increase their development, growth and meat production. The gene for ovine growth hormone is usually placed under control of metallothionein promoter. Such transgenic sheep has shown considerable improvement in their body weight, feed efficiency, meat/fat ratio and fat composition. Recently, Pursei et al 2004 successfully expressed IGF-1 in transgenic swine. In this study insulin-like growth factor I (IGF-I) transgene was specifically targeted into striated muscle of the transgenic swine. Transgenic pigs were produced with a fusion gene composed of avian skeletal alpha-actin regulatory sequences and a cDNA encoding human IGF-I. However the growth rate was not much encouraging in terms of feed efficiency, birth weight, weaning weight and proportion of pig survival. Similarly other studies indicated that transgenic pigs have suffered from less libido and gametogenesis in the transgenic boar therefore several other genes have been introduced such as **chicken 'ski' oncogene,** however this also resulted in limited success in the transgenic pigs.

Palmiter et al. 1982 and 1983 that transgenic mice with either rat or human GH genes had increased growth rates. Hammer et al. (1985) first demonstrated the hGH gene expression in transgenic pigs. Pursel et al. (1990) showed that bovine GH (bGH) transgenic pigs had higher plasma levels of insulin-like growth factor (IGF) with improved feed efficiency, decreased body fat content and in some pigs, enhanced growth rate. Gene constructs for GH, and insulin-like growth factor-I (IGF-I) in pigs and sheep resulted in expressing these constructs that were leaner and more feed-efficient. But as a result of high, unregulated levels of circulating GH, they also suffered a number of complications, such as increased gastric ulceration and decreased fertility, such as joint problems, indicating the need for tight control of hormone secretion.

Recently, two research groups reported preliminary data on the development of GH and IGF-I transgenic pigs with enhanced growth-performance traits. In both experiments, desirable effects on growth and body composition traits were achieved without apparent abnormalities, suggesting that someday useful animals could become available to swine breeders. Potentially useful GH-transgenic fish also have been produced, but biological containment of the transgene is great concern in species with existing wild fish populations. More recently, Nottle et al. (1997) produced transgenic pigs containing a GH construct consisting of a modified human metallothionein IIA (MT) promoter fused to the cDNA sequence for the porcine GH gene. Increased of transgene expression was achieved by feeding 1000 ppm zinc in the diet for 3 week. Although there were too few transgenic pigs to measure significant changes in growth parameters, this study did demonstrate that control of GH gene expression could be achieved using the MT promoter. Thus, in principle, it should be possible to develop a line of transgenic pigs in which expression of GH is not excessive.

22.12 ENVIRO-PIG

Pollution of waterways with phosphate and nitrate called as Eutrophication. Eutrophication is recognized as a national and international environmental problem and livestock account for 34% of the phosphate pollution in the European Union. Pigs contribute significantly to this problem as they cannot digest phosphorous as phytate - the form it takes in plants so the plant phosphorous in their diets is excreted as phosphate.

Fig. 22.5 Belgian blue

The level of pollution has been further exacerbated by the increasing practice of feeding pigs with mineral phosphate supplements in order to maximize growth. In an attempt to solve this problem, researchers at the University of Guelph in Ontario have introduced a gene from the bacteria *Escherichia coli* (coding for the phytase enzyme) into the salivary glands of pigs so that they can digest plant phytate. A reduction of approximately 65% in the phosphate content of pig manure was reported in transgenic pigs expressing phytase and virtual removal of the need for phosphate supplement. No health effects were reported in the G1 pigs (the generation bred from the transgenic founder), although slightly elevated levels of phytase were found in tissues otherthan the salivary gland. Further investigation on several more generations is required to determine whether there are adverse effects on the pig or on humans consuming the pork.

22.13 XENOTRANSPLANTATION

Xenotransplantation is a means of transplantation of organs between animals of different species. Demand of organ is increasing day by day and its supply is 4:1 (demand : supply). Patients die every year for lack of a organ heart, liver, or kidney. For example, about 5,000 organs are needed each year in the United Kingdom alone. Transgenic pigs may provide the (transplant) organs needed to alleviate the shortfall. Currently, xenotransplantation is hampered by a pig protein that can cause donor rejection. The main obstacle in organ transplant was *HAR hyperacute vascular rejection* which occur because of cellular and potential rejection due to some complement protein like *CD59,* and *CD46.*

Various knock out have been produced to avoid *HAR* which basically targets **1,3 α-gal epitopes** absent in human beings. One way around this problem is to produce transgenic pigs with a targeted deletion of the gene encoding the enzyme that produces the epitope, a -1,3 galactosyl transferase. Another approach is to preventing rejection of xenografts is to produce transgenic pigs expressing various human complement regulatory proteins, such as CD59, decay accelerating factor (*DAF or CD55*) and membrane cofactor protein, which protect cells from injury caused by autologous complement. In an *ex-vivo* model, hearts from transgenic pigs expressing human DAF resisted complement-mediated damage to the endothelial cells, thus preventing endothelial activation and consequent myocardial damage. In pig to baboon heterotopic heart transplants, Byrne et al. (1997) showed that expression of CD59 and DAF was sufficient to **block complement-mediated damage.** Many organs and tissues are now routinely transplanted from one human to another except for the rare cases where the donor and recipient are monozygotic "identical" twins, such grafts are called **allografts.**

The first xenografts were performed by Dr. Keith Reemtsma between 1963 and 1964, in which thirteen chimpanzee kidneys were transplanted into humans. The first cross-species heart transplant was performed in 1964, in which a 68-year-old man received a chimpanzee heart. He survived only two hours. Several other attempts have been made to transplant primate hearts into humans, with no patients surviving more than twenty-one days. While whole-organ xenografts have so far not been successful, but at the same time less radical transplants have shown success to some extent. Hundreds of thousands of patients have received pig heart valves since 1975, when the procedure first became commercially available. While primates are our closest relatives and seem to be the most logical organ donors, they can not be bred in captivity in large numbers. Therefore, pigs may be the major organ donors to the humans in the future. Procedure for producing transgenic lamb production by transfer of nucleus from somatic cell into unfertilized egg. Pig is useful for organtransplant because. 1. Their organ is similar in size, anatomy physiology is almost same. 2. They grow rapidly and high level of hygienic condition can be maintained at low cost. (See Fig. 22.6)

22.14 PRODUCTION OF HEMOGLOBIN

Functional human hemoglobin has been produced in transgenic swine. The transgenic protein could be purified from the porcine blood and showed oxygen-binding characteristics similar to natural human hemoglobin. The main obstacle was that only a small proportion of porcine red blood cells contained the human hemoglobin. Alternative approaches is to produce human blood substitutes focused on the chemical cross-linking of haemoglobin to the superoxide-dismutase system.

22.15 TRANSGENIC CHICKENS

Chickens grow faster than sheeps and can synthesize several grams of protein in the "white" of their eggs. The hen has been a potential candidate to produce human biopharmaceuticals at low-cost with high-yield. The reason for this is simple (1) The yolk and white of the egg are sterile. (2) The technology for fractionating egg yolk and egg white proteins is available and highly automated systems for efficiently producing and collecting thousands of eggs per day are well established. The egg white contains 4 g of protein, more than half of which comes from the expression of a single gene i.e. ovalbumin gene (OV). Hence, the OV promoter, combined with other expression elements, could yield significant amounts of protein with a great amount of purity and recovery of the protein. (3) Typically a hen lays >300 eggs per year, therefore a single hen could potentially produce 300 g of raw product annually. Above all they show posttranslational modifications that can be compared with humans. Recently Rapp et al 2003, demonstrated successful synthesis of glycosylated human interferon alpha-2b (hIFN), in the egg white of transgenic hens. The secreted hIFN was tested for its biological activity using viral inhibition assays and the results are strengthening the possibility of using hen as a bioreactor to produce commercially valuable and biologically active proteins.

22.16 IMMUNOCASTRATION

Immunization against gonadotropin releasing hormone (GnRH) has been recognized as a potential means of castration in domestic animals. However, to use this peptide as synthetic vaccines, they should be coupled to a carrier protein to make them more immunogenic. Manns and Robbins (1997) recently demonstrated the efficacy of a recombinant-based GnRH vaccine for immunocastration of boars. Using an *E. coli* for antigenic GnRH construct, it was shown that immunocastrated boars were leaner and had improved weight gain compared to surgically castrated barrows. Immunocastration also reduces fat in roster (one to levels similar to those in barrows). Since it is possible to obtain highly purified construct protein, this approach may be more commercially feasible than other approaches to immunocastration.

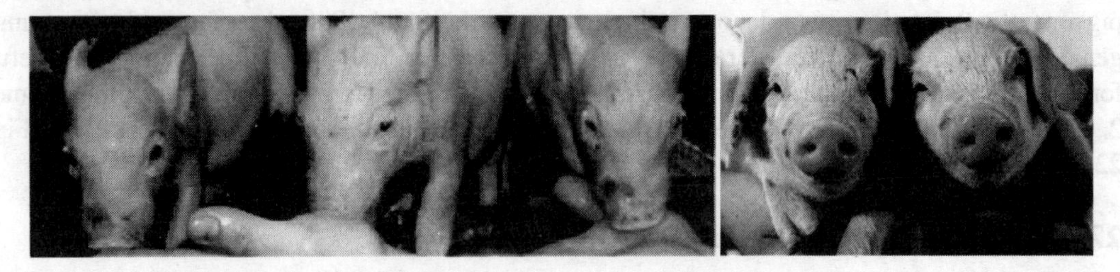

Fig. 22.6 Pig for Xenotransplantation

22.17 THE FUTURE

Research is presently limited to traits involving one or a few genes. It will probably require many years of research before scientists can manipulate complex traits (such as meat quality or animal behaviour) that are influenced by many genes. Much current research focuses on the understanding and developing useful promoter sequences to control transgenes and establishing more precise ways to insert and place the transgene in the recipient. Still much work is needed to improve our knowledge of specific genes and their actions and of the potential side effects of adding foreign DNA and of manipulating genes within an organism.

22.18 ETHICAL ISSUES IN TRANSGENIC ANIMALS

1. In most of the religions chimeras have been mentioned like for example makara rasi or an astrological sign, kamadhenu lady's face with the body of cow in Hindu mythology, sphinx of Egypt and Capricon in western astrological sign. Such chimeras may no longer be myth with the current advancement in science and technology. With the help of genetic engineering one can insert a piece of DNA from an unrelated species to another species. Such animal suffers from, various hierarchies in the public acceptance and this hierarchy depends on the animal, which is being experimented on and the purpose of doing so.

2. Another debate is over interest of people in different transgenic animals such debate is called as 'interest-sensitive speciesism'. For example about 65% of Japanese approve for using genetically modified bacteria for cleaning up the environment, whilst 42% approve and 40% disapprove when it came to cows' producing more milk, and only 19% supported the idea of producing large 'sport fish'. Even the European patent office adapts such a position while accepting /rejecting the patent applications. It rejected a patent application in which a transgenic mouse was designed to screen hair growth stimulants, while the same office granted a patent for 'Onco-mouse'. Another example, is where about 81% Americans accepted the idea of using plants to produce biopharmaceuticals, whilst 49% agreed for the same in animals.
3. Another area of debate is about either they should be considered as 'natural' and 'unnatural'. Introduction of human genes into animals and the vice-versa is seen as blurring the definition of 'humanness'. But what many do not understand is that comparable genes to the humans do occur in other animals, and therefore human genes are not unique to themselves. Such integration is also seem to be happening in the nature, where we have examples of genome of retrovirus integrating into the human genome and has not caused any devaluation of humanness as such. Nevertheless such integration is not only a matter of concern with general public but with scientists too.
4. People argue that attempting to genetically modify the species is an insult and every species has a right to exist as a separate identifiable entity. People sentiments can be further strengthened with the recent outbreaks of severe acute respiratory syndrome and avian influenza where it was seen the animal viruses causing diseases in humans, such an event may soon acquire the global epidemic. It is also true that some attempts in producing transgenic animals have resulted in producing animals which are more disease susceptible; suffer from arthritis and malfunctioning of other organs. So the question is what rights we have over the animals. In the case of patenting of transgenic animals, again the hierarchy plays an important role and most do not accept the patenting of higher life forms and there seems to be general acceptance when it comes to patenting of genes and gene sequences.

Table 22.8 Mutant Mouse Databases

1.	International Mouse Strain Resource -www.informatics.jax.org/imsr/index.jsp
2.	Frontier in Biosciences Gene Knockout Database -www.bioscience.org/knockout/knochome.htm
3.	Mouse Models of Human Cancers Database -cancermodels.nci.nih.gov/mmhcc/index.jsp
4.	Neuromice - www.neuromice.org
5.	Mouse Models of Infertility - reprogenomics.jax.org
6.	Mouse Phenome Database - aretha.jax.org/pubcgi/phenome/mpdcgi?rtn=docs/home
7.	Pathbase - eulep.anat.cam.ac.uk
8.	Mutant Mouse Regional Resource Centers (MMRRC) - www.mmrrc.org
9.	European Mouse Mutant Archive (EMMA) - www.emma.rm.cnr.it
10.	Trans-NIH Knockout Mouse Project (KOMP) -www.knockoutmouse.org/
11.	Mammalian Gene Collection mgc.nci.nih.gov
12.	Mouse cDNA Encyclopedia - genome.gsc.riken.jp

BRAIN QUEST

1. Describe different applications of transgenic animal.
2. Describe the benefit of using of knockouts mice model.
3. Describe the objections the transgenic animals (GMO).
4. Describe why gene is not regulated without promoter.
5. Why wool production has met with number of difficulties?
6. Why growth hormone gene construct was having number of disadvantages?
7. How egg can be utilised for the purpose of nonoclonal antibodies?
8. How pig can be better option for organ transplantation?
9. What are the ethical issue in production of transgenic animals?
10. What is the future scope of artifical blood production?

ASSISTED REPRODUCTIVE TECHNOLOGY (ART)

23

In this chapter we will learn about assisted reproductive technologie (ART) viz. In vitro fertilization (IVF), gamete intrafallopian transfer (GIFT), zygote intrafallopian transfer (ZIFT), donor insemination, egg donation, embryo cryopreservation, intracytoplasmic sperm injection, (ICSI), tubal embryo stage transfer, and intrauterine insemination (IUI) that are medically important to overcome infertility or to increase vigor.

23.1 INTRODUCTION

Reproductive technology has made significant strides over the past fifty years in improving cattles breed and now is being applied in infertile humans. **Assisted reproductive technology (ART)** is generally referring to methods used to achieve pregnancy by artificial or partially artificial means. Therefore, ART includes fertility treatments in which both eggs and sperm are handled seperately through artificial insemination, estrus synchronization, estrus induction, and superovulation, embryo collection, embryo transfer and embryo cryopreservation. In general, ART involve surgically removing eggs from a woman's ovaries, combining them with sperm in the laboratory, and returning them to the woman's body or surgically implanting to another woman. They do not include treatments in which only sperm are handled (i.e., intrauterine—or artificial—insemination) or procedures in which a woman takes medicine only to stimulate egg production without the intention of having eggs retrieved. The types of ART involves in vitro fertilization, gamete intrafallopian transfer, and zygote intrafallopian transfer.

23.2 CAUSES OF INFERTILITY

According to an estimates of the WHO, 13-19 million couples in India are infertile. Infertility may be due to reproductive tract infections or genital tuberculosis. Female infertility is most often caused by problems with ovulation (40%) or fallopian tubes (40%). Other possible causes include endometriosis, in which the uterine lining grows outside the uterus, in which a woman's ovaries stop functioning before she reaches the age of forty, and uterine fibroids. Sexually transmitted infections (STIs) also play a major role in infertility e.g. Chlamydia, has one of the highest numbers of reported cases in the United States (930,000 in 2004, with three times higher rate for women than men). However, since the symptoms are mild, it often goes untreated and, in women, can develop into pelvic inflammatory disease (PID). PID is an infection of the uterus, fallopian tubes, and other reproductive organs, and thus damage reproductive tissues and causes infertility.

It is believed that certain women develop **endometriosis** due to deficiencies in their immune system. Certain auto-antibodies may destroy the healthy endometrium in side the uterus, but are ineffective in destroying ectopic implants. Following symptoms appear after endometriosis such as pelvic pain and cramping before and during periods, pain during intercourse, Inability to conceive, Fatigue, Painful urination during periods, Gastrointestinal symptoms such as diarrhea, constipation, and nausea.

The other causes of infertility is believed due to result from the **scarring and adhesions** in the reproductive tract as a result of inflammation. Scar tissue and adhesions may reduce fertility by either obstructing or distorting the shape of the fallopian tubes, which in turn impedes the passage of sperm to the egg. In the event if sperm do reach the egg, they may encounter a hostile environment unfavorable to fertilization. Other issues known to contribute to infertility in women include stress, diet, exercise, and weight. Obesity contributes to infertility because it can cause irregular menstrual cycles and affect ovulation. Male infertility is most often attributed to low sperm count or abnormal sperm shape/structure. These conditions may be caused by health and lifestyle choices, including smoking, drinking alcohol, or taking recreational drugs or certain medications. Cancer treatments involving radiation and certain drugs can cause infertility in men and women as well.

23.3 *IN-VITRO* FERTILIZATION (IVF)

In vitro fertilization is generally used in cases of severe endometriosis, severe cases of low sperm count or sperm motility, blocked fallopian tubes and in cases where IUI has failed. It could be effective in age-related infertility or where there are problems with the eggs.

23.3.1 IVF-ET and Test Tube Baby

IVF is an acronym for In vitro fertilization ('in vitro' meaning 'in glass' hence also called test tube baby). In 1971, Edwards and colleagues reported the in vitro growth of human oocytes to the blastocyst stage. The first "test tube" baby delivered after in vitro fertilization treatment – IVF – was **Louise Brown,** born in July 1978 (Steptoe and Edwards, 1978). Scientists took 10 years to develop the complete process. In-vitro fertilization (IVF) involves removal of unfertilized eggs (oocytes) from the donor animal ovaries, (6-8 oocytes) and fertilization with sperm. The resulting zygotes are incubated and developed in the laboratory before being placed into the recipient. The world's second and India's first IVF baby, **Kanupriya** alias Durga, was born in Calcutta on October 3, 1978, about two months after the world's first IVF baby, Louise Brown. Since then, the field of assisted reproduction has developed rapidly. Newer techniques was the modifications of existing ones, and new approaches characterise this specialisation. In vitro fertilization was an option for many couples who cannot conceive through conventional therapies. IVF is performed through a **laparoscopy,**

Fig. 23.1 Louise Brown

which is eqipped with a small thin tube and a viewing lens and is inserted through an incision in the navel and the needle goes to the location in the ovaries with the help of an ultrasound machine. This allows the physician to see on a video monitor inside the uterus to locate the ovaries. Once the eggs are removed, they are mixed with sperm in a laboratory dish or test tube. The eggs are monitored for several days. Once there is evidence that fertilization has occurred and the cells begun to divide, they are returned to the woman's uterus.

23.3.2 Steps in IV

There are five steps in the IVF treatment. Ovarian stimulation, Egg retrieval, Insemination, Fertilization, embryo culture, and Embryo transfer.

23.3.2.1 Ovarian Stimulation

For super-ovulation fertility drugs, are given to boost egg production. Since normally, a woman produces one egg per month. Superovulation can also be achieved by administrating hormone follicle stimulating gonadotropin in the end or near to luteal phase of the cycle. This helps in producing multiple egg in same stage. The most common fertility drug are taken orally to help women who are not ovulating or who ovulate irregularly is **clomiphene citrate** (brand name Clomid or Serophene), which is aimed to produce one or more mature eggs. Clomiphene citrate and gonadotropins can be used on their own with intercourse, or combined with AI or IVF. Success rates depend on many factors, especially maternal age and the quality of the accompanying sperm.

23.3.2.2 Egg Retrieval/Embryo collection

Women usually produce one mature egg per menstrual cycle in phases- The first phase is called as 'down regulation', the second one is called as 'stimulation'. Down regulation involves 'switching off the pituitary gland'. This is done to stop the pituitary supply of hormones (FSH and LH) so that the ovarian production of endogenous hormones (which could interfere with follicle production) also get switched off.

In the stimulation phase **gonadotrophin drugs** are given over a 10-14 day period, to stimulate the ovaries to produce several follicles. Ideally each ovary should produce 5-10 follicles giving a total of 10-20 follicles. Successful IVF requires the fertilization of multiple eggs - some may not fertilize or develop normally after fertilization. The growth of the ovarian follicles and the development of the endometrium are usually monitored by serial ultrasound scanning. When the follicles are considered to be mature enough and the endometrium appropriately developed, the patient is given an injection of Human chorionic gonadotrophin - hCG to cause further maturation of the eggs. The egg recovery is planned some 36 hours after the hCG injection.

To allow fertilization, a single egg and about 100,000 sperm are placed together in special culture medium and incubated for about twenty-four hours. Before this viable sperm are collected from the man and washed in a special solution that activates them, for fertilization of the egg. The process of sperm activation is called **"capacitation"** and normally occurs when sperm are ejaculated and enter the female reproductive tract. **Capacitation** involves activating enzymes in the sperm's acrosomal cap , allowing the sperm head, (which contains the sperm's genetic material), to penetrate the outer and inner membranes of the egg (zona pellucida and vitelline membrane).

For males with **azoospermia**, microsurgical or aspiration techniques can directly extract sperm from either the epididymis or the testicles. **Azoospermia** is the most severe form of male infertility, caused by obstructions in the genital tract or by testicular failure. There are three method used successfully by doctors and researchers to micro-fertilize the egg. The first is **"zona drilling,"** in which a hole is punched into the zona pellucida, letting sperm penetrate the egg. The second method is called "**subzonal sperm insertion**," (SUZI) in which a sperm is injected directly under the zona pellucida. A third, related method is "intracytoplasmic sperm injection," in which a sperm is injected directly into the egg cytoplasm.

23.3.2.3 Artificial Insemination (AI)

The man's sperm is placed together with the best quality eggs and stored in an environmentally controlled chamber. The mixing of the sperm and egg is called insemination. The sperm usually enters (fertilizes) an egg a few hours after insemination. If the chance of fertilization is low, the laboratory staff may directly inject the sperm into the egg. This is called intracytoplasmic sperm injection (ICSI). Many fertility programs routinely do ICSI on some of the eggs even if everything is normal. AI can also be combined with hormonal drugs to stimulate production of multiple eggs to increase likelihood that one of them will be fertilized. Specifically, freshly ejaculated sperm, or sperm which has been frozen and thawed, is placed in the cervix (intracervical insemination) (ICI)) or in the female's uterus (intrauterine insemination) (IUI) by artificial means. In 1987, **Smith** at the University of Guelph introduced the concept of multiple ovulation and embryo transfer (MOET).

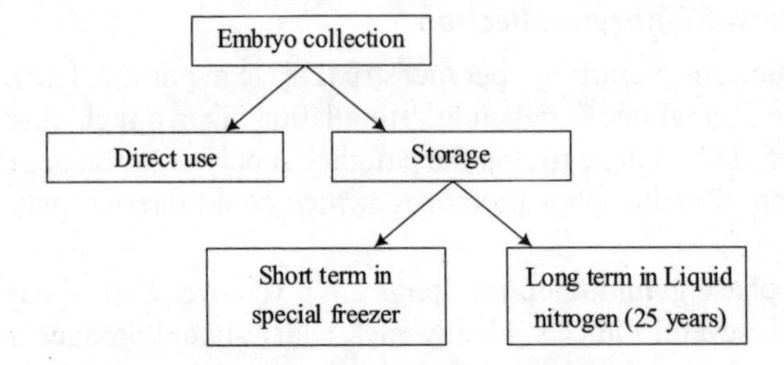

Fig. 23.2 Steps in embryo storage

(a) **Intracervical insemination** (ICI), the easiest way to inseminate, where semen is injected high into the cervix with a needle-less syringe
(b) **Intrauterine Insemination** (IUI), where sperm is injected directly into a woman's uterus.
(c) **Intratubal Insemination:** IUI can furthermore be combined with intratubal insemination (ITI), into the Fallopian tube although this procedure is no longer generally regarded as having any beneficial effect compared with IUI ITI however, should not be confused with gamete intrafallopian transfer, where both eggs and sperm are mixed outside the woman's body and then immediately inserted into the Fallopian tube where fertilization takes place.

23.3.2.4 Embryo Freezing

Cryopreservation is the process of freezing materials slowly, so that they can be used in future. Embryo preservation involves slow freezing, and vitrification. Both sperm or embryos, can be successfully frozen but Egg cryopreservation, has proven to be much more difficult because eggs have high water content and the freezing process often leads to the formation of ice crystals, resulting in burst of the egg cells. (Fig. 23.3)

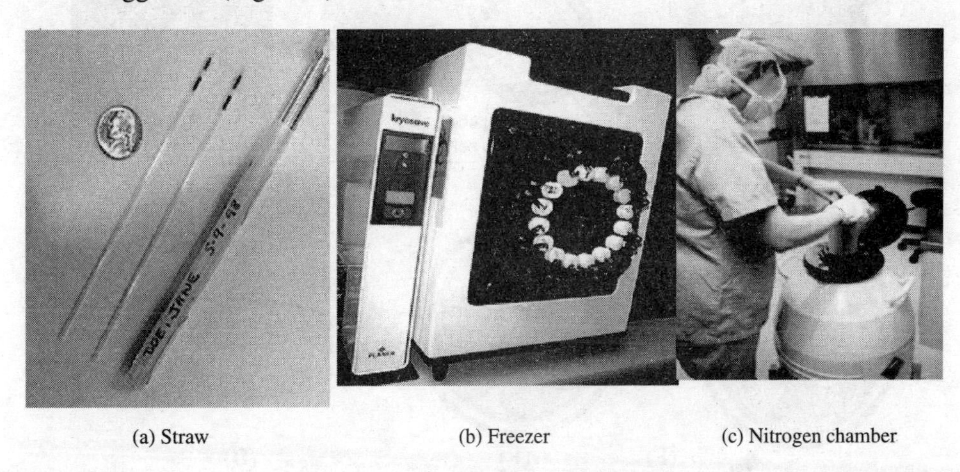

(a) Straw (b) Freezer (c) Nitrogen chamber

Fig. 23.3 The embryos are frozen in special straws in a controlled rate freezer. "Seeding" a straw at a critical temperature helps to avoid formation of intracellular ice. For long term storage liquid nitrogen can be used. (Ice formation can damage embryos)

Vitrification uses a high concentration of antifreeze (Viz DMSO and ethylene glycol), and it drops the temperature so rapidly that the water inside the cell never becomes ice. It is just instantaneously supercools into a solid with no ice crystal formation at all. Therefore, freezing is done in presence of some medium such as saline or saccharide (but for semen milk or egg yolk is used; from 0.2-1.5 M) and an cryoprotectant such as ethylene glycol, glycerol, DMSO. By using high cooling rates, vitrification is possible even in complete absence of cryoprotective agents.

23.3.5 Embryo Transfer (ET)

Embryo is transferred to the uterus or oviduct of recipients by laparotomy using laparoscopic techniques. Usually laparoscopic methods give high rate of pregnancy. This is normally carried out 48 hours after the egg collection. For this procedure a fine tube (catheter) is passed through the cervix and the embryos are injected high into the uterus in a minute amount of culture medium. This technique does not normally require sedation. A urine pregnancy test (BhCG) can be carried out 15 days following the embryo transfer.

A number of factors play a role in embryo transfer leading to a baby being born. Success rate is higher if embryo transfer takes place between forty-eight and seventy-two hours after oocyte collection. When more than one embryo is transferred at the same time, the success rate increases, but so does the chance for multiple pregnancies. Probably the single most important factor determining whether or not a successful embryo implantation will take place is the donated egg's age. Embryos formed from eggs donated by younger women have a higher implantation success rate than do embryos formed from eggs donated by older women. The age of the host uterus appears to have little or no effect on the outcome. The developmental stage of embryo is important during transfer; they usually ranged from follicular oocyte and zygote to elongated blastocysts.

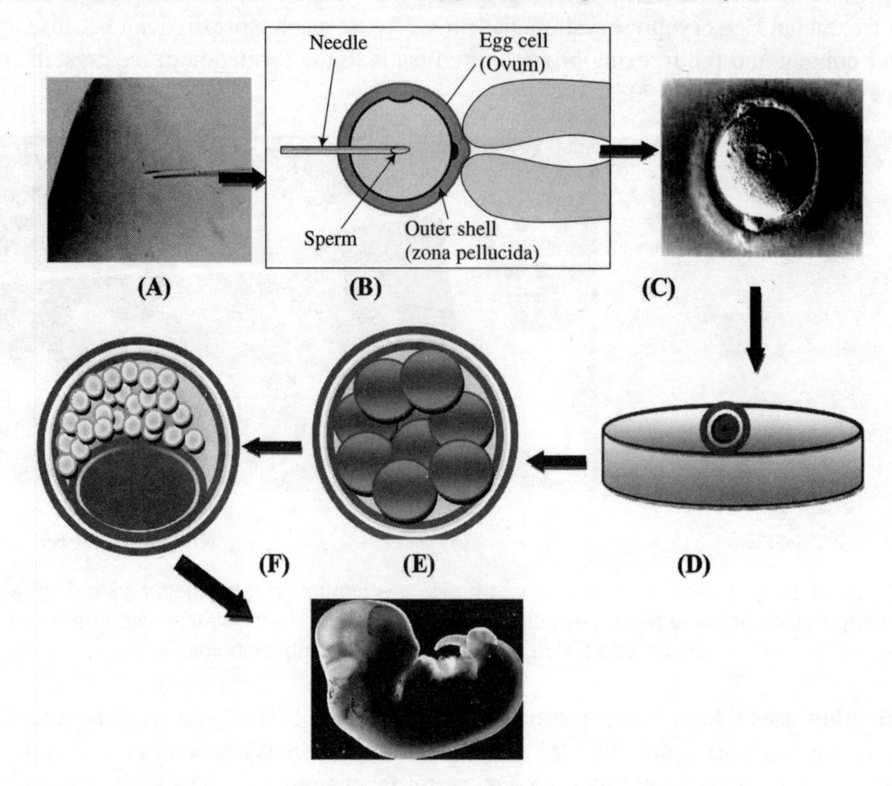

Fig. 23.4 Process details in In-vitro production of embryo

23.4 EXPANSIONS OF IVF

The following are the techniques involved in in-vitro fertilisation.

23.4.1 Transvaginal Ovum Retrieval (OCR)

It is the process whereby a small needle is inserted through the back of the vagina and guided via ultrasound into the ovarian follicles to collect the fluid that contains the eggs.

23.4.2 Assisted Zona Hatching (AZH)

It is performed shortly before the embryo is transferred to the uterus. A small opening is made in the outer layer surrounding the egg in order to help the embryo hatch out and aid in the implantation process of the growing embryo.

23.4.3 Intracytoplasmic Sperm Injection (ICSI)

It is beneficial in the case of male factor infertility where sperm counts are very low or failed fertilization occurred with previous IVF attempt(s). The ICSI procedure involves a single sperm carefully injected into the center of an egg using a microneedle. This method is also sometimes employed when donor sperm is used. (see Figs. 23.5, 23.6 and 23.7)

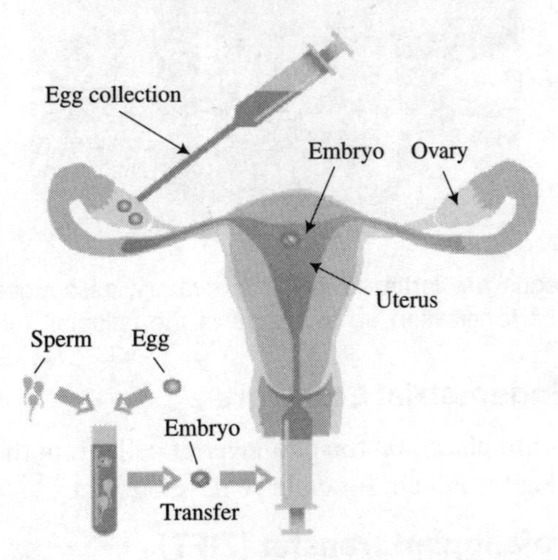

Fig. 23.5 Steps in IVF in humans from embryo collection to transfer in surrogate

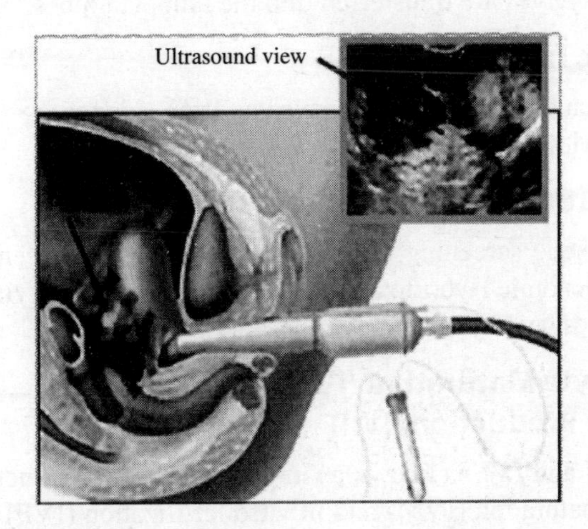

Fig. 23.6 Intracytoplasmic Sperm Injection (ICSI)

23.4.3.1 Steps in ICSI

(i) The mature egg are held with a specialized holding pipette.

(ii) A very delicate, sharp and hollow needle is used to immobilize and pick up a single sperm.

(iii) This needle is then carefully inserted through the zona (shell of egg) and in to the cytoplasm of the egg.

(iv) The sperm is injected in to the cytoplasm and the needle is carefully removed.

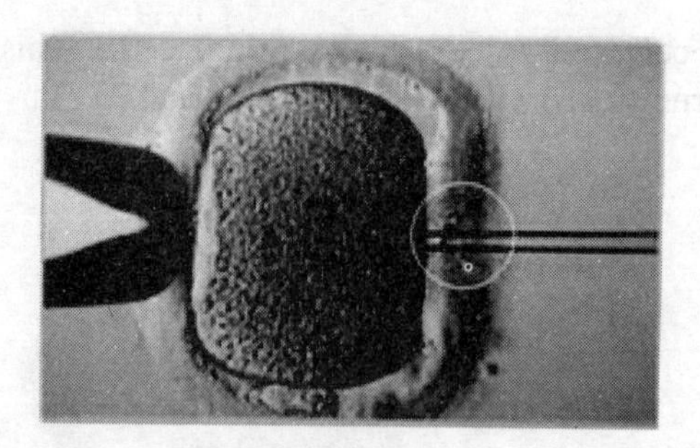

Fig. 23.7 With IVF, the eggs are fertilized in the laboratory, as opposed to GIFT, which simply leaves fertilization up to nature in the fallopian tubes.

23.4.4 Autologous Endometrial Coculture

The patient's fertilized eggs are placed on top of a layer of cells from the patient's own uterine lining, creating a more natural environment for embryo development.

23.4.5 In zygote Intrafallopian Transfer (ZIFT)

Zygote Intrafollopian Transfer or ZIFT is basically a variation of IVF. Eggs are fertilized in the laboratory and resulting zygotes are transferred into the fallopian tubes.

23.4.6 Cytoplasmic Transfer

It is the technique in which the contents of a fertile egg from a donor are injected into the infertile egg of the patient along with the sperm.

23.4.7 Preimplantation Genetic Diagnosis (PGD)

It involves the use of genetic screening mechanisms such as Fluorescent In Situ Hybridization (FISH) or Comparative Genomic Hybridization (CGH) to help in identifying genetically abnormal embryos and thus improves healthy outcomes.

23.4.8 In Vitro Oocyte Maturation (IVM), Fertilization and Embryo Production (IVP)

The combination of IVM and IVF is known as **in-vitro embryo production (IVP)**. The oocyte then undergoes in vitro maturation (IVM) and in vitro fertilization (IVF). The resultant embryos are cultured (*in vitro* culture or IVC) before being transferred to recipients. If IVP is performed in lambs, the process is called **juvenile in vitro embryo transfer (JIVET)**. Semen would provide germplasm for undertaking important future mating regimes flexibility in the conservation programme.

The following Assisted Reproduction techniques don't necessarily involve IVF

23.4.9 Ingamete Intrafallopian Transfer (GIFT)

Gamete Intrafallopian Transfer or GIFT was developed in 1984 and requires that the female partner should have at least one open fallopian tube. GIFT involves the placement of sperm and eggs into an unblocked fallopian tube through a laparoscope for fertilization inside the body. Two to four such unfertilized eggs are mixed with the sperm and then the mixture is placed in the woman's fallopian tubes. The main difference between GIFT and IVF is that in GIFT conception actually occurs in the woman's body, while in IVF conception occurs in the laboratory. (see Fig. 23.7).

23.5 SURROGATE MOTHERHOOD

There are basically two forms of surrogacy:

23.5.1 Genetic Surrogacy

In genetic surrogacy, a surrogate is contracted who is artificially inseminated with the husband's semen. The surrogate provides both an ovum and a uterus for a couple to use. According to Scott Rae, the surrogate "conceives, carries, and gives birth to the child and turns over her rights to the child to the contracting couple." Here surrogate has genetic relationship with child.

23.5.2 Gestational Surrogacy

In this case, the surogate has no genetic relationship with the child because both the gametes are provided by the couple.

23.6 PREIMPLANTATION GENETIC DIAGNOSIS (PGD)

PGD is a therapeutic and genetic testing tool (e.g. Amniocentesis)and has gained greater recognition for its ability to analyze the cells of a developing embryo via biopsy for genetic and chromosomal abnormalities. The objectives of these tests are the prevention of genetic diseases transmission (such as Down's syndrome or spina bifida) the prediction of phenotypic characteristics, and sex determination, prior to the embryo transfer or freezing. It can be generally divided into three phases (1) pre-implantation methods (2) post-implantation methods (3) Post-birth methods.

Amniocentesis and/or ultrasound is used to determine sex of an offspring, leading to subsequent abortion of any offspring of the unwanted sex. The more recent technique of fetal blood it possible to test the sex of the fetus from the sixth week of pregnancy.

23.6.1 Embryo Biopsy

When an embryo reaches the third day of development, it normally has eight cells. One or two of these cells, called "blastomeres," can be removed from the embryo with micromanipulation technique. One of the remarkable facts about mammalian development is that all the cells in the early (e.g., 8-cell) embryo are not needed to produce a healthy fetus (which is why a single fertilized egg can on occasions produce identical twins, triplets, etc.). So couples using in vitro fertilization (IVF) also can take advantage of genetic screening. While the embryo is in culture, a cell or two can safely be removed and tested for its genotype. For example: The sex of the embryo can be determined with a probe for Y-specific DNA. This permits prospective mothers carrying a severe X-linked trait like hemophilia A to choose a female rather than a male embryo for attempted implantation. Fluorescent probes specific for the DNA of particular chromosomes can detect (by FISH) if there is an abnormal number (aneuploidy) such as the three Ch. 21 chromosomes of Down syndrome.

In fact the entire karyotype of the embryo can be determined. Random fragments of DNA is prepared by the polymerase chain reaction (PCR) of all the DNA of a cell from the embryo that can be given a fluorescent label applied to the metaphase chromosomes of a standard reference cell that has a normal karyotype along with DNA fragments from the reference cell labelled with a different color. Comparing the intensity of the two colors from each chromosome shows whether the embryo has the normal amount of DNA for that chromosome or is aneuploid containing either: too much (e.g. 3 copies of Ch. 21 - trisomy) or too little (only a single copy of Ch. 14 - monosomy).

23.7 POST-IMPLANTATION METHODS

23.7.1 Post-birth Methods

1. ***Sex-selective infanticide*** Killing children of the unwanted sex. Though illegal in most parts of the world, it is still practiced.
2. ***Sex-selective child abandonment*** Abandoning children due to unwanted sex. Though illegal in most parts of the world, it is still practiced.
3. ***Sex-selective adoption*** Placing children of the unwanted sex up for adoption. Less commonly viewed as a method of social sex selection, adoption affords families that have a gender reference a legal means of choosing offspring of a particular sex.

23.8 AMNIOCENTESIS

Amniocentesis is an antenatal diagnostic test done in the pregnant mother to rule out the possible presence of birth defects in the fetus whereby the amniotic fluid (fluid present in the womb) is removed and tested for the various birth defects. It is usually performed between the sixteenth and twentieth week of pregnancy. during early stages of development.The word "amniocentesis" literally means "puncture of the amnion," the fluid-filled sac that encloses the fetus during pregnancy. Amniocentesis can detect a number of disorders that will affect babies, while they are still a small fetus in the uterus. For detecting genetic disorders DNA from amniotic fluid samples is used to identify a wide range of genetic disorders, including Fragile X syndrome, phenylketonuria, Tay-Sachs disease sickle cell disease, Down's syndrome and Neural tube defects, such as spina bifida Cystic fibrosis. (See Fig. 23.8)

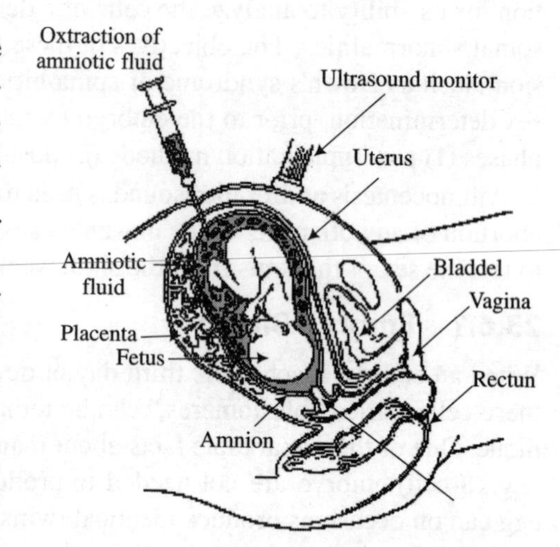

Fig. 23.8 Methods in amniocentesis

23.9 ECTOPIC PREGNANCY

Ectopic pregnancy is also known as a tubal pregnancy for condition when the pregnancy implants outside the womb. It occur at several places, e.g. the ovary, the abdomen and the cervix, at the joint between the tube and the womb (cornua), but the most common place is within the fallopian tube. Pregnancy can even occur in both the womb and the tube at the same time (heterotopic pregnancy), but this is rare and occur in only about 1/10,000 pregnancies. An abnormal rise in blood β-hCG levels may indicate an ectopic pregnancy. The threshold for discrimination of intrauterine pregnancy is around 1500 IU/ml of β-human chorionic gonadotropin (β–hCG). A high resolution, vaginal ultrasound scan showing no intrauterine pregnancy is presumptive evidence that an ectopic pregnancy is present, if the threshold of discrimination for β–hCG has been reached. The fetus produces enzymes that allow it to implant in varied types of tissues, and thus an embryo implanted elsewhere other than the uterus which cause high tissue damage in its efforts to reach a sufficient supply of blood. Thus an ectopic pregnancy is a medical emergency, and, if not treated properly, can lead to the death of the woman.

The advent of **methotrexate treatment** for ectopic pregnancy have reduced the need for surgery; however, surgical intervention is still required in cases where the Fallopian tube has ruptured or is in danger of doing so. This intervention may be laparoscopic or through a larger incision, known as a **laparotomy**.

23.9.1 Causes of the Ectopic Pregnancy

Many factors are known to increase the risk of having an ectopic pregnancy. Anything that alters the tubal function may affect further pregnancies. Fallopian tubes aren't like a hollow pipe that sits there with the egg rolling down. They have little hairs on the inside (cilia) which move with a wavelike motion to encourage the egg toward the womb. If the tube is blocked or the cilia get damaged then ectopic is more likely.

Women with pelvic inflammatory disease (PID) have a high occurrence of ectopic pregnancy. This results due to buildup of scar tissue in the Fallopian tubes, causing damage to cilia. If however both tubes were occluded by PID, pregnancy would not occur and this would be protective against ectopic pregnancy. Tubal surgery for damaged tubes could remove this protection and increase the risk of ectopic pregnancy. However tubal ligation can predispose to ectopic pregnancy. Seventy percent of pregnancies after tubal cautery are ectopic, while 70% of pregnancies after tubal clips are intrauterine. Reversal of tubal sterilization (Tubal reversal) carries a risk for ectopic pregnancy.

23.10 PROBLEMS AND CONCERNS

23.10.1 Age

Age is an important factor when a couples are considering for assisted reproductive technology. In humans, the age of the oocyte, not the age of the uterus, is the main cause of reproductive failure in IVF and embryo transfer techniques. Embryos formed from older oocytes demonstrated an increased incidence of aneuploidy.

Oocytes must reach full maturity before they can be ovulated normally and before they could fertilized, even artificially, to form embryos. If immature oocytes was artificially forced to mature *in vitro,* follicles should be taken from the ovaries of dying or dead women, or from cancer patients planning on undergoing chemotherapy treatments, which can damage oocytes. Unlike immature oocytes, immature sperm can effectively be used in fertilization. Additional research is needed in this area for assisted reproductive technology.

23.10.2 Legal, Ethical, and Moral Considerations

The use of such powerful techniques to facilitate reproduction in both humans and animals (the techniques can be used in cattle and pigs, and in the conservation of endangered wildlife) must be balanced against legal, ethical, and moral concerns. For example, would it be permissible to revive extinct animal species? Although a Jurassic Park-like scenario to reanimate extinct dinosaurs is not scientifically credible at this time, what if it became possible to use this technology to form embryos and clone an extinct mammoth, or the passenger pigeon? And what if we can do this for extinct humans? Just because we can develop the capability, would it be acceptable? What are the ethics involved?

Other concerns includes how long embryos should remain frozen and who owns frozen embryos not used by the parents. What happens if the parents separate, divorce, or die? What about the legal entanglements involved with surrogacy? Already in the media there have been a number of such cases reported. With the expected increase of these procedures in the future, it is likely that such complex questions will only escalate. Finally, there are basic concerns about helping people sidestep the natural birth process to bring into the world a new human.

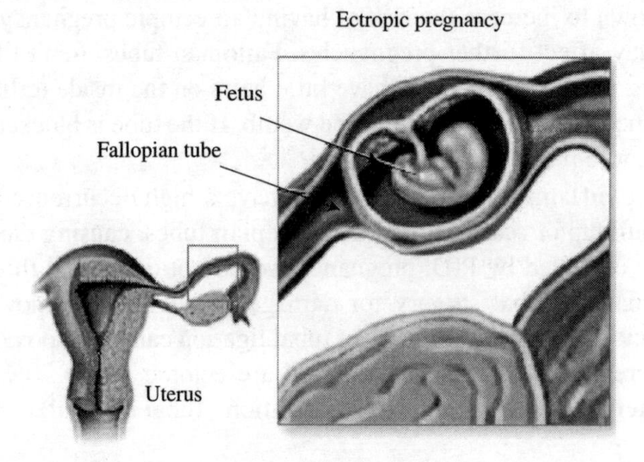

Fig. 23.9 Ectopic pregnancy

23.11 TERMINOLOGY

ICSI (Intracytoplasmic Sperm Injection). A procedure in which a single sperm is injected directly into an egg; this procedure is most commonly used to overcome male infertility problems.

IUI (Intrauterine Insemination). A medical procedure that involves placing sperm into a woman's uterus to facilitate fertilization. IUI is not considered an ART procedure because it does not involve the manipulation of eggs.

Tubal Factor. Structural or functional damage to one or both fallopian tubes that reduces fertility

ZIFT (Zygote Intrafallopian Transfer). An ART procedure in which eggs are collected from a woman's ovary and fertilized outside her body. A laparoscope is then used to place the resulting zygote (fertilized egg) into the woman's fallopian tube through a small incision in her abdomen.

GIFT (Gamete Intrafallopian Transfer). An ART procedure that involves removing eggs from the woman's ovary, combining them with sperm, and using a laparoscope to place the unfertilized eggs and sperm into the woman's fallopian tube through small incisions in her abdomen.

Artificial Insemination (AI) when sperm is placed into a female's uterus (intrauterine) or cervix (intracervical) using artificial means rather than by natural copulation.

Surrogacy, where a woman agrees to become pregnant and deliver a child for a contracted party. It may be her own biological child, or a child conceived through in vitro fertilization or embryo transfer using another woman's ova.

Reproductive Surgery, treating e.g. fallopian tube obstruction and vas deferens obstruction, or reversing a vasectomy by a reverse vasectomy.

In Surgical Sperm Retrieval (SSR) the reproductive urologist obtains sperm from the vas deferens, epididymis or directly from the testis in a short outpatient procedure.

BRAIN QUEST

1. What is the different type of female reproductive technology to improve the infertility?
2. Describe the benefit of ET technology.
3. Describe the technique of ICSI and GIFT.
4. Write down the method and benefits of Embryo cryopreservation.
5. Describe Technique involved in In-Vitro fertilization.
6. How an animal of specific sex can be obtained by ART technology.
7. Write short notes on amniocentesis.
8. What is Preimplantation Genetic Diagnosis.

INDEX

D

N

O

P